AF581784

Transition Metals in Catalysis

Transition Metals in Catalysis

The Functional Relationship of Fe–S Clusters and Molybdenum or Tungsten Cofactor-Containing Enzyme Systems

Editors

Silke Leimkühler
Axel Magalon
Oliver Einsle
Carola Schulzke

MDPI • Basel • Beijing • Wuhan • Barcelona • Belgrade • Manchester • Tokyo • Cluj • Tianjin

Editors
Silke Leimkühler
University of Potsdam
Germany

Axel Magalon
Aix-Marseille University &
CNRS
France

Oliver Einsle
Albert-Ludwigs-University
Freiburg
Germany

Carola Schulzke
University Greifswald
Germany

Editorial Office
MDPI
St. Alban-Anlage 66
4052 Basel, Switzerland

This is a reprint of articles from the Special Issue published online in the open access journal *Inorganics* (ISSN 2304-6740) (available at: https://www.mdpi.com/journal/inorganics/special_issues/MoTEC_Iron_Sulfur).

For citation purposes, cite each article independently as indicated on the article page online and as indicated below:

LastName, A.A.; LastName, B.B.; LastName, C.C. Article Title. *Journal Name* **Year**, *Volume Number*, Page Range.

ISBN 978-3-0365-0608-1 (Hbk)
ISBN 978-3-0365-0609-8 (PDF)

Contents

About the Editors

Silke Leimkühler obtained a Ph.D in microbiology from the Ruhr-University of Bochum, Germany, in 1998. After a post-doctoral position in the Department of Biochemistry at the Duke University Medical Center (Durham, NC, USA) from 1999 to 2001, she returned to Germany with an Emmy-Noether grant from the DFG to establish her own research group at the Technical University of Braunschweig, where she stayed until 2004. In 2005, she accepted a Junior Professor position at the University of Potsdam, Germany. Since 2009, she has had a full professor position in Molecular Enzymology at the University of Potsdam, Germany. Her major research interests focus on molybdenum cofactor biosynthesis, molybdoenzyme enzymology, cellular sulfur transfer mechanisms for sulfur-containing biomolecule synthesis and the role of Fe-S cluster assembly on molybdoenzyme maturation.

Axel Magalon obtained his Ph.D in Microbiology at Aix-Marseille University in 1997. After a two-year position in the Department of Microbiology at the Ludwig-Maximilians-Universität (Munich, Germany) as an Alexander von Humboldt postdoctoral fellow, he obtained a permanent position as a CNRS research scientist at the Laboratoire de Chimie Bactérienne (Marseille, France) in 2001. Since 2008, as research director, he has headed a group in this laboratory. His research focusses on the molecular and cell biology of respiration in enterobacteriaceae, with an emphasis on molybdoenzymes ranging from maturation to functionality and cellular organization questioning.

Oliver Einsle is the director of the Insitute for Biochemistry and dean of the faculty of Chemistry and Pharmacy at the University of Freiburg, Germany. He studied Biology at the University of Konstanz, Germany, and received a doctorate in Biochemistry and Biophysics from the same university in 2000, for a thesis work conducted at the Max Planck Institute for Biochemistry in Martinsried, Germany. During a post-doctoral fellowship as a Howard Hughes fellow at the California Institute of Technology in Pasadena, USA, he was appointed a Junior Professor for Protein Crystallography at the University of Göttingen in 2002 and full professor of Biochemistry in Freiburg in 2008. His research focuses on the structure and function of complex metalloenzymes, particularly in Biological Nitrogen Fixation. He is a member of the American Chemical Society (ACS), the Society of Biological Chemistry (SBIC) and the German National Academy of Sciences Leopoldina.

Carola Schulzke has been the Chair of Bioinorganic Chemistry at the University of Greifswald since 2012. She obtained her Dr. rer. nat. at the University of Hamburg in 2000. After post-doctoral positions in Ottawa Canada and Kiel, Germany, she became junior professor at the Georg-August-University in Göttingen, followed by an Assistant Professorship at Trinity College Dublin. Her research focusses on the synthesis of molybdenum and tungsten cofactor models and biologically active sulfur-rich compounds, synthetic and spectroscopic coordination chemistry, single crystal X-ray structural analysis and catalysis. She is a fellow of the Royal Society of Chemistry (RSC), the German Chemical Society (GDCh) and the Society of Biological Inorganic Chemistry (SBIC).

 MDPI

Editorial

Transition Metals in Catalysis: The Functional Relationship of Fe–S Clusters and Molybdenum or Tungsten Cofactor-Containing Enzyme Systems

Silke Leimkühler

Department of Molecular Enzymology, Institute of Biochemistry and Biology, University of Potsdam, 14469 Potsdam, Germany; sleim@uni-potsdam.de

Citation: Leimkühler, S. Transition Metals in Catalysis: The Functional Relationship of Fe–S Clusters and Molybdenum or Tungsten Cofactor-Containing Enzyme Systems. *Inorganics* **2021**, *9*, 6. https://doi.org/10.3390/inorganics9010006

Received: 26 October 2020
Accepted: 9 January 2021
Published: 13 January 2021

Following the "Molybdenum and Tungsten Enzyme conference—MoTEC2019" and the satellite meeting on "Iron–Sulfur for Life", we wanted to emphasize the link between iron–sulfur clusters and their importance for the biosynthesis, assembly, and activity of complex metalloenzymes in this Special Issue of *Inorganics*, entitled "Transition Metals in Catalysis: The Functional Relationship of Fe–S Clusters and Molybdenum or Tungsten Cofactor-Containing Enzyme Systems".

Iron–sulfur (Fe–S) centers are essential protein cofactors in all forms of life. They are involved in many of the key biological processes, including respiration, photosynthesis, metabolism of nitrogen, sulfur, carbon and hydrogen, biosynthesis of antibiotics, gene regulation, protein translation, replication and DNA repair, protection from oxidizing agents, and neurotransmission [1]. In particular, Fe–S centers are not only involved as enzyme cofactors in catalysis and electron transfer, but they are also indispensable for the biosynthesis of complex metal-containing cofactors. A prominent example is represented by the family of radical/*S*-adenosylmethionine-dependent enzymes, which were discovered in 2001 [2]. Members of this family play essential roles in the biosynthesis of metal centers as complex as the iron-molybdenum cofactor (FeMoco) of nitrogenase, the molybdenum cofactor (Moco) of various molybdoenzymes, the active sites of [Fe–Fe]- and [Fe]-hydrogenases, and the tetrapyrrole cofactors of hemes, corrins, and chlorins. In spite of the recent fundamental breakthroughs in metalloenzyme research, it has become evident that studies on single enzymes have to be transformed into the broader context of a living cell, where biosynthesis, function, and disassembly of these fascinating metal cofactors are coupled in a dynamic fashion. The various biosynthetic pathways were found to be tightly interconnected through a complex crosstalk mechanism that involves the dependence on the bioavailability of distinct metal ions, in particular, molybdenum, iron, and tungsten. The current lack of knowledge of such interaction networks is due to the sheer complexity of the metal cofactor biosynthesis with regard to both the (genetic) regulation and (chemical) metal center assembly.

This special issue intends to combine our recent knowledge on innovative model complexes and biogenesis pathways by emphasizing how they are interconnected by putting the focus on the metals, molybdenum, tungsten, and iron. In this issue, nine contributions, including four original research articles and five critical reviews, will update the reader on the broad spectrum of the role of molybdenum, tungsten, and iron in biology.

The understanding of the biological role of iron and the assembly of Fe–S clusters is reviewed in detail by Srour et al. [3]. The connection and requirement of Fe–S cluster assembly for the biosynthesis of the molybdenum cofactors is reviewed by Mendel et al. [4], while tungsten-containing enzymes and their assembly are reviewed by Seelmann et al. [5]. The review by Yang et al. [6] highlights the past work on metal–dithiolene interactions and how the unique electronic structure of the metal–dithiolene unit contributes to both the oxidative and reductive half reactions in pyranopterin molybdenum and tungsten enzymes. A more specific review focuses on the Moco and Fe–S cluster containing protein formate dehydrogenase. The review by Hille et al. [7] reports on the recent progress in the understanding of the maturation and reaction mechanism of the cytosolic and NAD^+-dependent enzyme from *Cupriavidus necator.* The review on formate dehydrogenase is complemented by an original research article on the formate-hydrogen-lyase (FHL) complex in *Escherichia coli* by Marloth et al. [8], which is composed of the molybdenum-containing formate dehydrogenase and type-4 [NiFe]-hydrogenase. The FHL complex is phylogenetically related to respiratory complex I, and it is suspected that it has a role in energy conservation similar to the proton-pumping activity of complex I. These results indicate a coupling not only between Na^+ transport activity and H_2 production activity, but also between the FHL reaction, proton import, and cation export. The original article by Huang et al. [9] focuses on the [Fe]-hydrogenase (Hmd) that catalyzes the reversible heterolytic cleavage of H_2, and hydride transfer to methenyl-tetrahydromethanopterin (methenyl-H_4MPT^+). The article reports on the crystal structure of an asymmetric homodimer of Hmd from *Methanolacinia paynteri* (pHmd), and the results suggest that Lys150 might be involved in the FeGP-cofactor incorporation into the Hmd protein in vivo.

The theoretical investigations by Rovaletti et al. [10] focus on the only binuclear molybdoenzyme, the Mo–Cu CO dehydrogenase from *Oligotropha carboxydovorans.* This original article studies the dihydrogen oxidation catalysis by this enzyme using QM/MM calculations. The study by Ahmadi et al. [11] introduces Ni to the topic and studies tetra-nuclear nickel dithiolene complexes.

In conclusion, we hope that these open-access contributions will serve as guiding lights for future research into the biological role of molybdenum, tungsten, and iron, and their interconnection at the cellular and enzymatic level. We thank the authors for their original contributions for the special issue, and we thank the reviewers for their insightful comments on each article.

Conflicts of Interest: The author declare no conflict of interest.

References

1. Beinert, H.; Holm, R.H.; Munck, E. Iron-sulfur clusters: Nature's modular, multipurpose structures. *Science* **1997**, *277*, 653–659. [CrossRef] [PubMed]
2. Sofia, H.J.; Chen, G.; Hetzler, B.G.; Reyes-Spindola, J.F.; Miller, N.E. Radical SAM, a novel protein superfamily linking unresolved steps in familiar biosynthetic pathways with radical mechanisms: Functional characterization using new analysis and information visualization methods. *Nucleic Acids Res.* **2001**, *29*, 1097–1106. [CrossRef] [PubMed]
3. Srour, B.; Gervason, S.; Monfort, B.; D'Autréaux, B. Mechanism of Iron–Sulfur Cluster Assembly: In the Intimacy of Iron and Sulfur Encounter. *Inorganics* **2020**, *8*, 55. [CrossRef]
4. Mendel, R.; Hercher, T.; Zupok, A.; Hasnat, M.; Leimkühler, S. The Requirement of Inorganic Fe–S Clusters for the Biosynthesis of the Organometallic Molybdenum Cofactor. *Inorganics* **2020**, *8*, 43. [CrossRef]
5. Seelmann, C.; Willistein, M.; Heider, J.; Boll, M. Tungstoenzymes: Occurrence, Catalytic Diversity and Cofactor Synthesis. *Inorganics* **2020**, *8*, 44. [CrossRef]
6. Yang, J.; Enemark, J.; Kirk, M. Metal–Dithiolene Bonding Contributions to Pyranopterin Molybdenum Enzyme Reactivity. *Inorganics* **2020**, *8*, 19. [CrossRef]
7. Hille, R.; Young, T.; Niks, D.; Hakopian, S.; Tam, T.; Yu, X.; Mulchandani, A.; Blaha, G. Structure: Function Studies of the Cytosolic, Mo- and NAD+-Dependent Formate Dehydrogenase from Cupriavidus necator. *Inorganics* **2020**, *8*, 41. [CrossRef]
8. Telleria Marloth, J.; Pinske, C. Susceptibility of the Formate Hydrogenlyase Reaction to the Protonophore CCCP Depends on the Total Hydrogenase Composition. *Inorganics* **2020**, *8*, 38. [CrossRef]
9. Huang, G.; Arriaza-Gallardo, F.; Wagner, T.; Shima, S. Crystal Structures of [Fe]-Hydrogenase from Methanolacinia paynteri Suggest a Path of the FeGP-Cofactor Incorporation Process. *Inorganics* **2020**, *8*, 50. [CrossRef]
10. Rovaletti, A.; Bruschi, M.; Moro, G.; Cosentino, U.; Greco, C.; Ryde, U. Theoretical Insights into the Aerobic Hydrogenase Activity of Molybdenum–Copper CO Dehydrogenase. *Inorganics* **2019**, *7*, 135. [CrossRef]
11. Ahmadi, M.; Correia, J.; Chrysochos, N.; Schulzke, C. A Mixed-Valence Tetra-Nuclear Nickel Dithiolene Complex: Synthesis, Crystal Structure, and the Lability of Its Nickel Sulfur Bonds. *Inorganics* **2020**, *8*, 27. [CrossRef]

Review

Mechanism of Iron–Sulfur Cluster Assembly: In the Intimacy of Iron and Sulfur Encounter

Batoul Srour, Sylvain Gervason, Beata Monfort and Benoit D'Autréaux *

Université Paris-Saclay, CEA, CNRS, Institute for Integrative Biology of the Cell (I2BC), 91198 Gif-sur-Yvette, France; Batoul.srour@i2bc.paris-saclay.fr (B.S.); Sylvain.GERVASON@i2bc.paris-saclay.fr (S.G.); beata.monfort@i2bc.paris-saclay.fr (B.M.)

* Correspondence: benoit.dautreaux@i2bc.paris-saclay.fr

Received: 18 July 2020; Accepted: 30 September 2020; Published: 3 October 2020

Abstract: Iron–sulfur (Fe–S) clusters are protein cofactors of a multitude of enzymes performing essential biological functions. Specialized multi-protein machineries present in all types of organisms support their biosynthesis. These machineries encompass a scaffold protein on which Fe–S clusters are assembled and a cysteine desulfurase that provides sulfur in the form of a persulfide. The sulfide ions are produced by reductive cleavage of the persulfide, which involves specific reductase systems. Several other components are required for Fe–S biosynthesis, including frataxin, a key protein of controversial function and accessory components for insertion of Fe–S clusters in client proteins. Fe–S cluster biosynthesis is thought to rely on concerted and carefully orchestrated processes. However, the elucidation of the mechanisms of their assembly has remained a challenging task due to the biochemical versatility of iron and sulfur and the relative instability of Fe–S clusters. Nonetheless, significant progresses have been achieved in the past years, using biochemical, spectroscopic and structural approaches with reconstituted system in vitro. In this paper, we review the most recent advances on the mechanism of assembly for the founding member of the Fe–S cluster family, the [2Fe2S] cluster that is the building block of all other Fe–S clusters. The aim is to provide a survey of the mechanisms of iron and sulfur insertion in the scaffold proteins by examining how these processes are coordinated, how sulfide is produced and how the dinuclear [2Fe2S] cluster is formed, keeping in mind the question of the physiological relevance of the reconstituted systems. We also cover the latest outcomes on the functional role of the controversial frataxin protein in Fe–S cluster biosynthesis.

Keywords: iron; sulfur; iron-sulfur cluster; persulfide; metallocofactor; ISC; SUF; NIF; frataxin; Friedreich's ataxia

---

1. Introduction

Iron–sulfur (Fe–S) clusters are small inorganic structures constituting the catalytic site of a multitude of enzymes. They are among the oldest biological cofactors on earth. Their antique role may date back to some four billion years ago, in the form of "Fe(Ni)–S" minerals catalyzing the first abiotic reactions at the origin of life [1–3]. Fe–S clusters then became integral part of living organisms to fulfil a wide range of biochemical reactions [4]. The use of Fe–S clusters by living organisms is now widespread among the prokaryotic, archaeal and eukaryotic worlds [5–9]. Even viruses use Fe–S proteins for their replication [10–12]. With 90 experimentally confirmed Fe–S cluster containing proteins and 60 based on predictions, about 7% of the proteins encoded by the genome of *Escherichia coli* are Fe–S proteins, which emphasizes the prominent roles of Fe–S clusters [13]. Fe–S clusters are composed of iron and sulfide ions , hold in specific protein-binding sites that ligate the iron ions usually via the thiolate of cysteines, but also the nitrogen of histidines and in some rare cases the nitrogen of arginines and oxygen of aspartates or serines [14–17]. The most common forms of Fe–S clusters are the [2Fe2S], [4Fe4S] and [3Fe4S] clusters which are the building blocks for more complex Fe–S clusters present in enzymes such as nitrogenase, hybrid cluster protein and carbon monoxide dehydrogenase [18–20]. Their biochemical functions can be divided into five main classes: electron transfer, redox catalysis, non-redox catalysis, DNA/RNA binding and maturation of Fe–S cluster proteins [13]. These biochemical functions cover a wide range of biological roles, including ATP production, protein synthesis, oxidative-stress defense and maintenance of genome integrity [7,9,21].

Even though the structures of the elementary Fe–S clusters are relatively simple, these inorganic compounds are not assimilated from the local environment, probably due to their instability. Thereby, the use of Fe–S clusters was concomitant with the development of multi-protein machineries by living organisms to support their biosynthesis [6–9,22]. Four machineries have been identified across species: the NIF (nitrogen fixation), the ISC (iron–sulfur cluster), the SUF (sulfur mobilization) and the CIA (cytosolic iron–sulfur cluster assembly) machineries that each have specialized functions [6–9,22]. In addition, carriers and accessory proteins achieve transport and insertion of Fe–S clusters in dedicated client proteins [6,7]. In eukaryotes, more than 30 proteins are needed to perform their synthesis, transport and insertion [6]. However, only a small subset of these proteins is required for their synthesis, which includes a scaffold protein on which Fe–S clusters are assembled, a cysteine desulfurase providing sulfur in the form of a cysteine bound persulfide (Cys-SSH) and a reductase to reduce the persulfide into sulfide (Figure 1) [23–26]. In a second step, Fe–S clusters are transferred to recipient apo-proteins with assistance of dedicated chaperones and accessory proteins (Figure 1) [6,7,27,28].

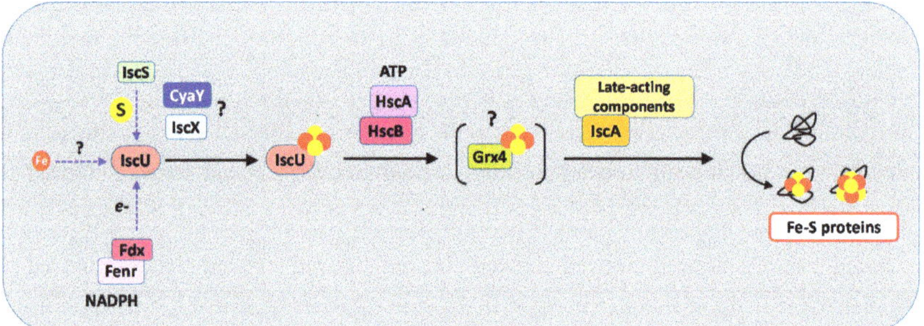

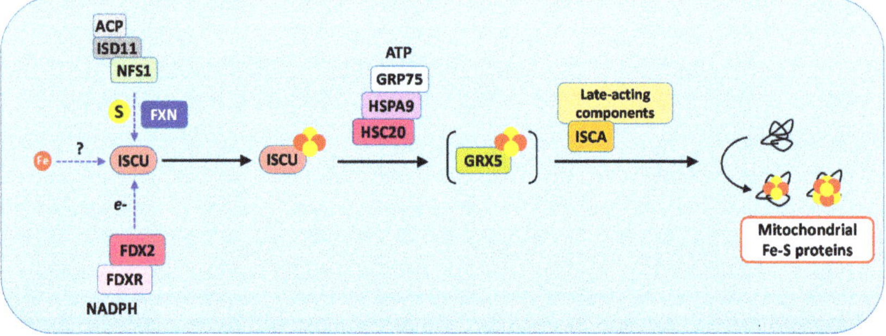

Figure 1. Main components of the bacterial and eukaryotic ISCmachineries. The bacterial ISC machinery encompasses proteins encoded by the ISC operon (IscU, IscS, IscX, Fdx, HscA, HscB and IscA), as well

as CyaY, Fenr, Grx4 and several late-acting components. Fe–S clusters are assembled on the IscU scaffold; iron is provided to IscU by a still ill-defined iron chaperone; sulfur is provided by the cysteine desulfurase IscS, and its activity is modulated by CyaY and IscX; and electrons are provided by Fdx that is reduced by the flavodoxin/ferredoxin NADP reductase (Fenr). The [2Fe2S] cluster formed on IscU is transferred to late-acting components by the ATP-dependent chaperone/co-chaperone HscA/HscB. Grx4 may function as a relay to late-acting components among which is the IscA proteins that have specialized functions in the assembly of [4Fe4S] clusters and/or insertion of [2Fe2S] and [4Fe4S] clusters into recipients' enzymes. In the eukaryotic ISC machinery, Fe–S clusters are assembled on the ISCU scaffold, iron is provided to ISCU by a still ill-defined iron chaperone, sulfur is provided by the cysteine desulfurase complex NFS1–ISD11–ACP, and electrons are provided by FDX2 that is reduced by the NADPH-dependent ferredoxin reductase FDXR. Binding of ISD11–ACP to NFS1 maintains NFS1 in a soluble form. FXN stimulates the whole process by acting on the sulfur donation step. The [2Fe2S] cluster formed on ISCU is transferred to late-acting components by the ATP-dependent chaperone/co-chaperone HSPA9/HSC20 with assistance from the nucleotide exchanger GRP75. GRX5 serves as a relay to late-acting components. The late-acting components, including ISCA, have specialized functions in the assembly of [4Fe4S] clusters and insertion of [2Fe2S] and [4Fe4S] clusters into recipients' enzymes. The dashed blue arrows describe the supply of each element needed to build a Fe–S cluster (iron, sulfur and electrons) and the proteins involved at these steps, without assumption on the sequence order and coordination between these singular steps. The solid black arrows describe actual sequence for Fe–S cluster biogenesis and transfer.

Despite huge progresses in the past 30 years to identify all the components of the assembly machineries, a central question remains how Fe–S clusters are assembled. While iron–sulfur clusters can form spontaneously in vitro by mixing iron and sulfide ions in the presence of scaffolding molecules, the biological processes of their synthesis rely on tightly orchestrated reactions to coordinate iron and sulfur insertions in the scaffold proteins. In recent years, combinations of biochemical, spectroscopic, structural and computational approaches with in vitro reconstituted machineries have significantly contributed to the understanding of the mechanism of Fe–S cluster assembly and the specific role of each component of these machineries. Besides, handling these reconstituted systems is a major challenge as reconstituted system can generate free sulfide, which contributes to Fe–S cluster formation in vitro. This process is not expected to be productive of Fe–S clusters in vivo as the sulfide ions can freely diffuse outside of the biosynthetic complexes. Moreover, iron can bind non-specifically to proteins thereby masking specific iron-binding sites. An additional level of complexity arises with the formation of the [4Fe4S] clusters. The current view is that [4Fe4S] clusters are synthesized by reductive coupling of two [2Fe2S] clusters, thus that the [2Fe2S] cluster is the elementary building block of all Fe–S clusters in the cell [6,25,29–33].

This review focuses on the subset of proteins that ensure the biosynthesis of the [2Fe2S] cluster, with a special emphasis on the questions of the synchronization of iron and sulfur supplies to the scaffold proteins, how persulfide is reduced into sulfide and the mechanism of nucleation of iron and sulfide ions leading to formation of the dinuclear [2Fe2S] center. We also review the latest data on the role of the frataxin protein, a key protein in the Fe–S cluster assembly process, the function of which has remained controversial until very recently.

2. Overview of the Fe–S Cluster Assembly Machineries

The NIF system was the first multi-protein machinery to be discovered, in the late 1980s, by Denis Dean's group [34–36]. In nitrogen-fixing bacteria, this machinery is dedicated to the biosynthesis of the M/V and P clusters of the nitrogenase enzyme [18,34–38]. This machinery encompasses two main components, NifU, the scaffold protein and NifS, a cysteine desulfurase that is the source of sulfur [35,36]. Cysteine desulfurases convert the sulfur from the amino acid L-cysteine into a cysteine bound persulfide intermediate (Cys-SSH) that is a source of sulfide ions upon reductive cleavage of its S–S bond. The discovery of cysteine desulfurase was a major advance since this mechanism now appears as a universal mode of sulfur delivery, not only for Fe–S clusters, but also for the production of a wide variety of sulfur containing biomolecules such as thiolated tRNA, thiamine and Moco [39,40].

In early 1990s, the ISC and SUF machineries were identified as the general providers of Fe–S clusters in most organisms, with the exception of some non-nitrogen fixing bacteria that lack the ISC and SUF machineries, in which the NIF machinery is the general source of Fe–S clusters [5,6,8,9,41–46]. Genome wide analysis show that the occurrence of the SUF system is much higher than the ISC one in bacteria and archaea, thus that the SUF system is the housekeeping assembly machinery in these organisms [47]. In contrast, nearly all eukaryotic organisms rely exclusively on the ISC system that was acquired from bacteria upon endosymbiosis [48], with the exception of plants that rely on both, the ISC and SUF systems [49,50]. In bacteria that encodes both the ISC and SUF machineries, the SUF machinery is expressed under oxidative-stress and iron-deprivation conditions, and in plants, it is expressed in the chloroplast, a compartment that is more exposed to oxidative conditions [8,45]. The underlying reason is that the SUF machinery is more resistant to oxidative stress [51] and handles iron apparently in a more efficient way than the ISC machinery. This raises important questions on the features of the mechanisms of Fe–S cluster assembly that provide such specificities to the ISC and SUF machineries. Some bacteria also express an incomplete machinery, the CSD (cysteine sulfinate desulfinase) system that includes only two components: the cysteine desulfurase CsdA and the sulfur acceptor CsdE that are homologous to the SufS and SufE components of the SUF machinery [9]. The CSD machinery apparently participates to Fe–S cluster biosynthesis by providing sulfur to the SUF machinery [52].

The CIA machinery is present in the cytoplasm of eukaryotic organisms, and in contrast to the other machineries, is not autonomous for sulfur acquisition. Although the cysteine desulfurase of the ISC machinery, NFS1, is also present in the cytoplasm, albeit at very low concentrations, it does not provide sulfur to the CIA machinery [53–56]. Instead, the CIA machinery uses a, as of yet, not identified compound synthesized by the mitochondrial ISC machinery, either a sulfur containing molecule or a preassembled [2Fe2S] cluster [7,25,57]. It is thus unclear whether the CIA machinery is able to synthesize de novo the [2Fe2S] building block. Consequently, we do not cover this topic here.

3. Mechanism of Assembly by the ISC Core Machinery

In bacteria, the genes encoding the components of the machineries are organized in operons. In *E. coli*, the ISC operon encodes eight proteins: **IscU**, the scaffold protein, **IscS**, the cysteine desulfurase, **Fdx**, a [2Fe2S] cluster ferredoxin, **HscA** and **HscB**, a chaperone/co-chaperone system involved in the transfer of Fe–S cluster from IscU to acceptor proteins, **IscA**, a scaffold and/or carrier protein, **IscX**, a putative regulator of the activity of IscS and **IscR** a transcriptional regulator of the whole operon (Figure 1) [22]. Homologs of IscU, IscS, Fdx, IscA, HscA and HscB were later found in yeast, mammals and other organisms with major contributions from Roland Lill's lab to these discoveries (Table 1) [6]. Thereafter, the nomenclature of the protein names of each species is used to describe particular experiments and a double nomenclature bacteria/mammal for more general considerations. Among the proteins of the ISC machinery, **IscS/NFS1**, **IscU/ISCU** and **Fdx/FDX2** together form the core complex for the biosynthesis of [2Fe2S] clusters, while the **IscA/ISCA** proteins have specialized functions in the synthesis of [4Fe4S] clusters from [2Fe2S] clusters and/or transfer of Fe–S clusters (Table 1 and Figure 1) [29–32]. Additional proteins that are not encoded by the ISC operon are needed for the biosynthesis of Fe–S clusters. A flavin-dependent ferredoxin reductase (**Fenr/FDXR**) that uses electrons from NAD(P)H to reduce the [2Fe2S] cluster of Fdx/FDX2 (Table 1 and Figure 1). The frataxin protein (**CyaY/FXN**) is also important for efficient Fe–S cluster synthesis (Table 1, Figure 1) [58,59]. Its exact role is a matter of controversy that we discuss later on [58]. In eukaryotes, **ISD11**, a protein belonging to the LYRM (Leu–Tyr–Arg motif) family, and the acyl carrier proteins (**ACP**) together form a complex with NFS1. The role of the ISD11–ACP complex is incompletely understood, it apparently controls the stability of NFS1 in response to the level of acetyl-CoA [60].

Table 1. Corresponding names of the components of the iron–sulfur cluster (ISC) machinery and accessory proteins in prokaryotes, yeast and mammals.

Functional Role	Prokaryote	Yeast	Mammal
Mitochondrial iron transporter	-	Mrs3/4	MFRN1/2
U-type scaffold	IscU	Isu1/2	ISCU
Cysteine desulfurase	IscS	Nfs1	NFS1
Desulfurase-interacting protein 11	-	Isd11, Lyrm4	ISD11, LYRM4
Acyl carrier protein	ACP	ACP	ACP
Frataxin	CyaY	Yfh1	FXN
IscX	IscX	-	-
Ferredoxin	Fdx	Yah1	FDX2
Ferredoxin reductase	Fenr	Arh1	FDXR
Hsp70 chaperone	HscA	Ssq1	HSPA9
J-type co-chaperone	HscB	Jac1	HSC20
Nucleotide exchanger	-	Mge1	GRPE
Glutaredoxin	Grx4	Grx5	GRX5
A-type scaffold	IscA	Isa1/2	ISCA1/2
Late-acting components	-	Iba57	IBA57
	Bola	Bola1	BOLA1
	-	Bola3	BOLA3
	NfuA	Nfu1	NFU1
	Mrp	Ind1	IND1

3.1. Step 1: Iron Insertion

3.1.1. A Mononuclear Ferrous Iron-Binding Site in IscU/ISCU

IscU/ISCU is a small highly conserved protein of 15 kDa that was identified as the scaffold protein based on its ability to bind a labile [2Fe2S] cluster in vivo when co-expressed with all the other ISC components and in vitro in Fe–S cluster reconstitution assays with IscS/NFS1 [61–63]. Spectroscopic and structural studies of bacterial, archaeal and eukaryotic IscU/ISCU proteins have provided evidence that the [2Fe2S] cluster is ligated in an asymmetric arrangement by well-conserved amino acids: three cysteines and a non-cysteinyl ligand most likely an aspartate [23,61,62,64–66]. This assembly site was thus proposed to be the entry point for iron.

However, metal titrations and tri-dimensional structures revealed that IscU/ISCU proteins purified from bacterial cells do not contain iron in the assembly site but a zinc ion instead [23,67–72]. The zinc ion was found coordinated in an overall tetrahedral geometry by the well-conserved amino acids of the assembly site in *Haemophilus influenza* [71] and *Mus musculus* (PDB code 1WFZ) IscU and the identity of the ligands was also assessed in *E. coli*, mouse and human IscU/ISCU using site directed mutagenesis experiments (Figure 2A) [23,68,69] Here and thereafter, to ease comparison, the mouse sequence is used for amino acid numbering in IscU/ISCU proteins. In these proteins, the Zn^{2+} ion is coordinated by the two cysteines Cys35 and Cys61 that are also ligands of the [2Fe2S] cluster, and the histidine His103, but the fourth ligand seems exchangeable. While the aspartate Asp37 is the fourth ligand in the vast majority of cases, it is exchanged by the cysteine Cys104 in the crystal structure of IscU from *H. influenza* [71]. This suggests structural plasticity of the metal-binding site. Analysis by quantum and molecular mechanics indeed indicate that small rearrangements, such as protonation of the ligands, could induce a ligand swapping at the zinc site [69]. This structural plasticity would facilitate the accommodation of several metal and sulfide ions.

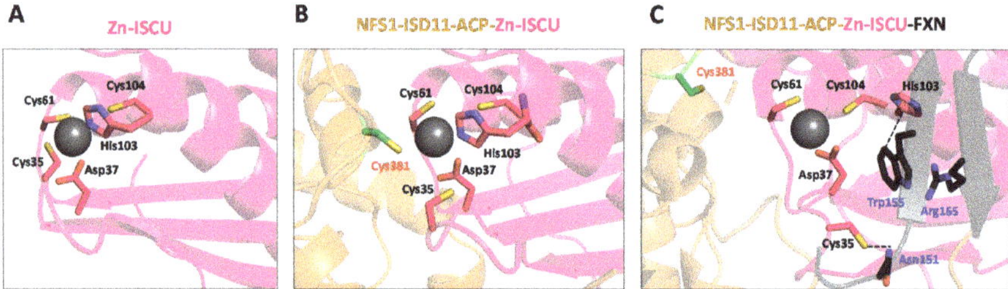

Figure 2. Structural rearrangement at the zinc site of ISCU upon binding of NFS1 and FXN. Zoom on the zinc site of (**A**) mouse ISCU (nuclear magnetic resonance (NMR) structure, PDB code 1WFZ) and in the human NFS1–ISD11–ACP–ISCU complex (**B**) without FXN (X-ray structure PDB code 5WLW) [67] and (**C**) with FXN (CryoEM structure PDB code 6NZU) [73]. ISCU and some of its key amino acids (Cys35, Asp37, Cys61 and His 103) are colored in pink, NFS1 is in yellow with its catalytic cysteine (Cys381) in green and FXN with its key amino acids (Asn151, Trp155 and Arg165) are in black. The dashed black line represents π staking between Trp155 and His103. The zinc ion is initially coordinated by Cys35, Asp37, Cys61 and His103 in ISCU. Upon binding to NFS1, the catalytic cysteine of NFS1, Cys381, binds to the zinc ion via exchange with Cys35. FXN binds to the NFS1–ISD11–ACP–ISCU complex and pushes aside His103 and Cys35 of ISCU via interactions with Trp155 and Asn151, respectively, thereby uncovering Cys104.

The plasticity of the IscU/ISCU protein also prevails at the level of its ternary structure and is directly connected to the metal-binding site. NMR studies revealed that IscU/ISCU exists in two inter-converting forms, a structured one and a de-structured one [69–71,74–76]. Interestingly, the coordination of the metal ion stabilizes the structured form. The ligands of the zinc ion belong to different parts of the protein: the cysteine Cys35 and the aspartate Asp37 are carried by a linker between two β-sheets, the cysteine Cys61 is at the tip of an α-helix and the histidine His103 along with the cysteine Cys104 are at the tip of another α-helix. Therefore, the coordination of the zinc ion connects distinct parts of the protein which stabilizes the ternary structure of the protein.

The first evidence suggesting that iron binds in the assembly site came from an NMR study on the *E. coli* IscU protein, which reported that incubation of apo-IscU with iron stabilizes the ordered form, as observed with zinc [77]. Our lab recently conducted an extensive study on mouse ISCU using several spectroscopic methods [23]. We found that the zinc ion hinders iron binding in the assembly site, but upon removal of the zinc ion, the monomeric form of ISCU binds Fe^{2+} in the assembly site. Site-directed mutagenesis identified Cys35, Asp37, Cys61 and His103 as the ligands of the iron center [23]. The Fe^{2+} ion thus adopts a similar arrangement as the Zn^{2+} ion in the assembly site of ISCU, which suggests interchangeable roles. Titration by circular dichroism (CD) indicated that ISCU binds a single Fe^{2+} ion and Mössbauer spectroscopies showed that it is a high spin Fe(II) center and confirmed the presence of several cysteines in the coordination sphere of the metal. Fe–S cluster assembly assays using the complete mouse ISC machinery with the NFS1–ISD11–ACP complex, FDX2 and FDXR show that iron-loaded ISCU (Fe–ISCU) is competent for Fe–S cluster assembly, while the zinc-loaded form (Zn–ISCU) is not, which indicates that binding of iron in the assembly site is the initial step in Fe–S cluster biosynthesis [23] Binding of iron in the assembly site was later on reported with *E. coli* IscU upon removal of the zinc ion, which suggests that the first step in the mechanism of Fe–S cluster synthesis is conserved [78].

Surprisingly, the investigations of iron-binding sites in IscU/ISCU proteins from *E. coli*, human, drosophila and yeast by X-ray absorption spectroscopy (XAS) led to the conclusion that iron does not initially bind in the assembly site but in a 6-coordinated site comprising only nitrogen and oxygen [79–81]. A similar species was detected by Mössbauer spectroscopy when mouse apo-ISCU was incubated with one equivalent of iron, but this species represents a minor fraction of iron (15%) [23]. Experiments conducted with sub-stoichiometric amounts of iron showed that only the assembly site is filled, thus that iron has a poor affinity for the putative secondary site. Moreover, reconstitution assays with ISCU containing iron exclusively in the assembly site showed that it is still fully competent for Fe–S cluster assembly, which rules out the idea that the minor fraction is important for Fe–S clusters synthesis, thus that there is a defined secondary site in ISCU [23]. This species most likely represents iron bound non-specifically to the protein. It must be noted that mouse ISCU is a difficult protein to handle as the disordered form that accumulates in the absence of metal forms higher order oligomers that do not bind iron. Thereby, the discrepancy with the XAS data may relate to protein preparations and buffer conditions if the other IscU/ISCU proteins behave similarly to mouse ISCU.

3.1.2. Iron Donor to IscU/ISCU

The fact that IscU/ISCU contains a zinc ion that hinders iron binding raises the question of the physiological mechanism of iron insertion. In vivo, the presence of the zinc might prevent accumulation of the disordered form. However, since zinc has a higher affinity than iron for IscU/ISCU [23,69], a still ill-defined metallochaperone might be required to exchange the metals. These data also raise the question of the physiological iron donor to IscU/ISCU.

Frataxin

The frataxin protein was initially proposed to function as an iron storage/donor for IscU/ISCU [82–87]. Although the iron-storage/chaperone function has been questioned by several studies [58], and the most recent data indicate that frataxin is an accelerator of persulfide transfer (see Section 3.5) [23,88,89], the iron chaperone hypothesis is still favored by a number of groups [90–96].

Eukaryotic cells that are defective or deleted for the gene encoding the frataxin protein display growth phenotypes associated with a deficit in Fe–S cluster biogenesis [97–101]. Indeed, the bacterial and eukaryotic frataxins interact with the core IscS–IscU and NFS1–ISCU complexes of the ISC machinery, indicating a direct role in Fe–S cluster biogenesis [86,102–105]. In humans, inherited mutations in the non-coding sequence of the frataxin locus lead to Friedreich's ataxia (FA), a neurodegenerative and cardiac disease [106]. This discovery has strongly fostered the research in the past years to elucidate the function of frataxin.

Based on the observations that (i) frataxin deficient cells accumulate iron in mitochondria and that (ii) bacterial CyaY and yeast Yfh1 frataxins bind ferric iron, which triggers auto-assembly into multimers, the hypothesis that frataxin functions as an iron storage has emerged [58,82,83,85,97,99,107,108]. However, the iron-storage function was challenged by several other studies. First, human frataxin (FXN) self-assembles into oligomers only under extreme conditions and in an iron-independent manner [109]. In yeast and mouse fibroblast, the self-assembly of frataxin is dispensable for its physiological function [110,111]. In yeast, overexpression of Yfh1 is unable to mitigate iron accumulation and binding of iron to Yfh1 could not be confirmed by immunoprecipitation [112,113]. In bacteria, the function of CyaY appears restricted to iron-rich conditions. Although this may be consistent with a role in iron storage, in yeast, there is no obvious effect of iron on the phenotype of cells lacking Yfh1 [59,97]. In mammals, a cardiac mouse model of FA showed that the accumulation of iron is not a primary defect but rather a consequence of the Fe–S cluster deficit [100]. Finally, the accumulation of iron is not a specific feature of frataxin deficiency since any defect in one of the core components of the ISC machinery also leads to dysregulated iron metabolism [114]. Altogether, these data have disqualified the iron-storage function for frataxin.

The frataxin protein was also proposed to function as an iron chaperone for IscU/ISCU, since monomeric frataxin from various organisms bind iron in vitro and several reconstitutions suggested that frataxin brings iron to the ISC machinery [86,87,92]. However, the number of iron-binding sites in frataxin varied from one study to another (from one to seven iron-binding sites), with both Fe^{2+} and Fe^{3+} binding with similar affinities, which rather suggest non-specific iron binding [82,87,108,115–118]. Indeed, FXN lacks a defined metal-binding site associated with well-conserved ligands such as cysteine, histidine and aspartate, but contains an acidic region that binds several metals non-specifically via electrostatic interactions. This may provide a mean for frataxin to bring several labile iron ions nearby the assembly site of IscU/ISCU, but with low specificity, which is difficult to rationalize with a specific role in iron donation. Moreover, data from our group indicated that mammalian FXN is not required for iron insertion in ISCU [23]. Altogether, these data argue against a role of frataxin in iron donation; instead, a function as an accelerator of persulfide transfer is now proposed (see Section 3.5).

ISCA

A-type proteins (IscA, NifIscA and SufA) from various organisms were shown to bind ferric iron with high affinity and to provide iron to IscU/ISCU in Fe–S cluster reconstitution assays through a cysteine mediated process [119–125]. The Fe(III) center of bacterial IscA can be reduced by L-cysteine and released as free or cysteine bound Fe^{2+} ions, which provide a source of iron for IscU [119,125]. However, several in vivo studies argue against a role of the IscA/ISCA proteins in iron donation. First, deletion of the genes encoding the Isa1 and Isa2 proteins in yeast led to a slight increase in the iron content of Isu1, instead of lowering it as expected for an iron chaperone [32]. Several groups found that A-type proteins do not co-purify with iron alone but with a Fe–S cluster [126–129]. Indeed, in vivo and in vitro data rather point to a role of the A-type proteins as scaffolds for the synthesis of [4Fe4S] clusters or in the transfer of [2Fe2S] and [4Fe4S] clusters [29,32,126,127,129,130]. Moreover, while no interaction with IscU/ISCU has been reported so far, the IscA/ISCA proteins interact with the late-acting components of the Fe–S cluster biogenesis pathway, which strengthen the idea that they are not involved at the early steps, i.e. iron insertion [29,131–133].

Connection with the Labile Iron Pool

If not frataxin or IscA/ISCA, then how does IscU/ISCU acquire iron? Only few iron metallochaperones have been identified in eukaryotes and prokaryotes [134–137]. Ferrochelatase is possibly the only attested iron chaperone in mitochondria [134]. Indeed, a general view of iron speciation in proteins is that non-heme iron proteins acquire their metal from a labile iron pool [135–137]. Due to the low structural diversity of metal-binding sites in proteins to discriminate one metal from another, insertion of the proper metal is controlled by its bioavailability and its intrinsic affinity, which follows the Irving–Williams series [135–137]. Thereby, metals with intrinsic high affinity are kept at very low concentration while weaker binders are present in the form of labile pools. Iron is among the weaker binders and the existence of labile iron pools is now corroborated by several studies. Recent investigations of cellular iron contents by Mössbauer spectroscopy suggest that it is composed of non-heme high-spin Fe(II) complex with primarily O and N donor ligands in concentrations estimated in the range of 1 to 10 μM, which is very close to the affinity of iron for most proteins [135–142]. Low-molecular-weight iron complexes with a mass-range of 500 to 1300 Da have been isolated from bacteria and mitochondria that may be central components of this labile iron pool [139–142]. Iron delivery to protein from these labile iron pools may involve small molecules, such as glutathione (GSH), instead of metallochaperones [137,143]. Finally, even though a chaperone is possibly dispensable for iron insertion, the most important step could be the release of the zinc ion from IscU/ISCU, which would require a chelatase.

3.2. *Step 2: Sulfur Insertion*

3.2.1. Two Different Classes of Cysteine Desulfurase

The second key step of Fe–S cluster synthesis is the insertion of sulfur. The cysteine desulfurases IscS and NFS1 are the source of sulfur for Fe–S cluster assembly in the ISC machinery. Cysteine desulfurases provide sulfur in the form of a persulfide bound to their catalytic cysteine (Cys-SSH) [39]. This persulfide is produced by desulfurization of L-cysteine catalyzed by their cofactor, a pyridoxal phosphate (PLP), which leads to formation of a persulfide on their catalytic cysteine. PLP-dependent enzymes form a huge family that perform various types of elimination and substitution reactions on amino acids and their derivatives. Several structures of cysteine desulfurase have been solved and all revealed an arrangement as homodimers, in which the PLP-binding pocket of one subunit is completed by the second one (Figure 3A,B) [39,67,144–150]. Cysteine desulfurases also display key differences allowing their categorization in two classes owing to the length of the mobile loop carrying their catalytic cysteine [39]. This loop is much longer in class I than in class II enzymes, which facilitates long-range sulfur delivery to acceptors for class I enzymes. IscS and NFS1 as well as NifS belong to class I, while SufS, the cysteine desulfurase of the SUF machinery, belongs to class II.

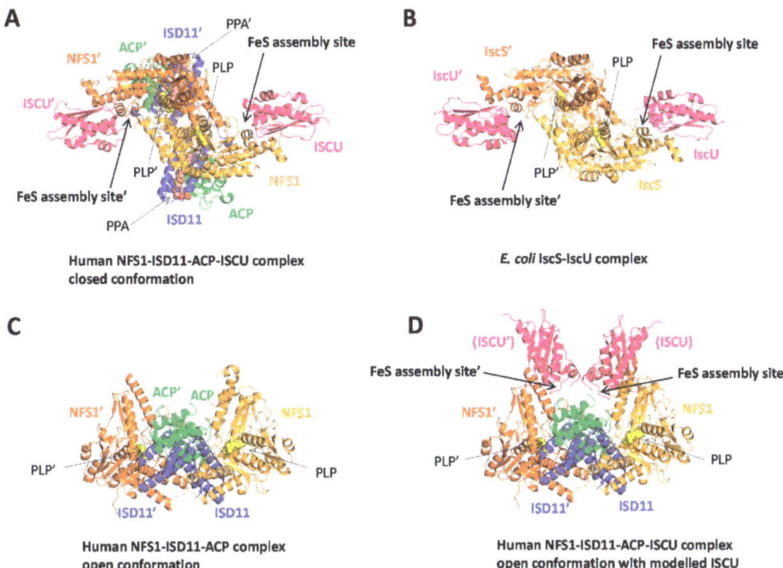

Figure 3. Structures of the human NFS1–ISD11–ACP–ISCU and *E. coli* IscS–IscU complexes. X-ray structures of (**A**) human NFS1–ISD11–ACP–ISCU complex in the closed conformation (PDB code 5WLW) [67], (**B**) *E. coli* IscS–IscU complex (PDB code 3LVL) [150], (**C**) human NFS1–ISD11–ACP complex in the open conformation (PDB code 5USR) [151] and (**D**) human NFS1–ISD11–ACP complex in the open conformation (PDB code 5USR) with the human ISCU proteins modeled according to the structure of the NFS1–ISD11–ACP–ISCU complex (PDB code 5WLW). The labels for the ACP protein have been omitted for clarity in D. IscS/NFS1 are colored in orange, IscU/ISCU in pink, ISD11 in blue and ACP in green. The pyridoxal phosphate (PLP)-binding sites, the phosphopantetheine with its fatty acyl chain (PPA) and the Fe–S cluster assembly sites of IscU/ISCU are indicated by arrows.

Moreover, in eukaryotes, NFS1 forms a tight complex with the ISD11–ACP complex that regulates the activity of the ISC system in response to the level of acetyl-coA [60,67,151]. The crystal structure of the human NFS1–ISD11–ACP complex shows that binding of ISD11–ACP to NFS1 is mediated by ISD11 and that ACP is anchored to ISD11 via its fatty acid chain (Figure 3A) [67]. The ISD11–ACP complex stabilizes NFS1 since in its absence, NFS1 is poorly soluble [152,153]. The crystal structure of the complex indeed shows that ISD11 binds to a hydrophobic patch composed of Leu78 and Ile82 at the surface of NFS1, which most likely helps maintain NFS1 in a soluble form [67].

Strikingly, another crystal structure of the human NFS1–ISD11–ACP complex was solved which displays a completely different arrangement of the NFS1 and ISD11 proteins relative to each other while keeping the overall structure of the ISD11–ACP complex (Figure 3C) [67,151]. In this structure, the dimer interface is created by contacts between the hydrophobic patch of each NFS1 and via ISD11 that connects the two subunits of NFS1. The PLP-binding pocket is incomplete and thus the protein is most likely catalytically inactive. This conformation is designated as open conformation [151] and the conformation in which the PLP-binding pocket is complete, is called the closed conformation [67]. It is still unclear whether the open conformation has a physiological relevance.

3.2.2. Mechanism Generating Persulfide in Class I Cysteine Desulfurase

While the mechanism of sulfur delivery clearly differentiates class I and II cysteine desulfurase, the mechanism of desulfurization is similar for both enzymes [39]. This mechanism relies on catalysis by the PLP cofactor which operates as an electron withdrawal group facilitating the elimination of sulfur on L-cysteine (Figure 4). The PLP cofactor is anchored to the enzyme via the side chain of a lysine through formation of an imine link, also called internal aldimine. The amino group of the L-cysteine substrate binds to the PLP by exchange with the amine of the lysine to form an external aldimine. This intermediate is converted into a ketimine in which a double bond is created between the enantiomeric α carbon of L-cysteine and its amino group. Then, the sulfur of L-cysteine is eliminated and "transferred" to the catalytic cysteine of the cysteine desulfurase, which leads to formation of a persulfide (-SSH) and creation of a double bond between the α and β carbon of the L-cysteine. Upon hydrolysis, alanine is released and a new cycle is initiated. Formally, a sulfanium cation (HS^+) is abstracted by the thiolate group of the catalytic cysteine. This formal description suggests that the reaction is an oxido-reduction process. The redox state of the terminal sulfur (also called sulfane sulfur) is commonly assumed to be "0" while the sulfur of the cysteine to which it is bound would stay at the -II redox state. However, this description is not correct since two identical atoms that are covalently attachedhave their redox state mutually affected and equals, as found in disulfides (RSSR') and hydroperoxides (ROOH) in which the sulfur and oxygen atoms are at the same -I redox state. Thereby, the formal redox states of the sulfurs in a persulfide are better described as -I for both of them, which indicates that the sulfur of L-cysteine is oxidized during the reaction and consequently the carbon carrying the sulfur of L-cysteine is reduced from -I to -III in L-alanine (Figure 4, red labels).

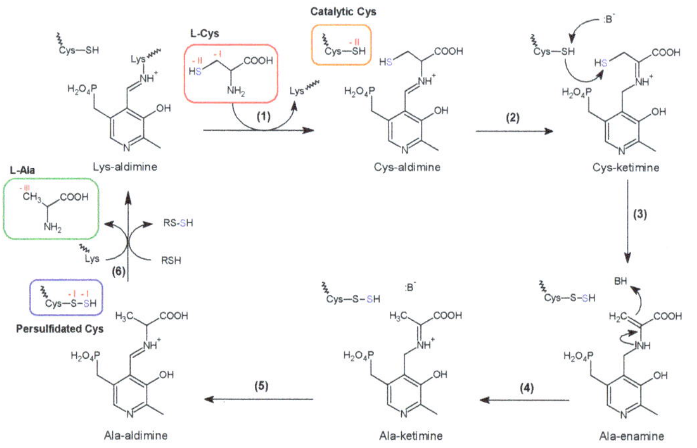

Figure 4. Mechanism of persulfide formation by cysteine desulfurases. The scheme presents the main intermediates in the catalytic cycle of cysteine desulfurases. The PLP cofactor is bound to the cysteine desulfurase backbone via the nitrogen of a lysine residue, in the form of a Lys-aldimine. The L-cysteine substrate exchanges with the lysine to form a Cys-aldimine (**1**), which is rearranged into a Cys-ketimine (**2**). A base (:B^-) deprotonates the thiol of the catalytic cysteine of the cysteine desulfurase and the resulting thiolate reacts with the thiol of the Cys-ketimine intermediate, which leads to formation of a persulfide on the catalytic cysteine (**3**). Concommitantly, adouble bond is created with the carbon of the former L-cysteine substrate in the form of an Ala-enamine (**3**). Upon protonation, the Ala-enamine is converted into Ala-ketimine (**4**) and then Ala-aldimine (**5**). The sulfane sulfur (blue) of the persulfide is transferred to a thiol acceptor and L-alanine is released by exchange with the internal lysine residue (**6**).

The formal oxidation states of the relevant atoms of the reaction (sulfur and adjacent carbons) are indicated in red for the starting and final material. The reaction starts with two sulfurs at the -II oxidation state in L-cysteine and the catalytic cysteine with a carbon at -I in L-cysteine to end up with two sulfurs at the -I oxidation state in the persulfidated catalytic cysteine and a carbon at -III in L-alanine.

3.2.3. Sulfur Transfer to the IscU/ISCU Scaffold, a Metal-Driven Process

The first studies on the process of sulfur supply to IscU/ISCU were reported almost 20 years ago, with *A. Vinelandii* and *E. coli* IscS and IscU [154–156]. These studies showed that IscS transfers several persulfides to IscU. A sulfur-first model was then proposed, in which the persulfides are transferred and later on reduced upon binding of Fe^{2+} ions. Surprisingly, several persulfides (2 to 4) accumulated on IscU, which was difficult to rationalize with the requirement of only two sulfurs for the formation of a [2Fe2S] cluster [154,156]. Another study with *T. maritima* IscU mentioned that the persulfides transferred to IscU were not reducible by iron, unless iron was already bound to IscU prior to persulfide transfer [157]. Although the experimental data were not shown, an iron-first model was proposed, in which the iron binds first then a persulfide is transferred and reduced by iron. Thereby, controversies on the number of persulfide transferred to IscU/ISCU and the sequence of iron and sulfur insertions have emerged with two opposite models: sulfur-first and iron-first. It is now becoming clear that iron comes first and triggers the transfer of a persulfide (Figure 5) [23,89].

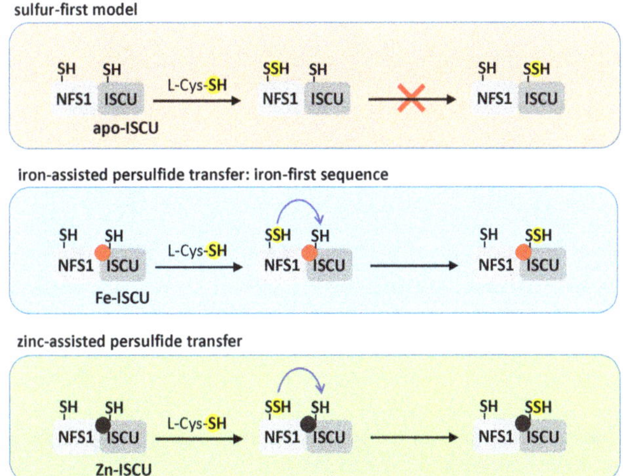

Figure 5. Metal-dependent persulfide transfer in the NFS1–ISCU complex. The sequence corresponding to the sulfur-first and iron-first models are depicted in the top two panels. First, a persulfide is formed on the catalytic cysteine of NFS1 (Cys381). Then, in the absence of a metal in ISCU (apo-ISCU), the persulfide of NFS1 is not transferred to ISCU (as shown by the red cross), which invalidates the sulfur-first model. In the presence of iron (red sphere, Fe–ISCU), the persulfide of NFS1 is transferred to the cysteine Cys104 of ISCU, which highlights the iron-first model. The third panel shows that, when zinc (dark gray sphere, Zn–ISCU) is present in the assembly site, instead of iron, it also triggers persulfide transfer. The blue arrows refer to movement of sulfur in trans-persulfuration reactions.

Persulfide transfer was also reported with the mouse and human NFS1–ISD11–ACP–ISCU complex [88,158]. Our analysis using mass spectrometry and an alkylation assay developed to directly visualize and quantify protein-bound persulfides, has established that a single persulfide is transferred to mouse ISCU [88]. The cysteine Cys104 was identified by mass spectrometry as the final persulfide receptor [88]. Furthermore, the transfer of persulfide to ISCU was correlated with Fe–S cluster assembly thus indicating that it is a relevant step in this process (see next section) [23]. The accumulation of more than one persulfide was also observed with the mouse system but at longer time course and was not correlated with Fe–S cluster assembly, which may explain the transfer of multiple persulfides previously observed with *A. Vinelandii* and *E. coli* IscU proteins [23,88].

Interestingly, with the mouse system it was shown that in the absence of iron, the transfer of persulfide is abolished, which points to a metal-dependent process [23]. A similar iron-dependent persulfide transfer process was recently reported with the *E. coli* system [78]. Thus, the presence of iron is required to allow the transfer of persulfide, which experimentally confirms the iron-first model that was earlier proposed and discards the sulfur-first hypothesis (Figure 5). Surprisingly, persulfide transfer to ISCU is also enabled when zinc is in the assembly site, which is probably related to the identical coordination mode of iron and zinc in ISCU (Figure 5). This may indicate a physiological role for zinc as proposed for the zinc ion of SufU, a closely related homolog of IscU/ISCU (see Section 5.3.3) [159]. Moreover, the iron-like effect of zinc may explain the initial controversy in the models of iron and sulfur insertion sequences. Since zinc was likely present in the preparations of IscU, although its presence was not assessed [154–156], it has probably allowed persulfide transfer in the absence of iron and misled the authors to the conclusion that sulfur comes first.

Trans-persulfuration reactions are common to several sulfur insertion processes but none of them have been described as iron-dependent [39,40]. Indeed, this feature is likely specific to Fe–S cluster assembly systems as it coordinates sulfur supply with the presence of iron in ISCU. This raises the question of the mechanism of this new type of reaction. Since both, iron and zinc, enable persulfide transfer, the examination of the structures of the NFS1–ISCU complexes solved with zinc may help elucidate this mechanism. The X-ray structure of the human NFS1–ISD11–ACP–ISCU complex shows that the catalytic cysteine of NFS1 binds to the zinc center through exchange with Cys35 of ISCU (Figure 2B) [67]. Thus, the metal ion may play a structural role by positioning the catalytic cysteine of NFS1 at a short distance of the receptor cysteine (Cys104) of ISCU and additionally may play the role of a Lewis acid catalyst by providing an electrophilic character to the sulfane sulfur of the persulfide to promote the nucleophilic attack by the receptor cysteine on the persulfide. However, Cys104 is shielded by the metal ion and its surrounding ligands (Asp37, Cys61 and His103), which seems to preclude a direct transfer of persulfide to this residue. Persulfide transfer could thus be indirect via Cys35 or Cys61 as relays. However, a mutation of the cysteine Cys104 to serine does not lead to accumulation of a persulfide on Cys35 or Cys61 as expected if they were functioning as relays, which tends to invalidate this hypothesis [23,88]. Therefore, in keeping with the structural plasticity of ISCU discussed in the previous section, a structural rearrangement uncovering Cys104 to facilitate persulfide transfer seems more likely.

In conclusion, the detailed study of the persulfide transfer process provides the first compelling evidence of an iron-first mechanism in which persulfide transfer is triggered by the presence of iron in IscU/ISCU [23,78]. This represents the first encounter of sulfur with iron in the Fe–S cluster assembly process.

3.3. Step 3: Persulfide Reduction

3.3.1. Reduction by Iron

To generate the sulfide ions that are the constituents of Fe–S clusters, the persulfide must be reduced into sulfide. As discussed in the previous section, Fe^{2+} ions were initially proposed to be the reductant of the persulfide thereby yielding Fe^{3+} and sulfide ions, to generate an oxidized $[2Fe2S]^{2+}$ cluster [154,155,157]. However, no reduction was observed with the mouse system after transfer of persulfide to ferrous iron-loaded ISCU [23]. Furthermore, the reduction of a persulfide by Fe^{2+} is chemically irrelevant since this reaction requires two electrons and Fe^{2+} provides only one electron by switching to the Fe^{3+} state.

3.3.2. Reduction by Thiols

Thiols, essentially dithiothreitol (DTT), but also GSH and cysteine were extensively used as reductants in Fe–S cluster reconstitution studies [61,78,158,160–164]. However, their mode of action was not thoroughly investigated until recently. Using the mouse ISC system, our group reported that thiols are able to release sulfide by direct reduction of the persulfide of NFS1, but are poorly efficient to reduce the persulfide transferred to ISCU, probably due to the presence of the metal that shields the receptor cysteine Cys104 [88]. The reduction of the persulfide of NFS1 by thiols occurs in reaction performed with excess of L-cysteine relative to the NFS1–ISCU complex (Figure 6A). Under these conditions, a persulfide is formed on NFS1 that is transferred to ISCU, then a new persulfide is formed on NFS1 that cannot be transferred to ISCU since it is already persulfidated, which leads to a complex persulfidated on both, NFS1 and ISCU. Since thiols react faster with NFS1 than with ISCU, this leads to preferential reduction of the persulfide sitting on NFS1. Thereby, the prominent source of sulfide ions in thiol-based reconstitution is NFS1 not ISCU. This reduction proceeds in two steps: (1) trans-persulfuration of the thiol leading to a persulfidated thiol (RSSH) and (2) cleavage of the S–S bond of the persulfidated thiol by another thiol molecule (Figure 6A). The sulfide ion then combined with iron to form a Fe–S cluster. The thiol-based reactions indiscriminately lead to formation of both [2Fe2S] and [4Fe4S] clusters with variable proportions [61,162,164–167]. DTT further enhances the proportion of [4Fe4S] cluster versus [2Fe2S] and also generates high-molecular-weight (HMW) Fe–S minerals [167]. This was attributed to the ability of DTT to stimulate the rate of sulfide release more strongly than L-cysteine, thereby creating a bulk of sulfide that reacts with [2Fe2S] clusters to generate [4Fe4S] clusters and eventually forms HMW Fe–S minerals. Surprisingly, the thiol-based reconstitution is also operative when zinc is present in the assembly site of ISCU, but is much slower probably due to hindrance of Fe–S cluster formation by the zinc ion [23].

Such thiol-based Fe–S cluster assembly reactions rely on "free" sulfide ions and their production is not coupled with the presence of iron. This type of reaction is probably inefficient in vivo for Fe–S cluster assembly as the sulfide ions will diffuse outside of the ISC complex and the probability of iron and sulfur encounter will be very low. Therefore, thiol-based reactions are not considered as physiologically relevant for Fe–S cluster assembly.

3.3.3. Ferredoxin and Iron Mediated Reduction of Persulfides

The [2Fe2S] ferredoxin Fdx/FDX2 was also proposed to function as a reductant of persulfides. The deletion of the gene encoding Fdx in *E. coli* and the depletion of Yah1 in the yeast *Saccharomyces cerevisiae*, both led to impaired Fe–S cluster biosynthesis [168–171]. Moreover, *yah1* is an essential gene in yeast, which points to a critical role of ferredoxins [169]. In human, two ferredoxins are present in mitochondria, the adrenodoxin (FDX1), and a closely related homolog named FDX2 (or FDX1-like) [172]. FDX1 is mainly expressed in adrenal glands where it is involved in steroid biosynthesis, while FDX2 is ubiquitously expressed and is essential for Fe–S cluster biogenesis in mitochondria [172]. Another study reported that both FDX1 and FDX2 function in Fe–S cluster biosynthesis [173]. However, the tissue specificity of FDX1 to adrenal glands most likely dictates its role. Cognate reductase partners were identified in *E. coli* (the ferredoxin–NADP reductase Fenr), human (FdxR) and yeast (adrenodoxin reductase Arh1) that provides electrons from NAD(P)H to Fdx/FDX2 (Table 1) [169,173–175].

Fe–S cluster reconstitution assays showed that *E. coli*, yeast and mouse ferredoxin/ferredoxin reductase systems are able to replace DTT [23,24,176]. NMR studies with mouse and bacterial proteins reported that Fdx binds IscS [177–179]. It was thus proposed that Fdx reduces the persulfide of IscS into sulfide. However, this mechanism is reminiscent of the thiol-based reconstitution in which sulfide is released irrespective of the presence of iron. Moreover, IscU was reported to displace Fdx, thereby preventing synchronization of sulfide production with the presence of IscU to uptake these sulfide ions [178]. In contrast, NMR and SAXS studies of the eukaryotic *Chaetomium thermophilum* ISC complex showed that FDX interacts with both NFS1 and ISCU [24,67]. The first experiments directly testing the effect of ferredoxin on the persulfide of NFS1 and ISCU were carried out with the mouse system by tracking the persulfide during the reaction, using mass spectrometry and alkylation assays [23]. The step-by-step analysis revealed that FDX2 selectively reduces the persulfide transferred to Fe–ISCU but does not reduce the persulfide of NFS1 (Figure 6B). Importantly, this reduction step leads to formation of a [2Fe2S] cluster in ISCU, thus indicating that it is a critical step of the assembly process. Titrations have further indicated that this reaction is highly efficient with about 90% of L-cysteine converted into a [2Fe2S] cluster whereas thiol-based reactions are poorly efficient with only 5% of L-cysteine actually incorporated as sulfide ions in the Fe–S cluster. Moreover, the FDX2-based reaction is about 100 times faster than the thiol-based one.

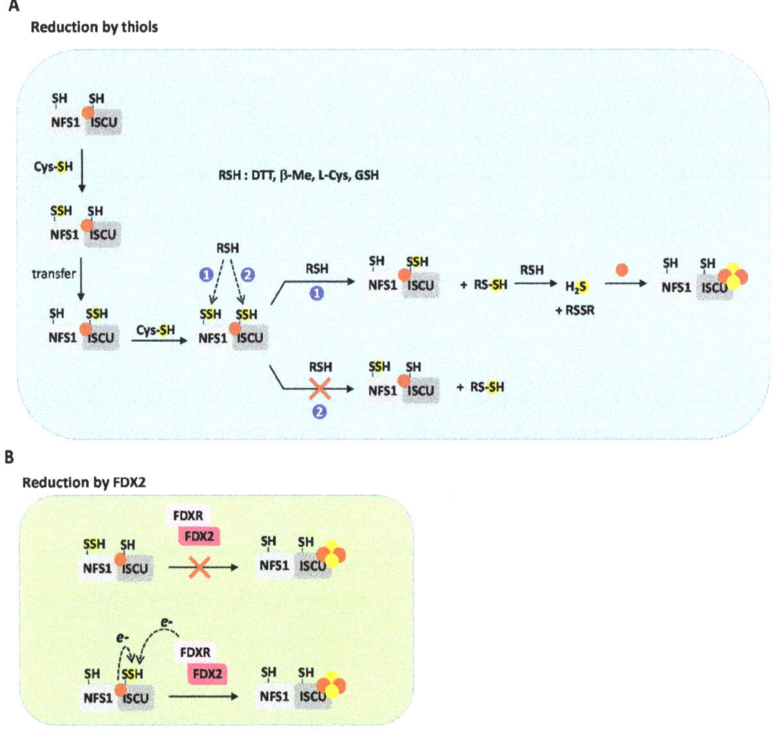

Figure 6. Selective reduction of the persulfide of NFS1 and ISCU by thiols and FDX2. (**A**) Thiols (RSH) such as dithiothreitol (DTT), L-cysteine, β-mercaptoethanol and GSH react with the persulfide of NFS1. Once a persulfide is formed it is transferred to ISCU, then a second persulfide is formed on NFS1. Thiols reduce the persulfide of NFS1 (**reaction 1**), which generates free persulfidated thiol (RSSH) that reacts with a second molecule of thiol to liberate sulfide ions in solution. Sulfide exists as H_2S, HS^- and S^{2-} depending on the pH, with HS^- the predominant form at physiological pH (pH ≈ 8 in mitochondria). The sulfide ions react with iron to generate [2Fe2S] and [4Fe4S] clusters in ISCU. In contrast, the persulfide of ISCU is poorly reactive with thiols (**reaction 2**, the red cross indicates that the reaction does not occur), most likely due to the presence of the metal ion that shields Cys104. Thereby, sulfide ions originate mainly from reduction of NFS1. The dashed black arrows describe the reaction of the thiols on the persulfide of either NFS1 or ISCU (**B**) FDX2 does not reduce the persulfide of NFS1 (top, the red cross indicates that the reaction does not occur), but selectively reduces the persulfide of ISCU when iron is present in the assembly site (bottom). A mechanism is proposed in which FDX2 provides one electron and the ferrous iron of ISCU the second one to reduce the S–S bond of the persulfide. The dashed black arrows show the origins of these two electrons.

Interestingly, the reduction of the persulfide of ISCU by FDX2 is abolished in the absence of metal and when zinc is present instead of iron in ISCU, which points to a specific role of iron in the reduction process. A mechanistic rationale could be that the iron ion participates as a redox partner. The iron ion is initially ferrous and ends-up as ferric in the $[2Fe2S]^{2+}$ cluster, thereby it might provide one of the two electrons needed to reduce the persulfide, the other one would be delivered by FDX2 (Figure 6B). Altogether, these data support the hypothesis that the reduction of the persulfide of ISCU is specifically coordinated with the presence of iron, which would ensure that the nascent sulfide is instantly trapped by the iron ion. In the SAXS-based model of the *C. thermophilum* ISC complex, the [2Fe2S] cluster of FDX is located nearby the assembly site of ISCU, thus in a suitable position to transfer electrons to the metal center for persulfide reduction [67]. Finally, the titrations of the number of persulfide reduced by FDX2 indicated that only one sulfide is produced per iron-loaded monomer of ISCU [23]. Hence, the most reasonable hypothesis is that the product of the reaction is a mononuclear iron–sulfide ([1Fe1S]) species. A number of non-heme mononuclear iron–sulfide inorganic complexes have been reported, thus providing credence to the formation of the same type of species in IscU/ISCU; however, a clear demonstration is still awaited [180–184].

In conclusion, the FDX2-driven reduction of the persulfide on iron-loaded ISCU appears as a concerted process that coordinates sulfide production with presence of iron, which concomitantly prevents escape of the sulfide ions. For these reasons, it has been proposed that the FDX2-based mechanism describes the physiological process of Fe–S cluster assembly [23]. This represents the second encounter of iron and sulfur in the Fe–S cluster assembly process. It is likely that a similar concerted mechanism occurs in the bacterial system, but it has not been demonstrated as of yet. Finally, these data highlight that thiols cannot be considered as surrogates of FDX2 in Fe–S cluster reconstitution assays, as they reduce the persulfide of NFS1, whereas FDX2 reduces the persulfide of ISCU; thus, the mechanisms of sulfide production are completely different. Furthermore, persulfide reduction by thiols is not coupled with the presence of iron.

3.4. Step 4: Fe–S Cluster Formation

The last step, after reduction of the persulfide, is the formation of the [2Fe2S] cluster. This requires formation of a dinuclear iron center with a di-μ-sulfido bridge. Since the titration of persulfides with the mouse system point to the formation of a mononuclear [1Fe1S] iron–sulfide complex, dimerization of IscU/ISCU is likely an obligatory step to form the dinuclear [2Fe2S] cluster (Figure 7). The dimerization would generate a bridged [2Fe2S] cluster, as observed with glutaredoxins that are involved in Fe–S cluster transfer [185–188]. In the reconstitutions performed with *C. thermophilum* Isu1, the final product was a [2Fe2S] cluster in a dimer of Isu1. However, this stoichiometry is also consistent with an asymmetric configuration in which only half of the subunits are occupied [24]. In contrast, a monomer of ISCU carrying one [2Fe2S] cluster has been identified by native mass spectrometry and Mössbauer spectroscopy as the final product in reconstitution assays performed with the mouse ISC system [23]. Therefore, if a bridged [2Fe2S] cluster is formed, it is a transient species that rapidly relocates on one of the subunits (Figure 7) [23]. Then, the dimer dissociates to release the monomer of ISCU holding a [2Fe2S] cluster.

An appealing hypothesis is that the dimeric structure of cysteine desulfurases assists the dimerization of IscU/ISCU within the complex. In the closed conformation of the human NFS1–ISD11–ACP–ISCU complex, the ISCU proteins are very distant (Figure 3A) [67,150], but in its open conformation, the predicted binding sites for ISCU are adjacent and the ISCU proteins are arranged along a C2 axis (Figure 3D) [151], thus in suitable positions for dimerization and formation of a dinuclear center. However, further tightening of the NFS1 subunits would be required to allow contact of the ISCU proteins, so it is not clear whether a dimer could actually form. In contrast, formation of dimers of holo-IscU/ISCU have been reported in several thiol-based reconstitutions [62,81,189,190]. Even a trimeric state has been crystalized with *Aquifex aeolicus* IscU, in which only one subunit harbors a [2Fe2S] cluster [64]. Fe–S reconstituted IscU proteins from *T. maritima* and *A. vinelandii* were shown to bind one [2Fe2S] cluster per dimer that could correspond to a bridged [2Fe2S] cluster [61,62]. However, the analysis by Mössbauer, electronic absorption and Raman spectroscopies led to the conclusion that the [2Fe2S] cluster in both species is coordinated in an asymmetric environment by the three cysteines Cys35, Cys61, Cys104 and most likely the aspartate Asp37. Since a bridged [2Fe2S] cluster is expected to be symmetric, this asymmetric arrangement indicates that the [2Fe2S] cluster is not bridged but hosted in one of the two subunits of the dimer. A similar signature was observed for the [2Fe2S] cluster in mouse ISCU, which is a key feature indicating that the [2Fe2S] is ligated by a monomer of ISCU [23].

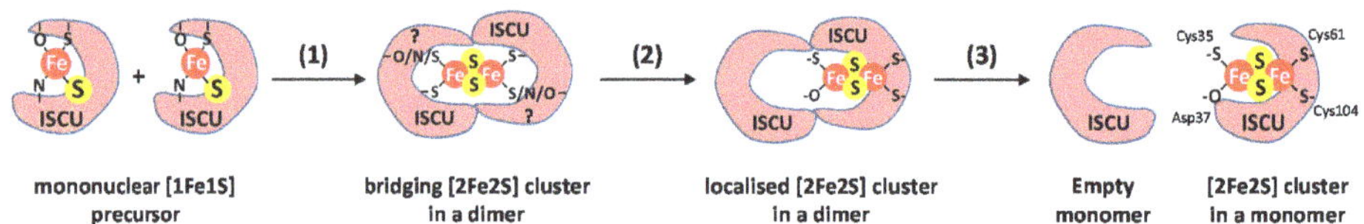

Figure 7. Hypothetic model for the formation of the [2Fe2S] cluster by dimerization of IscU/ISCU. The model for the formation of a [2Fe2S] cluster in Iscu/ISCU is decomposed in three steps: (**1**) the dimerization of two ISCUs coordinating a mononuclear [1Fe1S] intermediate generates a bridged [2Fe2S] cluster in a dimer of ISCU, (**2**) the [2Fe2S] segregates on one monomer and (**3**) the dimer dissociates to yield an empty monomer and a monomer containing the [2Fe2S] cluster. In the monomer, the [2Fe2S] cluster is coordinated by Cys35, Asp37, Cys61 and Cys104 of ISCU.

Interestingly, the kinetic analysis of the thiol-based Fe–S cluster assembly process with *A. Vinelandii* IscU showed that the dimer loaded with one [2Fe2S] cluster is a precursor of the dimer loaded with two [2Fe2S] clusters [61]. Therefore, the second [2Fe2S] cluster seems to be assembled by gathering of two iron and two sulfide ions in the empty subunit of the dimer loaded with one [2Fe2S] cluster, thus without requirement for a bridged [2Fe2S] cluster. A similar mechanism was recently proposed for GSH-based Fe–S cluster assembly with *E. coli* IscU [78]. It thus seems that the non-physiological DTT-based route proceeds via sequential accumulation of iron and sulfide ions on a single subunit of dimeric IscU/ISCU while the physiological FDX2-based route involves formation of a still ill-defined transient bridged [2Fe2S] cluster.

In conclusion, although a bridged [2Fe2S] cluster has not been isolated, the ability of IscU/ISCU to dimerize provides credence to a mechanism based on merging of two mononuclear [1Fe1S] precursors. The conformational flexibility of the NFS1–ISD11–ACP complex further raises the possibility that NFS1 could promote dimerization of ISCU, but it is unclear whether bacterial IscS are also capable of such a structural change.

3.5. Regulation of Fe–S Cluster Assembly by Frataxin

3.5.1. Effect of Frataxin on Cysteine Desulfurase Activity: The Key Question of the Rate-Limiting Step

The Pastore group reported that the bacterial frataxin homolog, CyaY, modulates the activity of IscS in the IscS–IscU complex [160]. CyaY decreases the rate of L-alanine production, the by-product of the desulfurization process and this was correlated with a concomitant decrease in the rate of Fe–S cluster production by the IscS–IscU complex in a DTT-based reconstitution. Moreover, the inhibitory effect of CyaY was more pronounced in the presence of iron. Thereby, a function as a gate-keeper preventing uncontrolled production of sulfide ions in the absence of iron, was proposed. In contrast, in DTT-based reconstitutions with the human and mouse ISC system, the eukaryotic frataxin protein (FXN) stimulates the cysteine desulfurase activity of NFS1 [88,164,191]. Moreover, it was shown that the stimulatory effect of FXN is only seen in the presence of ISCU [88,164,191]. These data thus indicate that frataxin affects the activity of cysteine desulfurases in the IscS/NFS1–IscU/ISCU complex, but in different ways depending on the organism.

Several groups have thus studied the cysteine desulfurase activity in more details to determine the kinetic parameters of this reaction [88,158,162–164,191–194]. Most of these assays were conducted in multiple turnover kinetics by monitoring the rates of sulfide or L-alanine productions in the presence of DTT to reduce the persulfide of NFS1 (Figure 8A) [88,158,162–164,191–194]. In multiple turnover kinetics, the rate-limiting step (i.e., the slowest step) dictates the global rate of the reaction. As the cysteine desulfurase assay encompasses two main reactions: persulfide formation (Figure 8A, reaction 1) and persulfide reduction (Figure 8A, reaction 2), either of these reactions could be rate-limiting. In some studies, saturation behaviors were apparently observed upon increase of L-cysteine, which was interpreted as formation of a complex between L-cysteine and NFS1 for persulfide formation (Figure 8A, reaction 1) [158,162–164,191–193]. Since L-cysteine was thought to be involved only at the first step, it was assumed that persulfide formation is the rate-limiting step. Thereby, the effects of FXN and CyaY were interpreted as a modulation of the rate of persulfide formation. The Michaelis–Menten formalism was applied and K_M and k_{cat} values were reported.

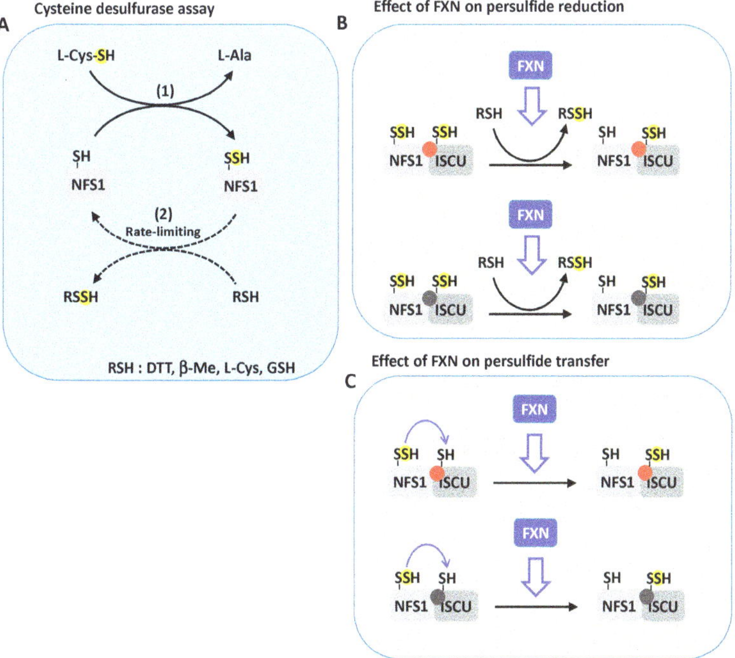

Figure 8. Effect of FXN on persulfide reduction and transfer. (**A**) The cysteine desulfurase assay encompasses two reactions: (1) formation of the persulfide and (2) its reduction by thiols (RSH). Reduction by thiols is the slowest step (rate-limiting step, dashed black arrows). (**B**) FXN accelerates the reduction of the persulfide of NFS1 by thiols in the NFS1–ISCU complex with ISCU metallated by iron (red ball) or zinc (gray ball). (**C**) FXN accelerates the transfer of the persulfide of NFS1 to ISCU in the NFS1–ISCU complex with ISCU metallated by iron (red ball) or zinc (gray ball). The solid blue arrow describes the transfer of persulfide. The blue arrows filled in white refer to acceleration by FXN.

However, a clear demonstration that persulfide formation is rate-limiting in those assays was not provided. Two kinetic studies have addressed the question of the rate-limiting step, with mouse NFS1 and *Synechocystis* IscS [88,195]. In both cases, the rate of L-alanine production was found proportional to the concentration of DTT and additionally to TCEP and β-mercaptoethanol in the case of *Synechocystis* IscS. Since DTT, TCEP and β-mercaptoethanol are not involved in the formation of the persulfide but in its reduction, this indicates that the reduction is the rate-limiting step. Second, it was shown that

L-cysteine is also able to reduce the persulfide of NFS1, thus that the saturation behaviors previously observed could also be interpreted as formation of a complex between L-cysteine and persulfidated NFS1 at the second step (Figure 8A, reaction 2) [88].

A definitive proof that reduction is the rate-limiting step was provided by monitoring persulfide formation and reduction independently in single turnover kinetics with the mouse NFS1–ISD11–ACP complex, using an alkylation assay to quantify protein bound persulfides [88]. Our group showed that persulfide formation is much faster than its reduction by DTT, hence that persulfide reduction, not its formation, is rate-limiting. Moreover, FXN was shown to directly accelerate persulfide reduction by DTT and to the same order of magnitude as its stimulatory effect on the global cysteine desulfurase activity, which further strengthens that persulfide reduction is the rate-limiting step (Figure 8B). Consequently, the ability of FXN to modulate the cysteine desulfurase activity of NFS1 indicates that it facilitates the reduction of the persulfide of NFS1 by thiols (Figure 8A reaction 2, Figure 8B). Even though other studies reported that FXN modulates the rate of persulfide formation, as this reaction is not rate-limiting, this effect of FXN is not expected to impact the cysteine desulfurase activity of NFS1 [89,194].

Moreover, no saturation behavior was observed with mouse NFS1 at low L-cysteine concentrations that could be consistent with the K_M and k_{cat} values reported in other studies [88]. Saturation curves were observed but at high, non-physiological, L-cysteine concentrations (above 50 mM). This was attributed to inhibitory effects by L-cysteine and/or DTT in other systems [43,195]. Anomalous behaviors were also reported for IscS and NifS from *E. coli* and *A. vinelandii*, which precluded determination of the kinetic parameters using the Michaelis–Menten formalism [41,43,196,197]. Altogether, this indicates that persulfide reduction does not proceed through formation of an enzyme-substrate complex but is a bimolecular reaction and that determination of K_M and k_{cat} for L-cysteine is not applicable [88].

Similar extensive studies have not been carried out yet with the *E. coli* system. However, the analysis of *Synechocystis* IscS affords evidence that persulfide reduction is also rate-limiting with a bacterial protein, which suggests that this could be a common feature of class I cysteine desulfurases. This would indicate that CyaY and FXN have opposite effects. However, this differential effect is rather due to the nature of the cysteine desulfurase, since exchanging IscS for the NFS1–ISD11–ACP complex turns CyaY into an activator [163]. Altogether, these data indicate that FXN, and probably CyaY too, modulates the rate of persulfide reduction by DTT and other thiols. However, as discussed in the previous sections, the thiol-based reaction is not physiologically relevant, which means that the effects of FXN and CyaY on the reduction of the persulfide of NFS1 by thiols are unrelated to Fe–S cluster biosynthesis in vivo.

3.5.2. Effect of Frataxin on Persulfide Transfer

Strikingly, assays of persulfide transfer and ^{35}S labeling experiments showed that mouse and human FXN have a strong impact on persulfide transfer to ISCU (Figure 8C) [23,88,158]. Mouse FXN was shown to enhance the rate of this reaction with both Zn–ISCU and Fe–ISCU (Figure 8C) [23,88]. The acceleration with Fe–ISCU was directly correlated to its stimulatory effect on Fe–S cluster assembly in FDX2-based reconstitution assays since persulfide transfer was identified as the rate-limiting step of the whole process [23]. Therefore, the sole function of mammalian FXN that appears relevant to the stimulation of the FDX2-based Fe–S cluster biosynthesis is the stimulation of persulfide transfer.

These data raise the question of the mechanism by which FXN stimulates persulfide transfer. Interestingly, the acceleration of persulfide transfer by FXN is metal-dependent as is persulfide transfer itself, with zinc and iron equally efficient [23]. This strengthens the idea that FXN accelerates the reaction, but does not modify its overall mechanism. The structure of the pentameric NFS1–ISD11–ACP–ISCU–FXN complex with zinc in the assembly site shows that FXN induces a rearrangement of the coordination sphere of the zinc ion (Figure 2C) [73]. The asparagine N141 and the Trp155–Arg165 pair of FXN tears out Cys35 and His103 from the metal ion through specific interactions. The movements induced by FXN thus seems to open the access to Cys104 by pushing aside Cys35 and

His103. Moreover, Cys104 comes closer to the metal ion as His103 is pulled out. Altogether, this may facilitate direct persulfide transfer to Cys104. However, FXN also interacts with the flexible loop of NFS1 and stabilizes the catalytic Cys381 halfway between the PLP and ISCU, which is difficult to rationalize with the stimulation of persulfide transfer and even persulfide formation on NFS1 [73]. However, this position should correspond to an intermediate state since the flexible loop needs to move back to the PLP pocket for persulfide loading.

Another important feature of the stimulation provided by FXN is that it does not modify the nature of the Fe–S cluster formed, i.e., a [2Fe2S] cluster [23]. This behavior fits the notion of a kinetic activator or an co-enzyme rather than an allosteric modulator as earlier proposed [162]. The term "allos" indeed comes from the ancient Greek "ἄλλος", meaning "other", and, thus, "allostery" applies to regulators that do not bind to the active site of the enzyme but at another site to modulate its activity, while FXN actually binds to the active site of the ISCU protein.

3.6. Toward a Model of [2Fe2S] Cluster Biosynthesis by the ISC Machinery

Altogether, the current data provide a more complete picture of the mechanism of Fe–S cluster assembly by the ISC machinery. Figures 9–11 depict the models based on these data. Since most mechanistic information originates from the eukaryotic systems, essentially the mouse and human ones, the mammalian nomenclature is used here to describe the mechanisms. Nonetheless, several features of the bacterial machineries suggest that the mechanism of Fe–S cluster assembly is conserved across species. A first striking conclusion is that Fe–S clusters can be assembled via two different routes: the FDX2-based reaction that is physiologically relevant (Figures 9 and 10) and the thiol-based one that is also productive of Fe–S clusters, but only in vitro and is thus not considered as physiologically relevant (Figure 11) [23]. The second breakthrough is the elucidation of the first steps of the FDX2-based reaction leading to production of the sulfide ions and its encounter with iron. The next steps are more elusive. The formation of a dinuclear [2Fe2S] cluster is supposed to rely on dimerization of ISCU and two mechanisms are postulated: either ISCU dimerizes within the complex with assistance from NFS1 (Figure 9) or the NFS1–ISCU complex dissociates and ISCU auto-dimerizes (Figure 10).

The first steps of the FDX2-based reaction are common to both hypothesis (Figures 9 and 10). The reaction is initiated by insertion of a ferrous iron in the assembly site of ISCU, which may require a still ill-defined metallochaperone to remove the zinc ion. It is unclear whether zinc removal and iron insertion in ISCU occur in the NFS1–ISCU complex or in free ISCU, or both. The iron site of ISCU is mononuclear and iron is stable as a high spin Fe(II) center. Upon formation of a complex with the cysteine desulfurase NFS1, a persulfide is transferred to the Cys104 of ISCU, in an iron dependent manner. This iron-assisted reaction allows synchronization of sulfur supply with the presence of iron. This mechanism is referred to the iron-first model since iron comes first then sulfur. Then, FDX2 reduces the persulfide bound to ISCU. The reduction is also iron-dependent, which probably ensures that the production of the sulfide ion is coordinated with the presence of iron, thereby allowing the iron center to trap the nascent sulfide ion.

The reduction by FDX2 would lead to a mononuclear [1Fe1S] species and the dimerization of ISCU would enable formation of a bridged dinuclear [2Fe2S] cluster, either within the NFS1–ISCU complex (Figure 9, internal dimerization) or as a free protein upon dissociation of the NFS1–ISCU complex (Figure 10, external dimerization). In both cases, the transient inter-subunit [2Fe2S] cluster would quickly segregate on one monomer and the chaperone/co-chaperone system HSPA9/HSC20 would enable the transfer of the [2Fe2S] cluster to client proteins. A last important feature of the FDX2-based process is that the rate of Fe–S cluster biosynthesis is controlled by frataxin that accelerates persulfide transfer, the rate-limiting step of the whole process.

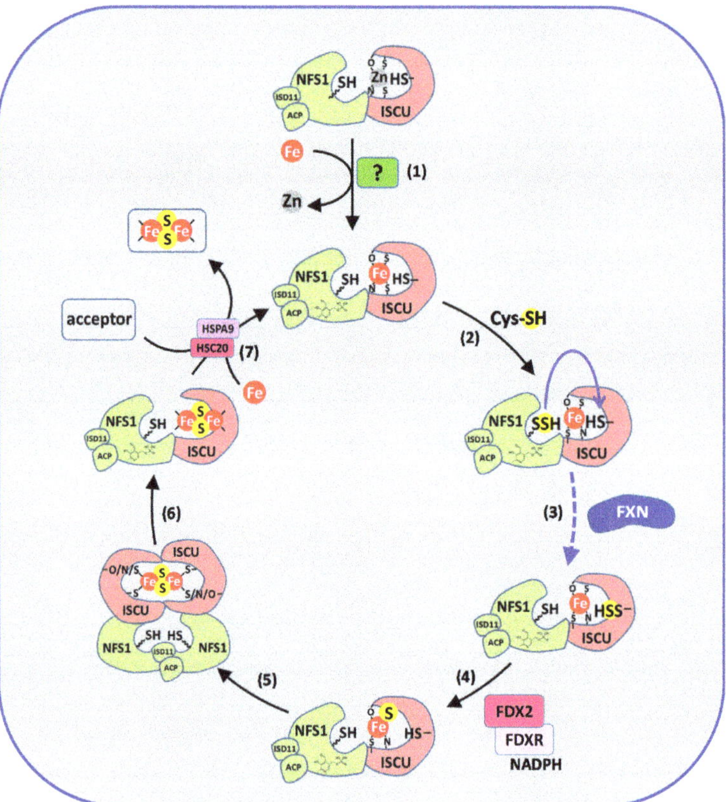

Figure 9. Model of FDX2-based Fe–S cluster biosynthesis by the eukaryotic ISC machinery relying on NFS1-assisted dimerization of ISCU. The reaction is initiated by insertion of iron in apo-ISCU, which may require a still ill-defined metallochaperone to remove the zinc ion (**1**). In the complex formed with NFS1–ISD11–ACP, a persulfide is formed on NFS1 (**2**) and transferred to ISCU (**3**). The persulfide of ISCU is reduced by FDX2, possibly into a mononuclear iron–sulfide intermediate (**4**). Then NFS1 in its open conformation enables the dimerization of ISCU, which leads to formation of a bridged [2Fe2S] cluster in ISCU (**5**). The bridged [2Fe2S] cluster segregates on one subunit (**6**) and is transferred to client proteins by the HSC20/HSPA9 chaperone system while a ferrous iron is inserted into apo-ISCU (**7**). FXN stimulates the whole reaction by accelerating persulfide transfer to ISCU (**3**). The solid blue arrow describes sulfur transfer to ISCU and the dashed blue arrow refers to the rate-limiting step that is accelerated by FXN.

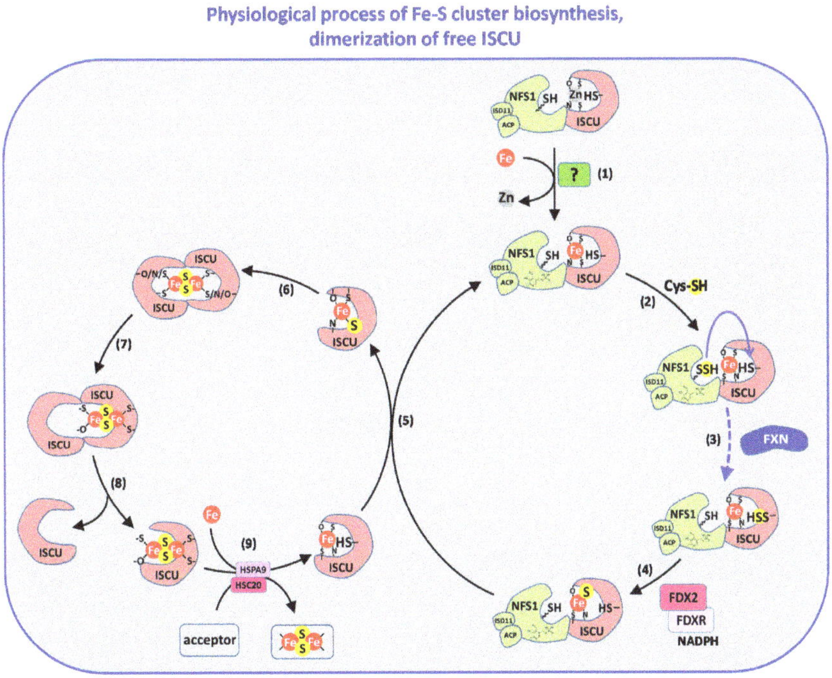

Figure 10. Model of FDX2-based Fe–S cluster biosynthesis by the eukaryotic ISC machinery relying on dimerization of free ISCU. Steps (**1**)–(**4**) are common to the mechanism described in Figure 9 until formation of the putative mononuclear iron–sulfide intermediate ([Fe–S]). Then [1Fe1S]-loaded ISCU dissociates from NFS1 either by exchange with Fe-loaded ISCU or spontaneously (**5**), which allows dimerization of ISCU and formation of a bridged [2Fe2S] cluster (**6**). The bridged [2Fe2S] cluster segregates on one subunit (**7**) and the dimer dissociates (**8**). The monomer of [2Fe2S]-loaded ISCU is recognized by the HSC20/HSPA9 chaperone system which enables the transfer of the [2Fe2S] cluster to client proteins while a ferrous iron is inserted into apo-ISCU (**9**). FXN stimulates the whole reaction by accelerating persulfide transfer to ISCU (**3**). The solid blue arrow describes sulfur transfer to ISCU and the dashed blue arrow refers to the rate-limiting step that is accelerated by FXN.

A distinct reaction occurs when thiols are used instead of Fdx/FDX2 as a reducing agent (Figure 11). Thiols such as DTT, GSH and L-cysteine reduce the persulfide of IscS/NFS1, which leads to release of sulfide ions in solution. The sulfide ions then combine with iron, either in solution or in the assembly site, to form [2Fe2S] and [4Fe4S] clusters. This process is not considered as physiologically relevant since production of sulfide is not confined to IscU/ISCU and is not synchronized with the presence of iron. In a biological context, the sulfide ions will diffuse outside of the ISC complex and the likelihood of iron and sulfur encounter will be very low.

However, the thiol-based reduction may compete with sulfur delivery to IscS/ISCU and thus could impede Fe–S cluster assembly. Moreover, in eukaryotes, FXN accelerates this reaction. Since the rate of the thiol-based reduction is proportional to the concentration of the thiol, the turning point at which this reaction starts to compete with Fe–S cluster assembly relies on its concentration. The relevant thiols that should be considered are the most abundant ones in the cells, i.e., GSH (5–10 mM) and L-cysteine (≈0.1 mM) [198–200]. Based on the kinetic constants determined for mouse NFS1 with GSH and L-cysteine, the thiol-based reduction would compete with transfer at concentrations of GSH and L-cysteine of 1 M and 15 mM, respectively, thus far from their physiological concentrations [88]. It is thus unlikely that persulfide reduction could compete with persulfide transfer in vivo.

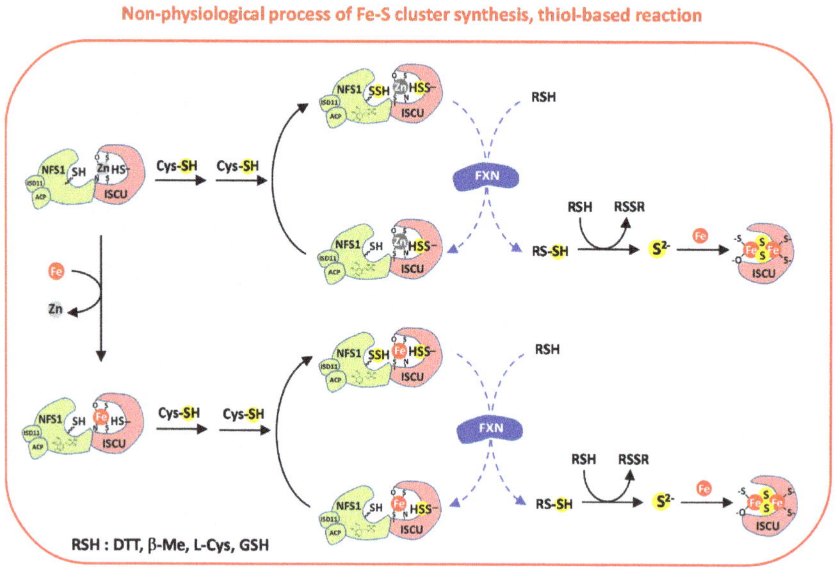

Figure 11. Thiol-based Fe–S cluster synthesis by the eukaryotic ISC machinery. The thiol-based reaction is described for the zinc- and iron-loaded ISCU. In both cases, the reaction is initiated by formation of an NFS1–ISD11–ACP–ISCU complex persulfidated on both NFS1 and ISCU, then thiols, (RSH) such as L-cysteine, DTT, GSH or β-mercaptoethanol, reduce the persulfide of NFS1, which leads to sulfide release. In the presence of iron, the sulfide ions generate Fe–S clusters in ISCU via a poorly defined mechanism. The dashed blue arrows refer to the rate-limiting steps that are accelerated by FXN.

4. Fe–S Assembly by the NIF Machinery

4.1. Iron Insertion in NifU

The NIF machinery comprises two main proteins, the scaffold protein NifU and the cysteine desulfurase NifS. NifU is a modular protein containing three distinct domains [36,41,201]. An N-terminal domain with very high homology to IscU/ISCU, a central domain harboring a permanent [2Fe2S] cluster homolog to ferredoxin and a C-terminal domain called NfU that has been proposed to function as a scaffold of [4Fe4S] clusters and a Fe–S cluster carrier.

The N-terminal domain contains all the well-conserved amino acids that bind the ferrous iron in IscU/ISCU: Cys35, Asp37, Cys61 and His103. It also contains the persulfide receptor Cys104 [23]. The N-terminal domain of NifU was shown to bind a mononuclear iron ion, most likely via Cys35, Cys61 and Cys104. However, in contrast to IscU/ISCU iron binds in the ferric state and Asp37 is not involved in its coordination [201]. Moreover, binding of Fe^{3+} to NifU is only observed below 2 °C, which questions its role in Fe–S cluster biosynthesis. In another study, time-course analysis by Mössbauer spectroscopy suggested that ferrous iron binds to NifU in a cysteine-rich site, which may correspond to the assembly site of the N-terminal IscU/ISCU domain [38]. Thereby an iron-first mechanism may also apply to NifU as reported for IscU/ISCU.

4.2. Still Elusive Steps Leading to Formation of the [2Fe2S] and [4Fe4S] Clusters

It is still unclear how NifU acquires sulfur. Although NifU and NifS form a complex, persulfide transfer to NifU has not been reported. Yet NifS-mediated Fe–S cluster reconstitutions in NifU were reported in DTT-based assays, which provide an idea of which domain binds Fe–S clusters. A truncated form of NifU containing only its N-terminal domain was shown to assemble a labile

[2Fe2S] cluster, as observed for IscU/ISCU [202]. Time-course experiments further suggested that this [2Fe2S] is converted into a [4Fe4S] cluster by reductive coupling [38]. However, as these reactions were performed with DTT as a reductant, it is still unclear whether assembly of [4Fe4S] clusters by the N-terminal domain is physiologically relevant. Moreover, no reductase has been identified in the NIF operon that could achieve the reduction of a persulfide transferred to NifU. An appealing hypothesis is that the permanent [2Fe2S] cluster of the central domain of NifU fills up this function, as described for FDX2. NifU may thus combine the functions of IscU/ISCU and Fdx/FDX2 within the same polypeptide for the biosynthesis of [2Fe2S] clusters.

Fe–S cluster reconstitution assays with the full-length NifU protein afforded evidence that the NfU domain binds [4Fe4S] clusters [38]. The formation of [4Fe4S] clusters is supposed to proceed via reductive coupling of two [2Fe2S] clusters and ferredoxin is able to catalyze this type of reaction [203]. As no reductase has been identified in the NIF operon, the permanent [2Fe2S] cluster of NifU may also fulfill this function. Thereby, the permanent [2Fe2S] cluster might play a bifunctional role, in persulfide reduction and in reductive coupling of the [2Fe2S] clusters. This process would require an external reductase to deliver electrons to the permanent [2Fe2S] cluster, but as for the ISC machinery, this reductase may not be encoded by the NIF operon. In analogy with NfU homologs that are involved in the assembly and/or transfer of [4Fe4S] clusters, the NfU domain of NifU might be the place for reductive coupling of the [2Fe2S] clusters leading to formation of the [4Fe4S] cluster [6].

5. Fe–S Assembly by the SUF Machinery

5.1. Overall Description of the SUF Machinery

There are several striking differences between the ISC and SUF machineries that are likely related to the higher resistance of the SUF machinery under stress conditions [26]. As the description of the SUF machinery has been extensively reviewed in other reports, the reader is referred to these reviews for more details [26,47,204]. We attempt, here, to focus on comparative elements between the ISC and SUF machineries.

Like the ISC machinery, the genes encoding the components of the SUF machinery are organized in operons [47]. The SUF system contains two to six genes depending on the organisms. The main components are SufS, the cysteine desulfurase providing sulfur and a scaffolding complex formed by the SufB, SufD and SufC proteins. Complexes containing various stoichiometries of the SufB, SufD and SufC proteins can be isolated by mixing the purified proteins in vitro, but the physiologically active complex is most likely a $SufBC_2D$ complex [26,205]. The crystal structure of the $SufBC_2D$ complex shows that SufB and SufD form a heterodimer with two SufC proteins bound to each subunit [206]. The site of Fe–S cluster assembly has been localized at the interface of the SufB–SufD heterodimer with Cys405, Glu434 or Glu432 and His433 from SufB and His360 from SufD as putative ligands of the Fe–S cluster [26]. SufC is an ATPase that drives structural changes to the SufB–SufD complex for proper Fe–S cluster assembly. The SUF operons also encode SufE or SufU, two sulfur relay proteins shuttling the persulfide generated by SufS to the scaffolding complex. SufE and SufU occur in a mutually exclusive manner in bacterial genomes and can complement each other, which indicates that they play very similar roles [47,207–210].

The persulfide transfer processes involving SufE or SufU have been particularly well studied. In contrast, it is still unclear how and where iron is inserted in the $SufBC_2D$ scaffold and whether or not persulfide supply is coupled with the presence of iron as observed with the ISC machinery. The mechanism of persulfide reduction is also still elusive.

5.2. Iron Insertion in the SufBC2D Scaffold

The few data available on the process of iron insertion in the $SufBC_2D$ complex suggest distinct mechanisms for the ISC and SUF machineries. Strains lacking SufC are impaired in the utilization of iron and a specific disruption of the ATPase activity of SufC lowers the iron content of the $SufBC_2D$

complex in vivo [211,212]. These data raise the possibility that SufC is important for iron acquisition by the SufBC_2D complex. An active process relying on the ATPase activity of SufC may provide a rationale to the higher ability of the SUF machinery to cope with iron-poor conditions, as compared to the ISC machinery for which a passive mode of iron insertion is postulated (see Section 3.1.2). Moreover, the SufBC_2D complex was proposed to function as a reductase of ferric iron [205,211]. The SufBC_2D complex isolated from *E. coli* co-purifies with one molecule of reduced FADH2 per complex. This FADH2 was shown to mobilize iron from ferric citrate and the ferric-loaded form of CyaY. This suggests that the SUF system is able to mobilize the poorly soluble ferric iron, which could provide a significant advantage under iron-poor conditions. However, the iron-binding site in the SufBC_2D complex has not been yet identified to test these hypotheses.

5.3. Sulfur Insertion in the SufBC_2D Complex

5.3.1. SufS, the Source of Sulfur

SufS belongs to the class II of the desulfurase family that are characterized by a shorter flexible loop carrying the catalytic cysteine [26]. The formation of the persulfide is also slower in class II than in class I, but accelerated by binding of SufU or SufE [26,213]. The reduction of the persulfide of SufS by thiols is apparently rate-limiting as for class I, which emphasizes that the persulfide is largely excluded from solvent [208,213–215]. Thereby, sulfur delivery is confined to the surrounding of the PLP-binding pocket and specific sulfur acceptors (SufE, SufU) are needed to mediate sulfur transfer to the SufBC_2D complex, most likely to ensure shielding of the persulfide.

5.3.2. Sulfur Transfer via SufE

SufE was shown to accept sulfur from SufS on its highly conserved cysteine residue, Cys51 [214,216]. The trans-persulfuration reaction relies on mutual changes in the structure of SufE and SufS [217–224]. The crystal structures of SufE show that the Cys51 is initially buried and protected from the solvent in a hydrophobic pocket [225,226]. Although there is still no tri-dimensional structure of the SufS–SufE complex, deuterium exchange experiments and comparison with the structure of the closely related homolog CsdA–CsdE complex have provided information on the mechanism of persulfide transfer [217–224]. Upon binding of CsdE to CsdA, the loop of CsdE carrying the receptor cysteine is twisted and comes in close proximity of the catalytic cysteine of CsdA, a similar remodeling is hypothesized for the SufS–SufE complex that would facilitate the nucleophilic attack by Cys51 of SufE. In addition, local changes in the SufS catalytic site and dimer interface may promote the desulfurase reaction and an outward-facing position of the persulfidated Cys364. This could provide a way to couple persulfide formation with its transfer. Upon dissociation of the SufS–SufE complex, the persulfide transferred to SufE is protected from the solvent most likely through a backward movement of the flexible to the hydrophobic cavity [225–227].

5.3.3. Sulfur Transfer via SufU

Due to its high homology with IscU/ISCU, SufU was initially proposed to function as a scaffold for Fe–S cluster assembly [209,228]. However, SufU cannot complement the lack of IscU/ISCU, which indicates that these proteins have different roles [159,207,210]. The main structural differences between SufU and IscU/ISCU are an insertion of 18–21 residues between the second and third cysteine residues and the lack of the well-conserved histidine H103 [23]. Since SufU accepts a persulfide from SufS, the current view is that SufU is a sulfurtransferase, as SufE, that conveys the persulfide from SufS to the SufBC_2D complex [208,228,229].

Interestingly, SufU hosts a tightly bound zinc ion that is essential for persulfide transfer [159,230,231]. The structure of SufU proteins revealed that the zinc ion is coordinated in a tetrahedral geometry by the three cysteines Cys41, Cys66, and Cys128 and the aspartate Asp43 (Figure 12A) [230,231]. These residues are equivalent to Cys35, Cys61, Cys104 and the aspartate Asp37 in IscU/ISCU. Upon formation of a complex with SufS, SufU and SufS undergo major structural changes (Figure 12B) [232,233]. The mobile loop of SufS carrying the catalytic cysteine Cys361 moves freely and the histidine residue His342 of SufS becomes a ligand of the zinc ion of SufU by swapping with the cysteine C41. Consequently, Cys41 is released and directed toward Cys361 of SufS. Soaking of the crystals of the SufS–SufU complex with L-cysteine leads to extra electronic densities at Cys361 and Cys41 consistent with cysteine persulfides, thus suggesting that persulfide transfer occurs between these residues [232].

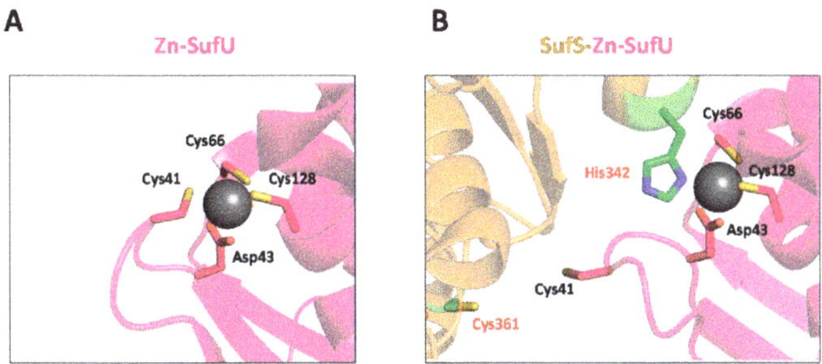

Figure 12. Structural rearrangement at the zinc site of SufU upon binding of SufS. Zoom on the zinc site of (**A**) *Bacillus subtilis* SufU (PDB 6JZV) and (**B**) *Bacillus subtilis* SufS–SufU complex (PDB 5XT5). SufU is colored in pink with key amino acids coordinating the zinc ion (Cys41, Asp43, Cys66 and Cys128) and SufS in orange with its catalytic cysteine Cys361 in green. The zinc ion is initially coordinated by Cys41, Asp43, Cys66 and Cys128 of SufU. Upon binding to SufS, His342 of SufS exchanges with Cys41, which comes close by Cys361, the catalytic cysteine of SufS.

This metal-dependent process is reminiscent of the persulfide transfer reaction reported for the NFS1–ISCU complex (Figure 2), but is mechanistically different. The main difference relates to the shorter length of the flexible loop carrying the catalytic cysteine in SufS that does not allow persulfide transfer to the distal cysteine (Cys128) but to a proximal one (Cys41) that is released upon binding of His342 to the zinc ion. This highlights two different ways of using a metal ion for persulfide transfer. In IscU/ISCU, the metal center anchors the persulfide of the desulfurase in close proximity of the receptor cysteine while in SufU, ligand exchange at the metal center expels the receptor cysteine and drives it close to the catalytic cysteine.

5.4. *Persulfide Transfer to SufBC$_2$D and Reduction via FADH$_2$?*

SufE interacts with the SufBC$_2$D complex and the cysteine Cys254 has been proposed as the primary acceptor in SufB [26,51,227,234]. Then this persulfide would be conveyed to the cysteine Cys405 of SufB through a hydrophobic tunnel involving non-cysteine amino acids as sulfur relays [26,206,234]. In contrast to the ISC machinery, there is, as of yet, no indication that iron is involved or required for persulfide transfer between SufE and SufB and within SufB. Moreover, the assembly site in the SufBC$_2$D complex is 20 A away from the entry point of sulfur on Cys254, it is thus still unclear how sulfur insertion could be coupled with presence of iron in the SufBC$_2$D complex. The sulfur transfer process form SufU to SufBC$_2$D has not been investigated yet, whether SufU transfers its persulfide to Cys254 as SufE is not known.

The SUF operon does not encode any obvious reductase that could play a role in persulfide reduction. Moreover, *E. coli* strains lacking Fdx that are thus defective in ISC-driven Fe–S cluster assembly, are viable. This indicates that the SUF machinery is active under these conditions and thus that Fdx is not the reductase of this machinery. Therefore, another reductase dedicated to the SUF operon should exist [168,170]. As mentioned above, the SufBC$_2$D complex isolated from *E. coli* co-purifies with one molecule of reduced FADH$_2$ per complex [205,211]. Although this cofactor is proposed to operate as a ferric reductase, it may also provide reducing equivalents for the reduction of the persulfide transferred to the $SufBC_2D$ complex.

6. Concluding Remarks

The recent advances in the field of Fe–S cluster biogenesis have shed new light on the process of their assembly. We are now close to provide a complete picture for the assembly of [2Fe2S] clusters by the ISC machinery. Two key primary steps have been uncovered with the mouse system: persulfide transfer to ISCU and its reduction by FDX2 that are both metal-dependent. This process is most likely conserved in bacteria, but this is still awaiting confirmation. In the first step, iron comes first and enables persulfide insertion in IscU/ISCU. In the second step, persulfide reduction is coupled to the presence of iron. Altogether, these metal dependent steps probably allow allocation of the proper amount of sulfur to the iron center of IscU/ISCU and coordinate sulfide release with the presence of iron to generate the first iron-sulfide link. However, a number of key questions remain to be solved, essentially the mechanism by which persulfide transfer and its reduction are coupled with iron and how the mononuclear center of IscU/ISCU is converted into a dinuclear center to form the [2Fe2S] cluster. By unravelling the primary steps of Fe–S cluster assembly, the functional role of frataxin is also becoming more clear. There is now compelling evidence pointing to a role of frataxin as a kinetic modulator of persulfide transfer.

Since the ISC and NIF systems are very similar in several ways, the data accumulated on the ISC system will likely help unravel the mechanism by which the NIF machinery assembles Fe–S clusters. In contrast, the comparisons of the ISC and SUF systems point to major differences. The persulfide transfer process in the SUF systems is split into discrete steps involving sulfur relay proteins and a trans-persulfuration reaction with the $SufBC_2D$ complex. In the first step, a buried area is created at the interface of the SufS–SufE/U complex, to achieve the trans-persulfuration reaction. This protected area probably shields the persulfide and the reactive cysteines from oxidants. The iron-insertion process also seems very different, but its description is still too elusive to allow a clear description. Several other questions are pending, essentially whether persulfide transfer is coupled with iron insertion, how the persulfide is reduced into sulfide and how the [2Fe2S] cluster is formed. Addressing these questions will improve our understanding of the mechanism by which the SUF machinery copes with iron-poor and oxidative-stress conditions.

Author Contributions: B.D., supervision and review-writing; B.S. writing contributions; S.G., writing contributions, bibliographic ressources; B.M., text corrections. All authors have read and agreed to the published version of the manuscript.

Funding: This research was funded by ANR grant (Frataxur, ANR-17-CE11-0021-01) and FARA grant (202212) awarded to B.D., B.M. was funded by FARA grant (202212).

Conflicts of Interest: The authors declare no conflict of interest.

References

1. Camprubi, E.; Jordan, S.F.; Vasiliadou, R.; Lane, N. Iron catalysis at the origin of life. *IUBMB Life* **2017**, *69*, 373–381. [CrossRef]
2. Huber, C.; Wachtershauser, G. Alpha-Hydroxy and alpha-amino acids under possible Hadean, volcanic origin-of-life conditions. *Science* **2006**, *314*, 630–632. [CrossRef]
3. Cody, G. Transition metal sulfides and the origin of metabolism. *Annu. Rev. Earth Planet. Sci.* **2004**, *32*, 569–599. [CrossRef]
4. Martin, W.; Russell, M.J. On the origins of cells: A hypothesis for the evolutionary transitions from abiotic geochemistry to chemoautotrophic prokaryotes, and from prokaryotes to nucleated cells. *Philos. Trans. R. Soc. Lond. B Biol. Sci.* **2003**, *358*, 59–83, discussion 55–83. [CrossRef]
5. Chahal, H.K.; Boyd, J.M.; Outten, F.W. Iron–Sulfur Cluster Biogenesis in Archaea and Bacteria. In *Encyclopedia of Inorganic and Bioinorganic Chemistry*; Scott, R.A., Ed.; Wiley: Hoboken, NJ, USA, 2013. [CrossRef]

6. Lill, R.; Freibert, S.A. Mechanisms of Mitochondrial Iron-Sulfur Protein Biogenesis. *Annu. Rev. Biochem.* **2020**, *89*, 471–499. [CrossRef]
7. Paul, V.D.; Lill, R. Biogenesis of cytosolic and nuclear iron-sulfur proteins and their role in genome stability. *Biochim. Biophys. Acta* **2015**, *1853*, 1528–1539. [CrossRef]
8. Przybyla-Toscano, J.; Roland, M.; Gaymard, F.; Couturier, J.; Rouhier, N. Roles and maturation of iron-sulfur proteins in plastids. *J. Biol. Inorg. Chem.* **2018**, *23*, 545–566. [CrossRef]
9. Roche, B.; Aussel, L.; Ezraty, B.; Mandin, P.; Py, B.; Barras, F. Iron/sulfur proteins biogenesis in prokaryotes: Formation, regulation and diversity. *Biochim. Biophys. Acta* **2013**, *1827*, 455–469. [CrossRef]
10. Martin, D.; Charpilienne, A.; Parent, A.; Boussac, A.; D'Autreaux, B.; Poupon, J.; Poncet, D. The rotavirus nonstructural protein NSP5 coordinates a [2Fe–2S] iron-sulfur cluster that modulates interaction to RNA. *FASEB J.* **2013**, *27*, 1074–1083. [CrossRef]
11. Tam, W.; Pell, L.G.; Bona, D.; Tsai, A.; Dai, X.X.; Edwards, A.M.; Hendrix, R.W.; Maxwell, K.L.; Davidson, A.R. Tail tip proteins related to bacteriophage lambda gpL coordinate an iron-sulfur cluster. *J. Mol. Biol.* **2013**, *425*, 2450–2462. [CrossRef]
12. Tsang, S.H.; Wang, R.; Nakamaru-Ogiso, E.; Knight, S.A.; Buck, C.B.; You, J. The Oncogenic Small Tumor Antigen of Merkel Cell Polyomavirus Is an Iron-Sulfur Cluster Protein That Enhances Viral DNA Replication. *J. Virol.* **2016**, *90*, 1544–1556. [CrossRef] [PubMed]
13. Estellon, J.; Ollagnier de Choudens, S.; Smadja, M.; Fontecave, M.; Vandenbrouck, Y. An integrative computational model for large-scale identification of metalloproteins in microbial genomes: A focus on iron-sulfur cluster proteins. *Metallomics* **2014**, *6*, 1913–1930. [CrossRef]
14. Liu, J.; Chakraborty, S.; Hosseinzadeh, P.; Yu, Y.; Tian, S.; Petrik, I.; Bhagi, A.; Lu, Y. Metalloproteins containing cytochrome, iron-sulfur, or copper redox centers. *Chem. Rev.* **2014**, *114*, 4366–4469. [CrossRef] [PubMed]
15. Stegmaier, K.; Blinn, C.M.; Bechtel, D.F.; Greth, C.; Auerbach, H.; Muller, C.S.; Jakob, V.; Reijerse, E.J.; Netz, D.J.A.; Schunemann, V.; et al. Apd1 and Aim32 Are Prototypes of Bishistidinyl-Coordinated Non-Rieske [2Fe–2S] Proteins. *J. Am. Chem. Soc.* **2019**, *141*, 5753–5765. [CrossRef]
16. Berkovitch, F.; Nicolet, Y.; Wan, J.T.; Jarrett, J.T.; Drennan, C.L. Crystal Structure of Biotin Synthase, an *S*-Adenosylmethionine-Dependent Radical Enzyme. *Science* **2004**, *303*, 76. [CrossRef]
17. Bak, D.W.; Elliott, S.J. Alternative FeS cluster ligands: Tuning redox potentials and chemistry. *Curr. Opinion Chem. Biol.* **2014**, *19*, 50–58. [CrossRef]
18. Hu, Y.; Ribbe, M.W. Biosynthesis of the Metalloclusters of Nitrogenases. *Annu. Rev. Biochem.* **2016**, *85*, 455–483. [CrossRef] [PubMed]
19. Dobbek, H.; Svetlitchnyi, V.; Gremer, L.; Huber, R.; Meyer, O. Crystal Structure of a Carbon Monoxide Dehydrogenase Reveals a [Ni–4Fe–5S] Cluster. *Science* **2001**, *293*, 1281. [CrossRef] [PubMed]
20. Cooper, S.J.; Garner, C.D.; Hagen, W.R.; Lindley, P.F.; Bailey, S. Hybrid-cluster protein (HCP) from Desulfovibrio vulgaris (Hildenborough) at 1.6 A resolution. *Biochemistry* **2000**, *39*, 15044–15054. [CrossRef] [PubMed]
21. Beinert, H. Iron-sulfur proteins: Ancient structures, still full of surprises. *J. Biol. Inorg. Chem.* **2000**, *5*, 2–15. [CrossRef]
22. Blanc, B.; Gerez, C.; Ollagnier de Choudens, S. Assembly of Fe/S proteins in bacterial systems: Biochemistry of the bacterial ISC system. *Biochim. Biophys. Acta* **2015**, *1853*, 1436–1447. [CrossRef] [PubMed]
23. Gervason, S.; Larkem, D.; Mansour, A.B.; Botzanowski, T.; Muller, C.S.; Pecqueur, L.; Le Pavec, G.; Delaunay-Moisan, A.; Brun, O.; Agramunt, J.; et al. Physiologically relevant reconstitution of iron-sulfur cluster biosynthesis uncovers persulfide-processing functions of ferredoxin-2 and frataxin. *Nat. Commun.* **2019**, *10*, 3566. [CrossRef] [PubMed]
24. Webert, H.; Freibert, S.A.; Gallo, A.; Heidenreich, T.; Linne, U.; Amlacher, S.; Hurt, E.; Muhlenhoff, U.; Banci, L.; Lill, R. Functional reconstitution of mitochondrial Fe/S cluster synthesis on Isu1 reveals the involvement of ferredoxin. *Nat. Commun.* **2014**, *5*, 5013. [CrossRef] [PubMed]
25. Ciofi-Baffoni, S.; Nasta, V.; Banci, L. Protein networks in the maturation of human iron–sulfur proteins. *Metallomics* **2018**, *10*, 49–72. [CrossRef]
26. Pérard, J.; Ollagnier de Choudens, S. Iron–sulfur clusters biogenesis by the SUF machinery: Close to the molecular mechanism understanding. *JBIC J. Biol. Inorg. Chem.* **2018**, *23*, 581–596. [CrossRef] [PubMed]
27. Rouault, T.A.; Maio, N. Biogenesis and functions of mammalian iron-sulfur proteins in the regulation of iron homeostasis and pivotal metabolic pathways. *J. Biol. Chem.* **2017**, *292*, 12744–12753. [CrossRef]

28. Delewski, W.; Paterkiewicz, B.; Manicki, M.; Schilke, B.; Tomiczek, B.; Ciesielski, S.J.; Nierzwicki, L.; Czub, J.; Dutkiewicz, R.; Craig, E.A.; et al. Iron-Sulfur Cluster Biogenesis Chaperones: Evidence for Emergence of Mutational Robustness of a Highly Specific Protein-Protein Interaction. *Mol. Biol. Evol.* **2016**, *33*, 643–656. [CrossRef]
29. Beilschmidt, L.K.; Ollagnier de Choudens, S.; Fournier, M.; Sanakis, I.; Hograindleur, M.A.; Clemancey, M.; Blondin, G.; Schmucker, S.; Eisenmann, A.; Weiss, A.; et al. ISCA1 is essential for mitochondrial Fe_4S_4 biogenesis in vivo. *Nat. Commun.* **2017**, *8*, 15124. [CrossRef]
30. Brancaccio, D.; Gallo, A.; Mikolajczyk, M.; Zovo, K.; Palumaa, P.; Novellino, E.; Piccioli, M.; Ciofi-Baffoni, S.; Banci, L. Formation of [4Fe–4S] clusters in the mitochondrial iron-sulfur cluster assembly machinery. *J. Am. Chem. Soc.* **2014**, *136*, 16240–16250. [CrossRef]
31. Brancaccio, D.; Gallo, A.; Piccioli, M.; Novellino, E.; Ciofi-Baffoni, S.; Banci, L. [4Fe–4S] Cluster Assembly in Mitochondria and Its Impairment by Copper. *J. Am. Chem. Soc.* **2017**, *139*, 719–730. [CrossRef]
32. Muhlenhoff, U.; Richter, N.; Pines, O.; Pierik, A.J.; Lill, R. Specialized function of yeast Isa1 and Isa2 proteins in the maturation of mitochondrial [4Fe–4S] proteins. *J. Biol. Chem.* **2011**, *286*, 41205–41216. [CrossRef] [PubMed]
33. Weiler, B.D.; Bruck, M.C.; Kothe, I.; Bill, E.; Lill, R.; Muhlenhoff, U. Mitochondrial [4Fe–4S] protein assembly involves reductive [2Fe–2S] cluster fusion on ISCA1-ISCA2 by electron flow from ferredoxin FDX2. *Proc. Natl. Acad. Sci. USA* **2020**, *117*, 20555–20565. [CrossRef] [PubMed]
34. Jacobson, M.R.; Cash, V.L.; Weiss, M.C.; Laird, N.F.; Newton, W.E.; Dean, D.R. Biochemical and genetic analysis of the nifUSVWZM cluster from Azotobacter vinelandii. *Mol. Gen. Genet.* **1989**, *219*, 49–57. [CrossRef]
35. Zheng, L.; White, R.H.; Cash, V.L.; Jack, R.F.; Dean, D.R. Cysteine desulfurase activity indicates a role for NIFS in metallocluster biosynthesis. *Proc. Natl. Acad. Sci. USA* **1993**, *90*, 2754–2758. [CrossRef] [PubMed]
36. Fu, W.; Jack, R.F.; Morgan, T.V.; Dean, D.R.; Johnson, M.K. nifU gene product from Azotobacter vinelandii is a homodimer that contains two identical [2Fe–2S] clusters. *Biochemistry* **1994**, *33*, 13455–13463. [CrossRef]
37. Johnson, D.C.; Dos Santos, P.C.; Dean, D.R. NifU and NifS are required for the maturation of nitrogenase and cannot replace the function of isc-gene products in Azotobacter vinelandii. *Biochem. Soc. Trans.* **2005**, *33*, 90–93. [CrossRef]
38. Smith, A.D.; Jameson, G.N.; Dos Santos, P.C.; Agar, J.N.; Naik, S.; Krebs, C.; Frazzon, J.; Dean, D.R.; Huynh, B.H.; Johnson, M.K. NifS-mediated assembly of [4Fe–4S] clusters in the N- and C-terminal domains of the NifU scaffold protein. *Biochemistry* **2005**, *44*, 12955–12969. [CrossRef]
39. Black, K.A.; Dos Santos, P.C. Shared-intermediates in the biosynthesis of thio-cofactors: Mechanism and functions of cysteine desulfurases and sulfur acceptors. *Biochim. Biophys. Acta* **2015**, *1853*, 1470–1480. [CrossRef]
40. Leimkühler, S.; Bühning, M.; Beilschmidt, L. Shared Sulfur Mobilization Routes for tRNA Thiolation and Molybdenum Cofactor Biosynthesis in Prokaryotes and Eukaryotes. *Biomolecules* **2017**, *7*, 5.
41. Zheng, L.; Cash, V.L.; Flint, D.H.; Dean, D.R. Assembly of iron-sulfur clusters. Identification of an iscSUA-hscBA-fdx gene cluster from Azotobacter vinelandii. *J. Biol. Chem.* **1998**, *273*, 13264–13272. [CrossRef]
42. Patzer, S.I.; Hantke, K. SufS is a NifS-like protein, and SufD is necessary for stability of the [2Fe–2S] FhuF protein in Escherichia coli. *J. Bacteriol.* **1999**, *181*, 3307–3309. [CrossRef] [PubMed]
43. Mihara, H.; Kurihara, T.; Yoshimura, T.; Esaki, N. Kinetic and mutational studies of three NifS homologs from Escherichia coli: Mechanistic difference between L-cysteine desulfurase and L-selenocysteine lyase reactions. *J. Biochem.* **2000**, *127*, 559–567. [CrossRef] [PubMed]
44. Nachin, L.; El Hassouni, M.; Loiseau, L.; Expert, D.; Barras, F. SoxR-dependent response to oxidative stress and virulence of Erwinia chrysanthemi: The key role of SufC, an orphan ABC ATPase. *Mol. Microbiol.* **2001**, *39*, 960–972. [CrossRef] [PubMed]
45. Takahashi, Y.; Tokumoto, U. A third bacterial system for the assembly of iron-sulfur clusters with homologs in archaea and plastids. *J. Biol. Chem.* **2002**, *277*, 28380–28383. [CrossRef] [PubMed]
46. Nyvltova, E.; Sutak, R.; Harant, K.; Sedinova, M.; Hrdy, I.; Paces, J.; Vlcek, C.; Tachezy, J. NIF-type iron-sulfur cluster assembly system is duplicated and distributed in the mitochondria and cytosol of Mastigamoeba balamuthi. *Proc. Natl. Acad. Sci. USA* **2013**, *110*, 7371–7376. [CrossRef]

47. Garcia, P.S.; Gribaldo, S.; Py, B.; Barras, F. The SUF system: An ABC ATPase-dependent protein complex with a role in Fe–S cluster biogenesis. *Res. Microbiol.* **2019**, *170*, 426–434. [CrossRef]
48. Freibert, S.A.; Goldberg, A.V.; Hacker, C.; Molik, S.; Dean, P.; Williams, T.A.; Nakjang, S.; Long, S.; Sendra, K.; Bill, E.; et al. Evolutionary conservation and in vitro reconstitution of microsporidian iron-sulfur cluster biosynthesis. *Nat. Commun.* **2017**, *8*, 13932. [CrossRef]
49. Couturier, J.; Touraine, B.; BRIAT, J.-F.; Gaymard, F.; Rouhier, N. The iron-sulfur cluster assembly machineries in plants: Current knowledge and open questions. *Front. Plant Sci.* **2013**, *4*, 1–22. [CrossRef]
50. Balk, J.; Schaedler, T.A. Iron Cofactor Assembly in Plants. *Annu. Rev. Plant Biol.* **2014**, *65*, 125–153. [CrossRef]
51. Dai, Y.; Outten, F.W. The E. coli SufS-SufE sulfur transfer system is more resistant to oxidative stress than IscS-IscU. *FEBS Lett.* **2012**, *586*, 4016–4022. [CrossRef]
52. Trotter, V.; Vinella, D.; Loiseau, L.; Ollagnier de Choudens, S.; Fontecave, M.; Barras, F. The CsdA cysteine desulphurase promotes Fe/S biogenesis by recruiting Suf components and participates to a new sulphur transfer pathway by recruiting CsdL (ex-YgdL), a ubiquitin-modifying-like protein. *Mol. Microbiol.* **2009**, *74*, 1527–1542. [CrossRef] [PubMed]
53. Fosset, C.; Chauveau, M.J.; Guillon, B.; Canal, F.; Drapier, J.C.; Bouton, C. RNA silencing of mitochondrial m-Nfs1 reduces Fe–S enzyme activity both in mitochondria and cytosol of mammalian cells. *J. Biol. Chem.* **2006**, *281*, 25398–25406. [CrossRef] [PubMed]
54. Kispal, G.; Csere, P.; Prohl, C.; Lill, R. The mitochondrial proteins Atm1p and Nfs1p are essential for biogenesis of cytosolic Fe/S proteins. *EMBO J.* **1999**, *18*, 3981–3989. [CrossRef] [PubMed]
55. Marelja, Z.; Stocklein, W.; Nimtz, M.; Leimkuhler, S. A novel role for human Nfs1 in the cytoplasm: Nfs1 acts as a sulfur donor for MOCS3, a protein involved in molybdenum cofactor biosynthesis. *J. Biol. Chem.* **2008**, *283*, 25178–25185. [CrossRef]
56. Muhlenhoff, U.; Balk, J.; Richhardt, N.; Kaiser, J.T.; Sipos, K.; Kispal, G.; Lill, R. Functional characterization of the eukaryotic cysteine desulfurase Nfs1p from Saccharomyces cerevisiae. *J. Biol. Chem.* **2004**, *279*, 36906–36915. [CrossRef]
57. Pandey, A.K.; Pain, J.; Dancis, A.; Pain, D. Mitochondria export iron-sulfur and sulfur intermediates to the cytoplasm for iron-sulfur cluster assembly and tRNA thiolation in yeast. *J. Biol. Chem.* **2019**, *294*, 9489–9502. [CrossRef]
58. Pastore, A.; Puccio, H. Frataxin: A protein in search for a function. *J. Neurochem.* **2013**, *126* (Suppl. S1), 43–52. [CrossRef]
59. Roche, B.; Huguenot, A.; Barras, F.; Py, B. The iron-binding CyaY and IscX proteins assist the ISC-catalyzed Fe–S biogenesis in Escherichia coli. *Mol. Microbiol.* **2015**, *95*, 605–623. [CrossRef]
60. Van Vranken, J.G.; Nowinski, S.M.; Clowers, K.J.; Jeong, M.Y.; Ouyang, Y.; Berg, J.A.; Gygi, J.P.; Gygi, S.P.; Winge, D.R.; Rutter, J. ACP Acylation Is an Acetyl-CoA-Dependent Modification Required for Electron Transport Chain Assembly. *Mol. Cell* **2018**, *71*, 567–580 e564. [CrossRef]
61. Agar, J.N.; Krebs, C.; Frazzon, J.; Huynh, B.H.; Dean, D.R.; Johnson, M.K. IscU as a scaffold for iron-sulfur cluster biosynthesis: Sequential assembly of [2Fe–2S] and [4Fe–4S] clusters in IscU. *Biochemistry* **2000**, *39*, 7856–7862. [CrossRef]
62. Mansy, S.S.; Wu, G.; Surerus, K.K.; Cowan, J.A. Iron-sulfur cluster biosynthesis. Thermatoga maritima IscU is a structured iron-sulfur cluster assembly protein. *J. Biol. Chem.* **2002**, *277*, 21397–21404. [CrossRef] [PubMed]
63. Raulfs, E.C.; O'Carroll, I.P.; Dos Santos, P.C.; Unciuleac, M.C.; Dean, D.R. In vivo iron-sulfur cluster formation. *Proc. Natl. Acad. Sci. USA* **2008**, *105*, 8591–8596. [CrossRef]
64. Shimomura, Y.; Wada, K.; Fukuyama, K.; Takahashi, Y. The asymmetric trimeric architecture of [2Fe–2S] IscU: Implications for its scaffolding during iron-sulfur cluster biosynthesis. *J. Mol. Biol.* **2008**, *383*, 133–143. [CrossRef] [PubMed]
65. Bonomi, F.; Iametti, S.; Morleo, A.; Ta, D.; Vickery, L.E. Facilitated transfer of IscU-[2Fe2S] clusters by chaperone-mediated ligand exchange. *Biochemistry* **2011**, *50*, 9641–9650. [CrossRef] [PubMed]
66. Marinoni, E.N.; de Oliveira, J.S.; Nicolet, Y.; Amara, P.; Dean, D.R.; Fontecilla-Camps, J.C. (IscS-IscU)2 complex structures provide insights into Fe2S2 biogenesis and transfer. *Angew. Chem. Int. Ed. Engl.* **2012**, *51*, 5439–5442. [CrossRef] [PubMed]
67. Boniecki, M.T.; Freibert, S.A.; Muhlenhoff, U.; Lill, R.; Cygler, M. Structure and functional dynamics of the mitochondrial Fe/S cluster synthesis complex. *Nat. Commun.* **2017**, *8*, 1287. [CrossRef]

68. Fox, N.G.; Martelli, A.; Nabhan, J.F.; Janz, J.; Borkowska, O.; Bulawa, C.; Yue, W.W. Zinc(II) binding on human wild-type ISCU and Met140 variants modulates NFS1 desulfurase activity. *Biochimie* **2018**, *152*, 211–218. [CrossRef]
69. Iannuzzi, C.; Adrover, M.; Puglisi, R.; Yan, R.; Temussi, P.A.; Pastore, A. The role of zinc in the stability of the marginally stable IscU scaffold protein. *Protein Sci.* **2014**, *23*, 1208–1219. [CrossRef]
70. Kim, J.H.; Fuzery, A.K.; Tonelli, M.; Ta, D.T.; Westler, W.M.; Vickery, L.E.; Markley, J.L. Structure and dynamics of the iron-sulfur cluster assembly scaffold protein IscU and its interaction with the cochaperone HscB. *Biochemistry* **2009**, *48*, 6062–6071. [CrossRef]
71. Ramelot, T.A.; Cort, J.R.; Goldsmith-Fischman, S.; Kornhaber, G.J.; Xiao, R.; Shastry, R.; Acton, T.B.; Honig, B.; Montelione, G.T.; Kennedy, M.A. Solution NMR structure of the iron-sulfur cluster assembly protein U (IscU) with zinc bound at the active site. *J. Mol. Biol.* **2004**, *344*, 567–583. [CrossRef]
72. Yan, R.; Kelly, G.; Pastore, A. The scaffold protein IscU retains a structured conformation in the Fe–S cluster assembly complex. *Chembiochem* **2014**, *15*, 1682–1686. [CrossRef] [PubMed]
73. Fox, N.G.; Yu, X.; Feng, X.; Bailey, H.J.; Martelli, A.; Nabhan, J.F.; Strain-Damerell, C.; Bulawa, C.; Yue, W.W.; Han, S. Structure of the human frataxin-bound iron-sulfur cluster assembly complex provides insight into its activation mechanism. *Nat. Commun.* **2019**, *10*, 2210. [CrossRef]
74. Bertini, I.; Cowan, J.A.; Del Bianco, C.; Luchinat, C.; Mansy, S.S. Thermotoga maritima IscU. Structural characterization and dynamics of a new class of metallochaperone. *J. Mol. Biol.* **2003**, *331*, 907–924. [CrossRef]
75. Cai, K.; Frederick, R.O.; Kim, J.H.; Reinen, N.M.; Tonelli, M.; Markley, J.L. Human mitochondrial chaperone (mtHSP70) and cysteine desulfurase (NFS1) bind preferentially to the disordered conformation, whereas co-chaperone (HSC20) binds to the structured conformation of the iron-sulfur cluster scaffold protein (ISCU). *J. Biol. Chem.* **2013**, *288*, 28755–28770. [CrossRef] [PubMed]
76. Mansy, S.S.; Wu, S.P.; Cowan, J.A. Iron-sulfur cluster biosynthesis: Biochemical characterization of the conformational dynamics of Thermotoga maritima IscU and the relevance for cellular cluster assembly. *J. Biol. Chem.* **2004**, *279*, 10469–10475. [CrossRef]
77. Kim, J.H.; Tonelli, M.; Markley, J.L. Disordered form of the scaffold protein IscU is the substrate for iron-sulfur cluster assembly on cysteine desulfurase. *Proc. Natl. Acad. Sci. USA* **2012**, *109*, 454–459. [CrossRef]
78. Lin, C.W.; McCabe, J.W.; Russell, D.H.; Barondeau, D.P. Molecular Mechanism of ISC Iron-Sulfur Cluster Biogenesis Revealed by High-Resolution Native Mass Spectrometry. *J. Am. Chem. Soc.* **2020**, *142*, 6018–6029. [CrossRef]
79. Lewis, B.E.; Mason, Z.; Rodrigues, A.V.; Nuth, M.; Dizin, E.; Cowan, J.A.; Stemmler, T.L. Unique roles of iron and zinc binding to the yeast Fe–S cluster scaffold assembly protein "Isu1". *Metallomics* **2019**, *11*, 1820–1835. [CrossRef]
80. Rodrigues, A.V.; Kandegedara, A.; Rotondo, J.A.; Dancis, A.; Stemmler, T.L. Iron loading site on the Fe–S cluster assembly scaffold protein is distinct from the active site. *Biometals* **2015**, *28*, 567–576. [CrossRef]
81. Dzul, S.P.; Rocha, A.G.; Rawat, S.; Kandegedara, A.; Kusowski, A.; Pain, J.; Murari, A.; Pain, D.; Dancis, A.; Stemmler, T.L. In vitro characterization of a novel Isu homologue from *Drosophila melanogaster* for de novo FeS-cluster formation. *Metallomics* **2017**, *9*, 48–60. [CrossRef]
82. Adamec, J.; Rusnak, F.; Owen, W.G.; Naylor, S.; Benson, L.M.; Gacy, A.M.; Isaya, G. Iron-dependent self-assembly of recombinant yeast frataxin: Implications for Friedreich ataxia. *Am. J. Hum. Genet.* **2000**, *67*, 549–562. [CrossRef] [PubMed]
83. Cavadini, P.; O'Neill, H.A.; Benada, O.; Isaya, G. Assembly and iron-binding properties of human frataxin, the protein deficient in Friedreich ataxia. *Hum. Mol. Genet.* **2002**, *11*, 217–227. [CrossRef] [PubMed]
84. Ding, H.; Yang, J.; Coleman, L.C.; Yeung, S. Distinct iron binding property of two putative iron donors for the iron-sulfur cluster assembly: IscA and the bacterial frataxin ortholog CyaY under physiological and oxidative stress conditions. *J. Biol. Chem.* **2007**, *282*, 7997–8004. [CrossRef]
85. Gakh, O.; Adamec, J.; Gacy, A.M.; Twesten, R.D.; Owen, W.G.; Isaya, G. Physical evidence that yeast frataxin is an iron storage protein. *Biochemistry* **2002**, *41*, 6798–6804. [CrossRef] [PubMed]
86. Layer, G.; Ollagnier-de Choudens, S.; Sanakis, Y.; Fontecave, M. Iron-sulfur cluster biosynthesis: Characterization of Escherichia coli CYaY as an iron donor for the assembly of [2Fe–2S] clusters in the scaffold IscU. *J. Biol. Chem.* **2006**, *281*, 16256–16263. [CrossRef] [PubMed]
87. Yoon, T.; Cowan, J.A. Iron-sulfur cluster biosynthesis. Characterization of frataxin as an iron donor for assembly of [2Fe–2S] clusters in ISU-type proteins. *J. Am. Chem. Soc.* **2003**, *125*, 6078–6084. [CrossRef]

88. Parent, A.; Elduque, X.; Cornu, D.; Belot, L.; Le Caer, J.P.; Grandas, A.; Toledano, M.B.; D'Autreaux, B. Mammalian frataxin directly enhances sulfur transfer of NFS1 persulfide to both ISCU and free thiols. *Nat. Commun.* **2015**, *6*, 5686. [CrossRef]
89. Patra, S.; Barondeau, D.P. Mechanism of activation of the human cysteine desulfurase complex by frataxin. *Proc. Natl. Acad. Sci. USA* **2019**, *116*, 19421–19430. [CrossRef]
90. Armas, A.M.; Balparda, M.; Terenzi, A.; Busi, M.V.; Pagani, M.A.; Gomez-Casati, D.F. Ferrochelatase activity of plant frataxin. *Biochimie* **2019**, *156*, 118–122. [CrossRef]
91. Bellanda, M.; Maso, L.; Doni, D.; Bortolus, M.; De Rosa, E.; Lunardi, F.; Alfonsi, A.; Noguera, M.E.; Herrera, M.G.; Santos, J.; et al. Exploring iron-binding to human frataxin and to selected Friedreich ataxia mutants by means of NMR and EPR spectroscopies. *Biochim. Biophys. Acta Proteins Proteom.* **2019**, *1867*, 140254. [CrossRef]
92. Cai, K.; Frederick, R.O.; Tonelli, M.; Markley, J.L. Interactions of iron-bound frataxin with ISCU and ferredoxin on the cysteine desulfurase complex leading to Fe–S cluster assembly. *J. Inorg. Biochem.* **2018**, *183*, 107–116. [CrossRef]
93. Castro, I.H.; Pignataro, M.F.; Sewell, K.E.; Espeche, L.D.; Herrera, M.G.; Noguera, M.E.; Dain, L.; Nadra, A.D.; Aran, M.; Smal, C.; et al. Frataxin Structure and Function. *Subcell Biochem.* **2019**, *93*, 393–438.
94. Gakh, O.; Ranatunga, W.; Galeano, B.K.; Smith, D.S.T.; Thompson, J.R.; Isaya, G. Defining the Architecture of the Core Machinery for the Assembly of Fe–S Clusters in Human Mitochondria. *Methods Enzymol.* **2017**, *595*, 107–160.
95. Galeano, B.K.; Ranatunga, W.; Gakh, O.; Smith, D.Y.; Thompson, J.R.; Isaya, G. Zinc and the iron donor frataxin regulate oligomerization of the scaffold protein to form new Fe–S cluster assembly centers. *Metallomics* **2017**, *9*, 773–801. [CrossRef]
96. Ranatunga, W.; Gakh, O.; Galeano, B.K.; Smith, D.Y.T.; Soderberg, C.A.; Al-Karadaghi, S.; Thompson, J.R.; Isaya, G. Architecture of the Yeast Mitochondrial Iron-Sulfur Cluster Assembly Machinery: The Sub-Complex Formed by the Iron Donor, Yfh1 Protein, and the Scaffold, Isu1 Protein. *J. Biol. Chem.* **2016**, *291*, 10378–10398. [CrossRef]
97. Babcock, M.; de Silva, D.; Oaks, R.; Davis-Kaplan, S.; Jiralerspong, S.; Montermini, L.; Pandolfo, M.; Kaplan, J. Regulation of mitochondrial iron accumulation by Yfh1p, a putative homolog of frataxin. *Science* **1997**, *276*, 1709–1712. [CrossRef] [PubMed]
98. Duby, G.; Foury, F.; Ramazzotti, A.; Herrmann, J.; Lutz, T. A non-essential function for yeast frataxin in iron-sulfur cluster assembly. *Hum. Mol. Genet.* **2002**, *11*, 2635–2643. [CrossRef] [PubMed]
99. Foury, F.; Cazzalini, O. Deletion of the yeast homologue of the human gene associated with Friedreich's ataxia elicits iron accumulation in mitochondria. *FEBS Lett.* **1997**, *411*, 373–377. [CrossRef]
100. Puccio, H.; Simon, D.; Cossee, M.; Criqui-Filipe, P.; Tiziano, F.; Melki, J.; Hindelang, C.; Matyas, R.; Rustin, P.; Koenig, M. Mouse models for Friedreich ataxia exhibit cardiomyopathy, sensory nerve defect and Fe–S enzyme deficiency followed by intramitochondrial iron deposits. *Nat. Genet.* **2001**, *27*, 181–186. [CrossRef]
101. Rotig, A.; de Lonlay, P.; Chretien, D.; Foury, F.; Koenig, M.; Sidi, D.; Munnich, A.; Rustin, P. Aconitase and mitochondrial iron-sulphur protein deficiency in Friedreich ataxia. *Nat. Genet.* **1997**, *17*, 215–217. [CrossRef] [PubMed]
102. Gerber, J.; Muhlenhoff, U.; Lill, R. An interaction between frataxin and Isu1/Nfs1 that is crucial for Fe/S cluster synthesis on Isu1. *EMBO Rep.* **2003**, *4*, 906–911. [CrossRef]
103. Ramazzotti, A.; Vanmansart, V.; Foury, F. Mitochondrial functional interactions between frataxin and Isu1p, the iron-sulfur cluster scaffold protein, in Saccharomyces cerevisiae. *FEBS Lett.* **2004**, *557*, 215–220. [CrossRef]
104. Schmucker, S.; Martelli, A.; Colin, F.; Page, A.; Wattenhofer-Donze, M.; Reutenauer, L.; Puccio, H. Mammalian frataxin: An essential function for cellular viability through an interaction with a preformed ISCU/NFS1/ISD11 iron-sulfur assembly complex. *PLoS ONE* **2011**, *6*, e16199. [CrossRef] [PubMed]
105. Prischi, F.; Konarev, P.V.; Iannuzzi, C.; Pastore, C.; Adinolfi, S.; Martin, S.R.; Svergun, D.I.; Pastore, A. Structural bases for the interaction of frataxin with the central components of iron-sulphur cluster assembly. *Nat. Commun.* **2010**, *1*, 95. [CrossRef]
106. Campuzano, V.; Montermini, L.; Molto, M.D.; Pianese, L.; Cossee, M.; Cavalcanti, F.; Monros, E.; Rodius, F.; Duclos, F.; Monticelli, A.; et al. Friedreich's ataxia: Autosomal recessive disease caused by an intronic GAA triplet repeat expansion. *Science* **1996**, *271*, 1423–1427. [CrossRef] [PubMed]

107. Sanchez-Casis, G.; Cote, M.; Barbeau, A. Pathology of the heart in Friedreich's ataxia: Review of the literature and report of one case. *Can. J. Neurol. Sci.* **1976**, *3*, 349–354. [CrossRef]
108. Cook, J.D.; Bencze, K.Z.; Jankovic, A.D.; Crater, A.K.; Busch, C.N.; Bradley, P.B.; Stemmler, A.J.; Spaller, M.R.; Stemmler, T.L. Monomeric yeast frataxin is an iron-binding protein. *Biochemistry* **2006**, *45*, 7767–7777. [CrossRef]
109. O'Neill, H.A.; Gakh, O.; Isaya, G. Supramolecular assemblies of human frataxin are formed via subunit-subunit interactions mediated by a non-conserved amino-terminal region. *J. Mol. Biol.* **2005**, *345*, 433–439. [CrossRef]
110. Aloria, K.; Schilke, B.; Andrew, A.; Craig, E.A. Iron-induced oligomerization of yeast frataxin homologue Yfh1 is dispensable in vivo. *EMBO Rep.* **2004**, *5*, 1096–1101. [CrossRef]
111. Schmucker, S.; Argentini, M.; Carelle-Calmels, N.; Martelli, A.; Puccio, H. The in vivo mitochondrial two-step maturation of human frataxin. *Hum. Mol. Genet.* **2008**, *17*, 3521–3531. [CrossRef]
112. Seguin, A.; Sutak, R.; Bulteau, A.L.; Garcia-Serres, R.; Oddou, J.L.; Lefevre, S.; Santos, R.; Dancis, A.; Camadro, J.M.; Latour, J.M.; et al. Evidence that yeast frataxin is not an iron storage protein in vivo. *Biochim. Biophys. Acta* **2010**, *1802*, 531–538. [CrossRef] [PubMed]
113. Muhlenhoff, U.; Richhardt, N.; Ristow, M.; Kispal, G.; Lill, R. The yeast frataxin homolog Yfh1p plays a specific role in the maturation of cellular Fe/S proteins. *Hum. Mol. Genet.* **2002**, *11*, 2025–2036. [CrossRef]
114. Stehling, O.; Wilbrecht, C.; Lill, R. Mitochondrial iron-sulfur protein biogenesis and human disease. *Biochimie* **2014**, *100*, 61–77. [CrossRef] [PubMed]
115. Adinolfi, S.; Nair, M.; Politou, A.; Bayer, E.; Martin, S.; Temussi, P.; Pastore, A. The factors governing the thermal stability of frataxin orthologues: How to increase a protein's stability. *Biochemistry* **2004**, *43*, 6511–6518. [CrossRef] [PubMed]
116. Bou-Abdallah, F.; Adinolfi, S.; Pastore, A.; Laue, T.M.; Dennis Chasteen, N. Iron binding and oxidation kinetics in frataxin CyaY of Escherichia coli. *J. Mol. Biol.* **2004**, *341*, 605–615. [CrossRef] [PubMed]
117. Yoon, T.; Dizin, E.; Cowan, J.A. N-terminal iron-mediated self-cleavage of human frataxin: Regulation of iron binding and complex formation with target proteins. *J. Biol. Inorg. Chem.* **2007**, *12*, 535–542. [CrossRef] [PubMed]
118. Huang, J.; Dizin, E.; Cowan, J.A. Mapping iron binding sites on human frataxin: Implications for cluster assembly on the ISU Fe–S cluster scaffold protein. *J. Biol. Inorg. Chem.* **2008**, *13*, 825–836. [CrossRef]
119. Ding, B.; Smith, E.S.; Ding, H. Mobilization of the iron centre in IscA for the iron-sulphur cluster assembly in IscU. *Biochem. J.* **2005**, *389*, 797–802. [CrossRef]
120. Ding, H.; Clark, R.J.; Ding, B. IscA mediates iron delivery for assembly of iron-sulfur clusters in IscU under the limited accessible free iron conditions. *J. Biol. Chem.* **2004**, *279*, 37499–37504. [CrossRef]
121. Lu, J.; Bitoun, J.P.; Tan, G.; Wang, W.; Min, W.; Ding, H. Iron-binding activity of human iron-sulfur cluster assembly protein hIscA1. *Biochem. J.* **2010**, *428*, 125–131. [CrossRef]
122. Lu, J.; Yang, J.; Tan, G.; Ding, H. Complementary roles of SufA and IscA in the biogenesis of iron-sulfur clusters in Escherichia coli. *Biochem. J.* **2008**, *409*, 535–543. [CrossRef] [PubMed]
123. Yang, J.; Bitoun, J.P.; Ding, H. Interplay of IscA and IscU in biogenesis of iron-sulfur clusters. *J. Biol. Chem.* **2006**, *281*, 27956–27963. [CrossRef] [PubMed]
124. Ding, H.; Clark, R.J. Characterization of iron binding in IscA, an ancient iron-sulphur cluster assembly protein. *Biochem. J.* **2004**, *379*, 433–440. [CrossRef]
125. Mapolelo, D.T.; Zhang, B.; Naik, S.G.; Huynh, B.H.; Johnson, M.K. Spectroscopic and functional characterization of iron-bound forms of Azotobacter vinelandii (Nif)IscA. *Biochemistry* **2012**, *51*, 8056–8070. [CrossRef]
126. Gupta, V.; Sendra, M.; Naik, S.G.; Chahal, H.K.; Huynh, B.H.; Outten, F.W.; Fontecave, M.; Ollagnier de Choudens, S. Native Escherichia coli SufA, coexpressed with SufBCDSE, purifies as a [2Fe–2S] protein and acts as an Fe–S transporter to Fe–S target enzymes. *J. Am. Chem. Soc.* **2009**, *131*, 6149–6153. [CrossRef] [PubMed]
127. Morimoto, K.; Yamashita, E.; Kondou, Y.; Lee, S.J.; Arisaka, F.; Tsukihara, T.; Nakai, M. The asymmetric IscA homodimer with an exposed [2Fe–2S] cluster suggests the structural basis of the Fe–S cluster biosynthetic scaffold. *J. Mol. Biol.* **2006**, *360*, 117–132. [CrossRef]

128. Wang, W.; Huang, H.; Tan, G.; Si, F.; Liu, M.; Landry, A.P.; Lu, J.; Ding, H. In vivo evidence for the iron-binding activity of an iron-sulfur cluster assembly protein IscA in Escherichia coli. *Biochem. J.* **2010**, *432*, 429–436. [CrossRef]
129. Zeng, J.; Geng, M.; Jiang, H.; Liu, Y.; Liu, J.; Qiu, G. The IscA from Acidithiobacillus ferrooxidans is an iron-sulfur protein which assemble the [Fe4S4] cluster with intracellular iron and sulfur. *Arch. Biochem. Biophys.* **2007**, *463*, 237–244. [CrossRef]
130. Mapolelo, D.T.; Zhang, B.; Randeniya, S.; Albetel, A.N.; Li, H.; Couturier, J.; Outten, C.E.; Rouhier, N.; Johnson, M.K. Monothiol glutaredoxins and A-type proteins: Partners in Fe–S cluster trafficking. *Dalton Trans.* **2013**, *42*, 3107–3115. [CrossRef] [PubMed]
131. Kim, K.D.; Chung, W.H.; Kim, H.J.; Lee, K.C.; Roe, J.H. Monothiol glutaredoxin Grx5 interacts with Fe–S scaffold proteins Isa1 and Isa2 and supports Fe–S assembly and DNA integrity in mitochondria of fission yeast. *Biochem. Biophys. Res. Commun.* **2010**, *392*, 467–472. [CrossRef] [PubMed]
132. Gelling, C.; Dawes, I.W.; Richhardt, N.; Lill, R.; Muhlenhoff, U. Mitochondrial Iba57p is required for Fe/S cluster formation on aconitase and activation of radical SAM enzymes. *Mol. Cell Biol.* **2008**, *28*, 1851–1861. [CrossRef]
133. Tokumoto, U.; Nomura, S.; Minami, Y.; Mihara, H.; Kato, S.; Kurihara, T.; Esaki, N.; Kanazawa, H.; Matsubara, H.; Takahashi, Y. Network of protein-protein interactions among iron-sulfur cluster assembly proteins in Escherichia coli. *J. Biochem.* **2002**, *131*, 713–719. [CrossRef] [PubMed]
134. Al-Karadaghi, S.; Franco, R.; Hansson, M.; Shelnutt, J.A.; Isaya, G.; Ferreira, G.C. Chelatases: Distort to select? *Trends Biochem. Sci.* **2006**, *31*, 135–142. [CrossRef] [PubMed]
135. Foster, A.W.; Osman, D.; Robinson, N.J. Metal preferences and metallation. *J. Biol. Chem.* **2014**, *289*, 28095–28103. [CrossRef]
136. Capdevila, D.A.; Edmonds, K.A.; Giedroc, D.P. Metallochaperones and metalloregulation in bacteria. *Essays Biochem.* **2017**, *61*, 177–200.
137. Philpott, C.C.; Jadhav, S. The ins and outs of iron: Escorting iron through the mammalian cytosol. *Free Radic Biol. Med.* **2019**, *133*, 112–117. [CrossRef]
138. Kruszewski, M. The role of labile iron pool in cardiovascular diseases. *Acta Biochim. Pol.* **2004**, *51*, 471–480. [CrossRef]
139. Wofford, J.D.; Bolaji, N.; Dziuba, N.; Outten, F.W.; Lindahl, P.A. Evidence that a respiratory shield in Escherichia coli protects a low-molecular-mass Fe(II) pool from O_2-dependent oxidation. *J. Biol. Chem.* **2019**, *294*, 50–62. [CrossRef]
140. Lindahl, P.A.; Moore, M.J. Labile Low-Molecular-Mass Metal Complexes in Mitochondria: Trials and Tribulations of a Burgeoning Field. *Biochemistry* **2016**, *55*, 4140–4153. [CrossRef]
141. McCormick, S.P.; Moore, M.J.; Lindahl, P.A. Detection of Labile Low-Molecular-Mass Transition Metal Complexes in Mitochondria. *Biochemistry* **2015**, *54*, 3442–3453. [CrossRef]
142. Moore, M.J.; Wofford, J.D.; Dancis, A.; Lindahl, P.A. Recovery of mrs3Δmrs4Δ Saccharomyces cerevisiae Cells under Iron-Sufficient Conditions and the Role of Fe580. *Biochemistry* **2018**, *57*, 672–683. [CrossRef] [PubMed]
143. Hider, R.C.; Kong, X.L. Glutathione: A key component of the cytoplasmic labile iron pool. *Biometals* **2011**, *24*, 1179–1187. [CrossRef] [PubMed]
144. Cupp-Vickery, J.R.; Urbina, H.; Vickery, L.E. Crystal structure of IscS, a cysteine desulfurase from Escherichia coli. *J. Mol. Biol.* **2003**, *330*, 1049–1059. [CrossRef]
145. Fujii, T.; Maeda, M.; Mihara, H.; Kurihara, T.; Esaki, N.; Hata, Y. Structure of a NifS homologue: X-ray structure analysis of CsdB, an Escherichia coli counterpart of mammalian selenocysteine lyase. *Biochemistry* **2000**, *39*, 1263–1273. [CrossRef] [PubMed]
146. Kaiser, J.T.; Clausen, T.; Bourenkow, G.P.; Bartunik, H.D.; Steinbacher, S.; Huber, R. Crystal structure of a NifS-like protein from Thermotoga maritima: Implications for iron sulphur cluster assembly. *J. Mol. Biol.* **2000**, *297*, 451–464. [CrossRef] [PubMed]
147. Nakamura, R.; Hikita, M.; Ogawa, S.; Takahashi, Y.; Fujishiro, T. Snapshots of PLP-substrate and PLP-product external aldimines as intermediates in two types of cysteine desulfurase enzymes. *FEBS J.* **2020**, *287*, 1138–1154. [CrossRef]
148. Roret, T.; Pegeot, H.; Couturier, J.; Mulliert, G.; Rouhier, N.; Didierjean, C. X-ray structures of Nfs2, the plastidial cysteine desulfurase from Arabidopsis thaliana. *Acta Crystallogr. F Struct. Biol. Commun.* **2014**, *70*, 1180–1185. [CrossRef]

149. Rybniker, J.; Pojer, F.; Marienhagen, J.; Kolly, G.S.; Chen, J.M.; van Gumpel, E.; Hartmann, P.; Cole, S.T. The cysteine desulfurase IscS of Mycobacterium tuberculosis is involved in iron-sulfur cluster biogenesis and oxidative stress defence. *Biochem. J.* **2014**, *459*, 467–478. [CrossRef]
150. Shi, R.; Proteau, A.; Villarroya, M.; Moukadiri, I.; Zhang, L.; Trempe, J.F.; Matte, A.; Armengod, M.E.; Cygler, M. Structural basis for Fe–S cluster assembly and tRNA thiolation mediated by IscS protein-protein interactions. *PLoS Biol.* **2010**, *8*, e1000354. [CrossRef]
151. Cory, S.A.; Van Vranken, J.G.; Brignole, E.J.; Patra, S.; Winge, D.R.; Drennan, C.L.; Rutter, J.; Barondeau, D.P. Structure of human Fe–S assembly subcomplex reveals unexpected cysteine desulfurase architecture and acyl-ACP-ISD11 interactions. *Proc. Natl. Acad. Sci. USA* **2017**, *114*, E5325–E5334. [CrossRef]
152. Turowski, V.R.; Busi, M.V.; Gomez-Casati, D.F. Structural and functional studies of the mitochondrial cysteine desulfurase from Arabidopsis thaliana. *Mol. Plant* **2012**, *5*, 1001–1010. [CrossRef]
153. Herrera, M.G.; Pignataro, M.F.; Noguera, M.E.; Cruz, K.M.; Santos, J. Rescuing the Rescuer: On the Protein Complex between the Human Mitochondrial Acyl Carrier Protein and ISD11. *ACS Chem. Biol.* **2018**, *13*, 1455–1462. [CrossRef] [PubMed]
154. Smith, A.D.; Agar, J.N.; Johnson, K.A.; Frazzon, J.; Amster, I.J.; Dean, D.R.; Johnson, M.K. Sulfur transfer from IscS to IscU: The first step in iron-sulfur cluster biosynthesis. *J. Am. Chem. Soc.* **2001**, *123*, 11103–11104. [CrossRef]
155. Urbina, H.D.; Silberg, J.J.; Hoff, K.G.; Vickery, L.E. Transfer of sulfur from IscS to IscU during Fe/S cluster assembly. *J. Biol. Chem.* **2001**, *276*, 44521–44526. [CrossRef] [PubMed]
156. Smith, A.D.; Frazzon, J.; Dean, D.R.; Johnson, M.K. Role of conserved cysteines in mediating sulfur transfer from IscS to IscU. *FEBS Lett.* **2005**, *579*, 5236–5240. [CrossRef]
157. Nuth, M.; Yoon, T.; Cowan, J.A. Iron-sulfur cluster biosynthesis: Characterization of iron nucleation sites for assembly of the $[2Fe–2S]^{2+}$ cluster core in IscU proteins. *J. Am. Chem. Soc.* **2002**, *124*, 8774–8775. [CrossRef]
158. Bridwell-Rabb, J.; Fox, N.G.; Tsai, C.L.; Winn, A.M.; Barondeau, D.P. Human frataxin activates Fe–S cluster biosynthesis by facilitating sulfur transfer chemistry. *Biochemistry* **2014**, *53*, 4904–4913. [CrossRef] [PubMed]
159. Selbach, B.P.; Chung, A.H.; Scott, A.D.; George, S.J.; Cramer, S.P.; Dos Santos, P.C. Fe–S Cluster Biogenesis in Gram-Positive Bacteria: SufU Is a Zinc-Dependent Sulfur Transfer Protein. *Biochemistry* **2014**, *53*, 152–160. [CrossRef] [PubMed]
160. Adinolfi, S.; Iannuzzi, C.; Prischi, F.; Pastore, C.; Iametti, S.; Martin, S.R.; Bonomi, F.; Pastore, A. Bacterial frataxin CyaY is the gatekeeper of iron-sulfur cluster formation catalyzed by IscS. *Nat. Struct. Mol. Biol.* **2009**, *16*, 390–396. [CrossRef]
161. Nuth, M.; Cowan, J.A. Iron-sulfur cluster biosynthesis: Characterization of IscU-IscS complex formation and a structural model for sulfide delivery to the [2Fe–2S] assembly site. *J. Biol. Inorg. Chem.* **2009**, *14*, 829–839. [CrossRef]
162. Tsai, C.L.; Barondeau, D.P. Human frataxin is an allosteric switch that activates the Fe–S cluster biosynthetic complex. *Biochemistry* **2010**, *49*, 9132–9139. [CrossRef] [PubMed]
163. Bridwell-Rabb, J.; Iannuzzi, C.; Pastore, A.; Barondeau, D.P. Effector role reversal during evolution: The case of frataxin in Fe–S cluster biosynthesis. *Biochemistry* **2012**, *51*, 2506–2514. [CrossRef] [PubMed]
164. Colin, F.; Martelli, A.; Clemancey, M.; Latour, J.M.; Gambarelli, S.; Zeppieri, L.; Birck, C.; Page, A.; Puccio, H.; Ollagnier de Choudens, S. Mammalian frataxin controls sulfur production and iron entry during de novo Fe4S4 cluster assembly. *J. Am. Chem. Soc.* **2013**, *135*, 733–740. [CrossRef] [PubMed]
165. Unciuleac, M.C.; Chandramouli, K.; Naik, S.; Mayer, S.; Huynh, B.H.; Johnson, M.K.; Dean, D.R. In vitro activation of apo-aconitase using a [4Fe–4S] cluster-loaded form of the IscU [Fe-S] cluster scaffolding protein. *Biochemistry* **2007**, *46*, 6812–6821. [CrossRef]
166. Iannuzzi, C.; Adinolfi, S.; Howes, B.D.; Garcia-Serres, R.; Clemancey, M.; Latour, J.M.; Smulevich, G.; Pastore, A. The role of CyaY in iron sulfur cluster assembly on the E. coli IscU scaffold protein. *PLoS ONE* **2011**, *6*, e21992. [CrossRef]
167. Fox, N.G.; Chakrabarti, M.; McCormick, S.P.; Lindahl, P.A.; Barondeau, D.P. The Human Iron-Sulfur Assembly Complex Catalyzes the Synthesis of [2Fe–2S] Clusters on ISCU2 That Can Be Transferred to Acceptor Molecules. *Biochemistry* **2015**, *54*, 3871–3879. [CrossRef]
168. Tokumoto, U.; Takahashi, Y. Genetic analysis of the isc operon in Escherichia coli involved in the biogenesis of cellular iron-sulfur proteins. *J. Biochem.* **2001**, *130*, 63–71. [CrossRef]

169. Lange, H.; Kaut, A.; Kispal, G.; Lill, R. A mitochondrial ferredoxin is essential for biogenesis of cellular iron-sulfur proteins. *Proc. Natl. Acad. Sci. USA* **2000**, *97*, 1050–1055. [CrossRef]
170. Djaman, O.; Outten, F.W.; Imlay, J.A. Repair of oxidized iron-sulfur clusters in Escherichia coli. *J. Biol. Chem.* **2004**, *279*, 44590–44599. [CrossRef]
171. Muhlenhoff, U.; Gerber, J.; Richhardt, N.; Lill, R. Components involved in assembly and dislocation of iron-sulfur clusters on the scaffold protein Isu1p. *EMBO J.* **2003**, *22*, 4815–4825. [CrossRef]
172. Sheftel, A.D.; Stehling, O.; Pierik, A.J.; Elsasser, H.P.; Muhlenhoff, U.; Webert, H.; Hobler, A.; Hannemann, F.; Bernhardt, R.; Lill, R. Humans possess two mitochondrial ferredoxins, Fdx1 and Fdx2, with distinct roles in steroidogenesis, heme, and Fe/S cluster biosynthesis. *Proc. Natl. Acad. Sci. USA* **2010**, *107*, 11775–11780. [CrossRef] [PubMed]
173. Shi, Y.; Ghosh, M.; Kovtunovych, G.; Crooks, D.R.; Rouault, T.A. Both human ferredoxins 1 and 2 and ferredoxin reductase are important for iron-sulfur cluster biogenesis. *Biochim. Biophys. Acta* **2012**, *1823*, 484–492. [CrossRef] [PubMed]
174. Li, J.; Saxena, S.; Pain, D.; Dancis, A. Adrenodoxin reductase homolog (Arh1p) of yeast mitochondria required for iron homeostasis. *J. Biol. Chem.* **2001**, *276*, 1503–1509. [CrossRef]
175. Wan, J.T.; Jarrett, J.T. Electron acceptor specificity of ferredoxin (flavodoxin): NADP+ oxidoreductase from Escherichia coli. *Arch. Biochem. Biophys.* **2002**, *406*, 116–126. [CrossRef]
176. Yan, R.; Adinolfi, S.; Pastore, A. Ferredoxin, in conjunction with NADPH and ferredoxin-NADP reductase, transfers electrons to the IscS/IscU complex to promote iron-sulfur cluster assembly. *Biochim. Biophys. Acta* **2015**, *1854*, 1113–1117. [CrossRef]
177. Cai, K.; Tonelli, M.; Frederick, R.O.; Markley, J.L. Human Mitochondrial Ferredoxin 1 (FDX1) and Ferredoxin 2 (FDX2) Both Bind Cysteine Desulfurase and Donate Electrons for Iron-Sulfur Cluster Biosynthesis. *Biochemistry* **2017**, *56*, 487–499. [CrossRef]
178. Kim, J.H.; Frederick, R.O.; Reinen, N.M.; Troupis, A.T.; Markley, J.L. [2Fe–2S]-ferredoxin binds directly to cysteine desulfurase and supplies an electron for iron-sulfur cluster assembly but is displaced by the scaffold protein or bacterial frataxin. *J. Am. Chem. Soc.* **2013**, *135*, 8117–8120. [CrossRef]
179. Yan, R.; Konarev, P.V.; Iannuzzi, C.; Adinolfi, S.; Roche, B.; Kelly, G.; Simon, L.; Martin, S.R.; Py, B.; Barras, F.; et al. Ferredoxin competes with bacterial frataxin in binding to the desulfurase IscS. *J. Biol. Chem.* **2013**, *288*, 24777–24787. [CrossRef]
180. Galardon, E.; Roger, T.; Deschamps, P.; Roussel, P.; Tomas, A.; Artaud, I. Synthesis of a Fe(II)SH complex stabilized by an intramolecular N-H...S hydrogen bond, which acts as a H2S donor. *Inorg. Chem.* **2012**, *51*, 10068–10070. [CrossRef]
181. Tsou, C.C.; Chiu, W.C.; Ke, C.H.; Tsai, J.C.; Wang, Y.M.; Chiang, M.H.; Liaw, W.F. Iron(III) bound by hydrosulfide anion ligands: NO-promoted stabilization of the [Fe(III)-SH] motif. *J. Am. Chem. Soc.* **2014**, *136*, 9424–9433. [CrossRef]
182. Dey, A.; Hocking, R.K.; Larsen, P.; Borovik, A.S.; Hodgson, K.O.; Hedman, B.; Solomon, E.I. X-ray absorption spectroscopy and density functional theory studies of $[(H_3buea)Fe^{III}{-}X]^{n-}$ ($X = S^{2-}, O^{2-}, OH^{-}$): Comparison of bonding and hydrogen bonding in oxo and sulfido complexes. *J. Am. Chem. Soc.* **2006**, *128*, 9825–9833. [CrossRef] [PubMed]
183. Conradie, J.; Tangen, E.; Ghosh, A. Trigonal bipyramidal iron(III) and manganese(III) oxo, sulfido, and selenido complexes. An electronic-structural overview. *J. Inorg. Biochem.* **2006**, *100*, 707–715. [CrossRef] [PubMed]
184. Larsen, P.L.; Gupta, R.; Powell, D.R.; Borovik, A.S. Chalcogens as terminal ligands to iron: Synthesis and structure of complexes with Fe(III)-S and Fe(III)-Se motifs. *J. Am. Chem. Soc.* **2004**, *126*, 6522–6523. [CrossRef] [PubMed]
185. Li, H.; Outten, C.E. Monothiol CGFS glutaredoxins and BolA-like proteins: [2Fe–2S] binding partners in iron homeostasis. *Biochemistry* **2012**, *51*, 4377–4389. [CrossRef]
186. Rouhier, N.; Couturier, J.; Johnson, M.K.; Jacquot, J.P. Glutaredoxins: Roles in iron homeostasis. *Trends Biochem. Sci.* **2010**, *35*, 43–52. [CrossRef]
187. Banci, L.; Brancaccio, D.; Ciofi-Baffoni, S.; Del Conte, R.; Gadepalli, R.; Mikolajczyk, M.; Neri, S.; Piccioli, M.; Winkelmann, J. [2Fe–2S] cluster transfer in iron-sulfur protein biogenesis. *Proc. Natl. Acad. Sci. USA* **2014**, *111*, 6203–6208. [CrossRef]

188. Shakamuri, P.; Zhang, B.; Johnson, M.K. Monothiol glutaredoxins function in storing and transporting [Fe2S2] clusters assembled on IscU scaffold proteins. *J. Am. Chem. Soc.* **2012**, *134*, 15213–15216. [CrossRef]
189. Agar, J.N.; Zheng, L.; Cash, V.L.; Dean, D.R.; Johnson, M.K. Role of the IscU Protein in Iron−Sulfur Cluster Biosynthesis: IscS-mediated Assembly of a [Fe2S2] Cluster in IscU. *J. Am. Chem. Soc.* **2000**, *122*, 2136–2137. [CrossRef]
190. Wu, G.; Mansy, S.S.; Wu Sp, S.P.; Surerus, K.K.; Foster, M.W.; Cowan, J.A. Characterization of an iron-sulfur cluster assembly protein (ISU1) from Schizosaccharomyces pombe. *Biochemistry* **2002**, *41*, 5024–5032. [CrossRef]
191. Tsai, C.L.; Bridwell-Rabb, J.; Barondeau, D.P. Friedreich's ataxia variants I154F and W155R diminish frataxin-based activation of the iron-sulfur cluster assembly complex. *Biochemistry* **2011**, *50*, 6478–6487. [CrossRef]
192. Bridwell-Rabb, J.; Winn, A.M.; Barondeau, D.P. Structure-function analysis of Friedreich's ataxia mutants reveals determinants of frataxin binding and activation of the Fe–S assembly complex. *Biochemistry* **2011**, *50*, 7265–7274. [CrossRef] [PubMed]
193. Fox, N.G.; Das, D.; Chakrabarti, M.; Lindahl, P.A.; Barondeau, D.P. Frataxin Accelerates [2Fe–2S] Cluster Formation on the Human Fe–S Assembly Complex. *Biochemistry* **2015**, *54*, 3880–3889. [CrossRef] [PubMed]
194. Pandey, A.; Gordon, D.M.; Pain, J.; Stemmler, T.L.; Dancis, A.; Pain, D. Frataxin directly stimulates mitochondrial cysteine desulfurase by exposing substrate-binding sites, and a mutant Fe–S cluster scaffold protein with frataxin-bypassing ability acts similarly. *J. Biol. Chem.* **2013**, *288*, 36773–36786. [CrossRef] [PubMed]
195. Behshad, E.; Parkin, S.E.; Bollinger, J.M., Jr. Mechanism of cysteine desulfurase Slr0387 from Synechocystis sp. PCC 6803: Kinetic analysis of cleavage of the persulfide intermediate by chemical reductants. *Biochemistry* **2004**, *43*, 12220–12226. [CrossRef]
196. Lacourciere, G.M.; Stadtman, T.C. The NIFS protein can function as a selenide delivery protein in the biosynthesis of selenophosphate. *J. Biol. Chem.* **1998**, *273*, 30921–30926. [CrossRef]
197. Zheng, L.; White, R.H.; Cash, V.L.; Dean, D.R. Mechanism for the desulfurization of L-cysteine catalyzed by the nifS gene product. *Biochemistry* **1994**, *33*, 4714–4720. [CrossRef]
198. Ponsero, A.J.; Igbaria, A.; Darch, M.A.; Miled, S.; Outten, C.E.; Winther, J.R.; Palais, G.; D'Autreaux, B.; Delaunay-Moisan, A.; Toledano, M.B. Endoplasmic Reticulum Transport of Glutathione by Sec61 Is Regulated by Ero1 and Bip. *Mol. Cell* **2017**, *67*, 962–973 e965. [CrossRef]
199. Furne, J.; Saeed, A.; Levitt, M.D. Whole tissue hydrogen sulfide concentrations are orders of magnitude lower than presently accepted values. *Am. J. Physiol. Regul. Integr. Comp. Physiol.* **2008**, *295*, R1479–R1485. [CrossRef]
200. Ross-Inta, C.; Tsai, C.Y.; Giulivi, C. The mitochondrial pool of free amino acids reflects the composition of mitochondrial DNA-encoded proteins: Indication of a post-translational quality control for protein synthesis. *Biosci. Rep.* **2008**, *28*, 239–249. [CrossRef]
201. Agar, J.N.; Yuvaniyama, P.; Jack, R.F.; Cash, V.L.; Smith, A.D.; Dean, D.R.; Johnson, M.K. Modular organization and identification of a mononuclear iron-binding site within the NifU protein. *J. Biol. Inorg. Chem.* **2000**, *5*, 167–177. [CrossRef]
202. Yuvaniyama, P.; Agar, J.N.; Cash, V.L.; Johnson, M.K.; Dean, D.R. NifS-directed assembly of a transient [2Fe–2S] cluster within the NifU protein. *Proc. Natl. Acad. Sci. USA* **2000**, *97*, 599–604. [CrossRef] [PubMed]
203. Chandramouli, K.; Unciuleac, M.C.; Naik, S.; Dean, D.R.; Huynh, B.H.; Johnson, M.K. Formation and properties of [4Fe–4S] clusters on the IscU scaffold protein. *Biochemistry* **2007**, *46*, 6804–6811. [CrossRef] [PubMed]
204. Outten, F.W. Recent advances in the Suf Fe–S cluster biogenesis pathway: Beyond the Proteobacteria. *Biochim. Biophys. Acta* **2015**, *1853*, 1464–1469. [CrossRef] [PubMed]
205. Wollers, S.; Layer, G.; Garcia-Serres, R.; Signor, L.; Clemancey, M.; Latour, J.M.; Fontecave, M.; Ollagnier de Choudens, S. Iron–sulfur (Fe–S) cluster assembly: The SufBC$_2$D complex is a new type of Fe–S scaffold with a flavin redox cofactor. *J. Biol. Chem.* **2010**, *285*, 23331–23341. [CrossRef] [PubMed]
206. Hirabayashi, K.; Yuda, E.; Tanaka, N.; Katayama, S.; Iwasaki, K.; Matsumoto, T.; Kurisu, G.; Outten, F.W.; Fukuyama, K.; Takahashi, Y.; et al. Functional Dynamics Revealed by the Structure of the SufBC$_2$D Complex, a Novel ATP-binding Cassette (ABC) Protein That Serves as a Scaffold for Iron-Sulfur Cluster Biogenesis. *J. Biol. Chem.* **2015**, *290*, 29717–29731. [CrossRef] [PubMed]

207. Riboldi, G.P.; Verli, H.; Frazzon, J. Structural studies of the Enterococcus faecalis SufU [Fe-S] cluster protein. *BMC Biochem.* **2009**, *10*, 3. [CrossRef]
208. Selbach, B.; Earles, E.; Dos Santos, P.C. Kinetic analysis of the bisubstrate cysteine desulfurase SufS from Bacillus subtilis. *Biochemistry* **2010**, *49*, 8794–8802. [CrossRef] [PubMed]
209. Albrecht, A.G.; Netz, D.J.; Miethke, M.; Pierik, A.J.; Burghaus, O.; Peuckert, F.; Lill, R.; Marahiel, M.A. SufU is an essential iron-sulfur cluster scaffold protein in Bacillus subtilis. *J. Bacteriol.* **2010**, *192*, 1643–1651. [CrossRef]
210. Yokoyama, N.; Nonaka, C.; Ohashi, Y.; Shioda, M.; Terahata, T.; Chen, W.; Sakamoto, K.; Maruyama, C.; Saito, T.; Yuda, E.; et al. Distinct roles for U-type proteins in iron-sulfur cluster biosynthesis revealed by genetic analysis of the Bacillus subtilis sufCDSUB operon. *Mol. Microbiol.* **2018**, *107*, 688–703. [CrossRef]
211. Saini, A.; Mapolelo, D.T.; Chahal, H.K.; Johnson, M.K.; Outten, F.W. SufD and SufC ATPase activity are required for iron acquisition during in vivo Fe–S cluster formation on SufB. *Biochemistry* **2010**, *49*, 9402–9412. [CrossRef]
212. Nachin, L.; Loiseau, L.; Expert, D.; Barras, F. SufC: An unorthodox cytoplasmic ABC/ATPase required for [Fe-S] biogenesis under oxidative stress. *EMBO J.* **2003**, *22*, 427–437. [CrossRef] [PubMed]
213. Tirupati, B.; Vey, J.L.; Drennan, C.L.; Bollinger, J.M., Jr. Kinetic and structural characterization of Slr0077/SufS, the essential cysteine desulfurase from Synechocystis sp. PCC 6803. *Biochemistry* **2004**, *43*, 12210–12219. [CrossRef] [PubMed]
214. Outten, F.W.; Wood, M.J.; Munoz, F.M.; Storz, G. The SufE protein and the $SufBC_2D$ complex enhance SufS cysteine desulfurase activity as part of a sulfur transfer pathway for Fe–S cluster assembly in Escherichia coli. *J. Biol. Chem.* **2003**, *278*, 45713–45719. [CrossRef] [PubMed]
215. Zafrilla, B.; Martinez-Espinosa, R.M.; Esclapez, J.; Perez-Pomares, F.; Bonete, M.J. SufS protein from Haloferax volcanii involved in Fe–S cluster assembly in haloarchaea. *Biochim. Biophys. Acta* **2010**, *1804*, 1476–1482. [CrossRef]
216. Ollagnier-de-Choudens, S.; Lascoux, D.; Loiseau, L.; Barras, F.; Forest, E.; Fontecave, M. Mechanistic studies of the SufS-SufE cysteine desulfurase: Evidence for sulfur transfer from SufS to SufE. *FEBS Lett.* **2003**, *555*, 263–267. [CrossRef]
217. Singh, H.; Dai, Y.; Outten, F.W.; Busenlehner, L.S. Escherichia coli SufE sulfur transfer protein modulates the SufS cysteine desulfurase through allosteric conformational dynamics. *J. Biol. Chem.* **2013**, *288*, 36189–36200. [CrossRef]
218. Dai, Y.; Kim, D.; Dong, G.; Busenlehner, L.S.; Frantom, P.A.; Outten, F.W. SufE D74R Substitution Alters Active Site Loop Dynamics To Further Enhance SufE Interaction with the SufS Cysteine Desulfurase. *Biochemistry* **2015**, *54*, 4824–4833. [CrossRef]
219. Kim, D.; Singh, H.; Dai, Y.; Dong, G.; Busenlehner, L.S.; Outten, F.W.; Frantom, P.A. Changes in Protein Dynamics in Escherichia coli SufS Reveal a Possible Conserved Regulatory Mechanism in Type II Cysteine Desulfurase Systems. *Biochemistry* **2018**, *57*, 5210–5217. [CrossRef]
220. Dunkle, J.A.; Bruno, M.R.; Outten, F.W.; Frantom, P.A. Structural Evidence for Dimer-Interface-Driven Regulation of the Type II Cysteine Desulfurase, SufS. *Biochemistry* **2019**, *58*, 687–696. [CrossRef]
221. Blahut, M.; Wise, C.E.; Bruno, M.R.; Dong, G.; Makris, T.M.; Frantom, P.A.; Dunkle, J.A.; Outten, F.W. Direct observation of intermediates in the SufS cysteine desulfurase reaction reveals functional roles of conserved active-site residues. *J. Biol. Chem.* **2019**, *294*, 12444–12458. [CrossRef]
222. Kim, S.; Park, S. Structural changes during cysteine desulfurase CsdA and sulfur acceptor CsdE interactions provide insight into the trans-persulfuration. *J. Biol. Chem.* **2013**, *288*, 27172–27180. [CrossRef] [PubMed]
223. Dunkle, J.A.; Bruno, M.R.; Frantom, P.A. Structural evidence for a latch mechanism regulating access to the active site of SufS-family cysteine desulfurases. *Acta Crystallogr. D Struct. Biol.* **2020**, *76*, 291–301. [CrossRef] [PubMed]
224. Fernández, F.J.; Ardá, A.; López-Estepa, M.; Aranda, J.; Peña-Soler, E.; Garces, F.; Round, A.; Campos-Olivas, R.; Bruix, M.; Coll, M.; et al. Mechanism of Sulfur Transfer Across Protein–Protein Interfaces: The Cysteine Desulfurase Model System. *ACS Catal.* **2016**, *6*, 3975–3984. [CrossRef]
225. Goldsmith-Fischman, S.; Kuzin, A.; Edstrom, W.C.; Benach, J.; Shastry, R.; Xiao, R.; Acton, T.B.; Honig, B.; Montelione, G.T.; Hunt, J.F. The SufE sulfur-acceptor protein contains a conserved core structure that mediates interdomain interactions in a variety of redox protein complexes. *J. Mol. Biol.* **2004**, *344*, 549–565. [CrossRef]

226. Liu, G.; Li, Z.; Chiang, Y.; Acton, T.; Montelione, G.T.; Murray, D.; Szyperski, T. High-quality homology models derived from NMR and X-ray structures of E. coli proteins YgdK and Suf E suggest that all members of the YgdK/Suf E protein family are enhancers of cysteine desulfurases. *Protein Sci.* **2005**, *14*, 1597–1608. [CrossRef]
227. Selbach, B.P.; Pradhan, P.K.; Dos Santos, P.C. Protected sulfur transfer reactions by the Escherichia coli Suf system. *Biochemistry* **2013**, *52*, 4089–4096. [CrossRef]
228. Riboldi, G.P.; de Oliveira, J.S.; Frazzon, J. Enterococcus faecalis SufU scaffold protein enhances SufS desulfurase activity by acquiring sulfur from its cysteine-153. *Biochim. Biophys. Acta* **2011**, *1814*, 1910–1918. [CrossRef]
229. Albrecht, A.G.; Peuckert, F.; Landmann, H.; Miethke, M.; Seubert, A.; Marahiel, M.A. Mechanistic characterization of sulfur transfer from cysteine desulfurase SufS to the iron-sulfur scaffold SufU in Bacillus subtilis. *FEBS Lett.* **2011**, *585*, 465–470. [CrossRef]
230. Kornhaber, G.J.; Snyder, D.; Moseley, H.N.; Montelione, G.T. Identification of zinc-ligated cysteine residues based on 13Calpha and 13Cbeta chemical shift data. *J. Biomol. NMR* **2006**, *34*, 259–269. [CrossRef]
231. Liu, J.; Oganesyan, N.; Shin, D.H.; Jancarik, J.; Yokota, H.; Kim, R.; Kim, S.H. Structural characterization of an iron-sulfur cluster assembly protein IscU in a zinc-bound form. *Proteins* **2005**, *59*, 875–881. [CrossRef]
232. Fujishiro, T.; Terahata, T.; Kunichika, K.; Yokoyama, N.; Maruyama, C.; Asai, K.; Takahashi, Y. Zinc-Ligand Swapping Mediated Complex Formation and Sulfur Transfer between SufS and SufU for Iron-Sulfur Cluster Biogenesis in Bacillus subtilis. *J. Am. Chem. Soc.* **2017**, *139*, 18464–18467. [CrossRef] [PubMed]
233. Blauenburg, B.; Mielcarek, A.; Altegoer, F.; Fage, C.D.; Linne, U.; Bange, G.; Marahiel, M.A. Crystal Structure of Bacillus subtilis Cysteine Desulfurase SufS and Its Dynamic Interaction with Frataxin and Scaffold Protein SufU. *PLoS ONE* **2016**, *11*, e0158749. [CrossRef] [PubMed]
234. Yuda, E.; Tanaka, N.; Fujishiro, T.; Yokoyama, N.; Hirabayashi, K.; Fukuyama, K.; Wada, K.; Takahashi, Y. Mapping the key residues of SufB and SufD essential for biosynthesis of iron-sulfur clusters. *Sci. Rep.* **2017**, *7*, 9387. [CrossRef] [PubMed]

Review

The Requirement of Inorganic Fe-S Clusters for the Biosynthesis of the Organometallic Molybdenum Cofactor

Ralf R. Mendel [1], Thomas W. Hercher [1], Arkadiusz Zupok [2], Muhammad A. Hasnat [2] and Silke Leimkühler [2,*]

[1] Institute of Plant Biology, Braunschweig University of Technology, Humboldtstr. 1, 38106 Braunschweig, Germany; r.mendel@tu-bs.de (R.R.M.); t.hercher@tu-bs.de (T.W.H.)

[2] Department of Molecular Enzymology, Institute of Biochemistry and Biology, University of Potsdam, Karl-Liebknecht-Str. 24-25, 14476 Potsdam, Germany; zupok@uni-potsdam.de (A.Z.); hasnat@uni-potsdam.de (M.A.H.)

* Correspondence: sleim@uni-potsdam.de; Tel.: +49-331-977-5603; Fax: +49-331-977-5128

Received: 18 June 2020; Accepted: 14 July 2020; Published: 16 July 2020

Abstract: Iron-sulfur (Fe-S) clusters are essential protein cofactors. In enzymes, they are present either in the rhombic [2Fe-2S] or the cubic [4Fe-4S] form, where they are involved in catalysis and electron transfer and in the biosynthesis of metal-containing prosthetic groups like the molybdenum cofactor (Moco). Here, we give an overview of the assembly of Fe-S clusters in bacteria and humans and present their connection to the Moco biosynthesis pathway. In all organisms, Fe-S cluster assembly starts with the abstraction of sulfur from L-cysteine and its transfer to a scaffold protein. After formation, Fe-S clusters are transferred to carrier proteins that insert them into recipient apo-proteins. In eukaryotes like humans and plants, Fe-S cluster assembly takes place both in mitochondria and in the cytosol. Both Moco biosynthesis and Fe-S cluster assembly are highly conserved among all kingdoms of life. Moco is a tricyclic pterin compound with molybdenum coordinated through its unique dithiolene group. Moco biosynthesis begins in the mitochondria in a Fe-S cluster dependent step involving radical/S-adenosylmethionine (SAM) chemistry. An intermediate is transferred to the cytosol where the dithiolene group is formed, to which molybdenum is finally added. Further connections between Fe-S cluster assembly and Moco biosynthesis are discussed in detail.

Keywords: Moco biosynthesis; Fe-S cluster assembly; L-cysteine desulfurase; ISC; SUF; NIF; iron; molybdenum; sulfur

1. Introduction

As one of the most abundant metals on earth, iron naturally is one of the prevalent metal ions in biological systems [1]. Iron is a major constituent of iron-sulfur (Fe-S) clusters and plays an important role in life on earth. Fe-S centers are essential protein cofactors in all forms of life [2]. They are involved in many key biological pathways including the metabolism of carbon, nitrogen and sulfur, photosynthesis, respiration, biosynthesis of antibiotics, protein translation, replication and DNA repair, gene regulation, protection from oxidizing agents, and neurotransmission. In particular, Fe-S centers are not only involved as enzyme cofactors in catalysis and electron transfer, but they have been revealed to be essential for the assembly of other metal-containing cofactors.

The most common clusters are [2Fe-2S], [3Fe-4S], and [4Fe-4S], and they are the most versatile and presumably oldest cofactors of proteins in the cell (Figure 1). Their synthesis and insertion into apo-proteins require the function of complex cellular machinery [2–4]. In addition to the roles named above, Fe-S cluster-containing proteins play critical roles in the assembly of other metal-containing enzymes or metal-containing cofactors such as the molybdenum cofactor (Moco). In this case, Fe-S cluster assembly has to precede the biosynthesis of this metal-dependent molecule.

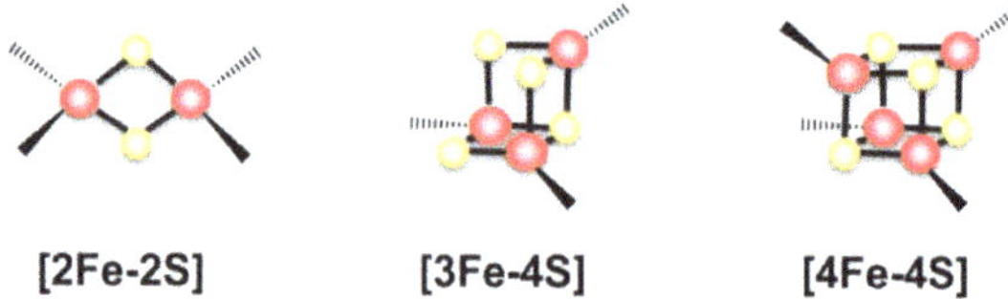

Figure 1. Three common types of Fe-S clusters. Shown are the structures of a rhombic [2Fe-2S] and a cubane [4Fe-4S] cluster. The [3Fe-4S] cluster can be generated by the loss of iron from a [4Fe-4S] cluster. Sulfur is symbolized in yellow and iron is represented in red.

Fe-S clusters were first discovered in the early 1960s by H. Beinert, R.H. Sands, and others, in photosynthetic organisms [5], nitrogen-fixing bacteria [6], and sub-mitochondrial fractions of mammalian origin [7]. To date, numerous different types of proteins or enzymes containing diverse Fe-S clusters have been identified [2]. Fe-S containing proteins are ubiquitous and unarguably constitute the oldest but the structurally-heterogeneous class of proteins in biology. Fe-S enzymes are quite diverse in function and many of them catalyze key redox reactions in central metabolism under both aerobic and anaerobic conditions. Fe-S clusters can form spontaneously in solution from Fe^{2+}/Fe^{3+}, R-S^{-}, S^{2-}; or from Fe^{2+}, R-SH, and sulfur [8], and it has been suggested that the first biologically meaningful reactions were catalyzed by Fe-S clusters [9]. Presumably, the first Fe-S proteins arose from the incorporation of preformed inorganic Fe-S clusters into polypeptides. In the world as we currently know it, the process of Fe-S cluster biosynthesis in living organisms turns out to be highly regulated and is catalyzed by numerous biogenesis factors that are remarkably conserved among prokaryotes and eukaryotes [4,10].

The complexity of Fe-S cluster biosynthesis became evident in 1998, when a complex gene cluster was discovered in bacteria that codes for proteins that are involved in their controlled assembly [11]. In eukaryotes, the mitochondria were identified as the primary compartment for Fe-S cluster assembly and were found to contain a very similar system as found in prokaryotes [12]. Since the late 1990s, the proteins involved in Fe-S cluster biosynthesis have been studied and characterized extensively. While these studies provided a general outline of in vitro and in vivo Fe-S cluster assembly, a number of major questions remain to be answered. Remaining gaps in our knowledge are: how Fe-S clusters are transferred to their target proteins, how specificity in this process is achieved and, in particular, how the iron for cluster assembly is provided in the cell.

Among the most recent additions to the field of Fe-S dependent enzymes was the discovery of the superfamily of radical/S-adenosylmethionine (radical/SAM) enzymes in 2001 [13]. These enzymes utilize a [4Fe-4S] cluster and SAM to initiate a diverse set of radical reactions, in most cases via the generation of a 5′-deoxyadenosyl radical intermediate. While this superfamily has been already identified in 2001 by studies that were mainly based on bioinformatics, the discovery of new Fe-S containing enzymes that employ SAM to initiate radical reactions still continues. Especially in recent years, it has become obvious that most reaction pathways for the synthesis of complex metal-containing cofactors have recruited radical/SAM chemistry [14]. One example has been the identification of the mechanism of the radical/SAM enzyme MoaA, a GTP 3′,8-cyclase in the biosynthesis of Moco of the diverse class of molybdoenzymes [15,16]. The number of known radical/SAM-dependent enzymes grew exponentially during the last years, with an initial identification of 600 members of the superfamily by Sofia et al. in 2001 that until today has increased to more than 113,000 members [14]. These enzymes are found across species and catalyze a diverse set of reactions, the vast majority of which have yet to be characterized. Due to their functional diversity, most cellular processes depend on this superfamily of [4Fe-4S]-containing enzymes.

2. The Assembly of Fe-S Clusters in Bacteria

Three main Fe-S cluster assembly pathways have been identified to date, namely the NIF (nitrogen fixation) system, the SUF (sulfur formation) system, and the ISC (iron sulfur cluster) system (reviewed in Reference [4]) (Figure 2). The three systems have different phylogenetic distributions throughout the three kingdoms of life. For example, in cyanobacteria, the SUF pathway is the major system for Fe-S cluster assembly, while in *Escherichia coli* the ISC has the predominant role, while the SUF pathway is more important under stress conditions [17]. Furthermore, in gram-positive pathogens such as *Mycobacteria* or *Clostridia*, as well as some Archaea, the SUF pathway is essential. In other bacteria like the plant-pathogenic bacterium *Dickeya dandatii*, all three ISC, SUF, and NIF systems are present [18]. In eukaryotes, the Fe-S cluster assembly pathway is further complicated by a different localization to specific organelles [19]. Homologues of the ISC pathway are present predominantly in mitochondria, while SUF homologues are restricted to the chloroplasts of some photosynthetic organisms (Figure 2). In addition, a cytosolic iron sulfur cluster (CIA) machinery is present that appears to be distinct from the SUF and ISC pathway.

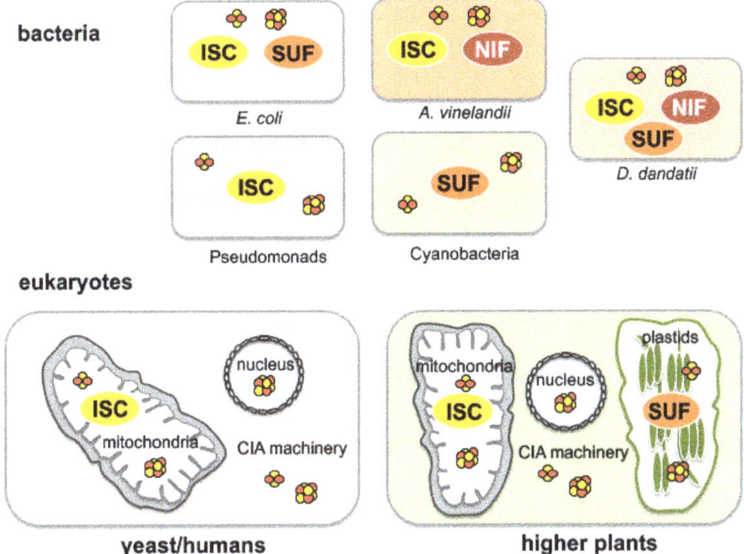

Figure 2. Possible links of Fe-S cluster assembly systems in prokaryotes and eukaryotes. Bacterial organisms harbor different complements of the NIF (nitrogen fixation), ISC (iron-sulfur cluster), and SUF (sulfur mobilization) systems. The NIF system is specialized for the assembly of nitrogenase in azototrophic bacteria. In bacteria like *E. coli*, both the ISC and SUF systems are present, while Pseudomonads contain only the ISC system and Cyanobacteria contain only the SUF system, and in *Dickeya dandatii* all three systems are present. The ISC assembly machinery of mitochondria is likely inherited from an ancestor of α-proteobacteria, the evolutionary origin of these organelles. The SUF machinery of plastids in higher plants has been likely inherited by endosymbiosis of a photosynthetic bacterium. The cytosolic iron-sulfur protein assembly (CIA) machinery for the maturation of cytosolic and nuclear Fe-S proteins depends on the mitochondrial ISC assembly machinery. These three systems are highly conserved in eukaryotes from humans to yeast and plants.

Fe-S clusters are mainly bound to proteins by cysteine or histidine residues either in the rhombic [2Fe-2S] or the cubic [4Fe-4S] forms (Figure 1) [20]. In model organisms, like *E. coli*, the Fe-S cluster assembly pathways have been well studied. *E. coli* possesses two systems for Fe-S cluster assembly, which share the same basic principles in cluster assembly. While the ISC machinery is transcribed by the *iscRSUA-hscBA-fdx-iscX* operon, the SUF machinery is organized in the *sufABCDSE* operon (Figure 3) [21–23]. As a starting point, the L-cysteine desulfurases IscS or SufS convert L-cysteine to L-alanine and provide the sulfur in the form of a protein-bound persulfide [23]. IscS was shown to act as housekeeping L-cysteine desulfurase [11], while SufS acts under conditions of iron-limitation and oxidative stress [24]. For house-keeping Fe-S cluster assembly, IscS interacts with IscU, thereby making IscU accessible to receive the persulfide sulfur from IscS. In this step, IscU serves as a scaffold protein for the initial assembly of [2Fe-2S] clusters and [4Fe-4S] clusters (Figure 3) [25,26]. The iron source for nascent Fe-S cluster formation has not been identified yet; however, several proteins have been discussed as candidates [27]. IscS and IscU proteins form a heterotetrameric complex together with CyaY, the bacterial homologue to frataxin (see below) [26,28]. During this step of the assembly of Fe-S clusters electrons are required for persulfide reduction. These electrons are most likely provided by ferredoxin (Fdx) [29–31]. Initially, one [2Fe-2S] cluster is formed per IscU monomer inducing a conformational change within the IscU protein that decreases the stability of the IscS–IscU interaction [23]. This step is expected to further enable the reductive coupling of two [2Fe-2S] clusters to form one single [4Fe-4S] cluster on IscU.

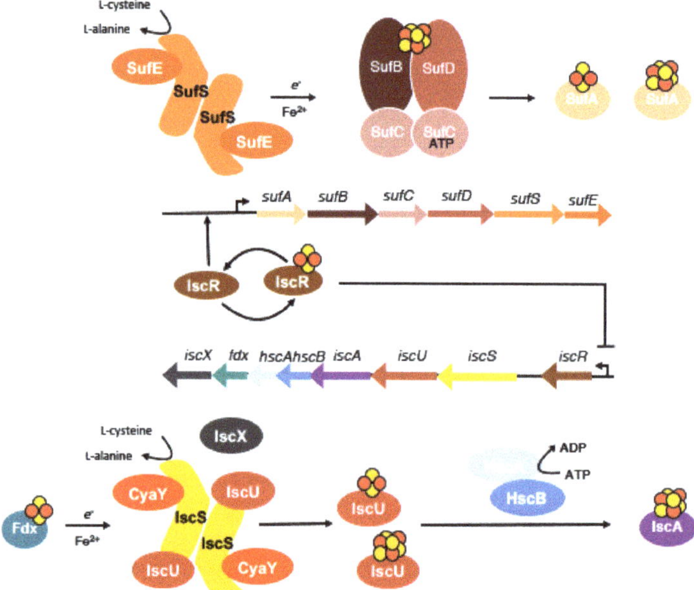

Figure 3. A model for the assembly of Fe-S clusters in *E. coli* by the ISC and SUF machinery. For each system, the proteins involved in each step are indicated in addition to their operon organization. The building of the Fe-S clusters in both systems is facilitated by the scaffold proteins IscU or SufB. After the formation of [2Fe-2S] or [4Fe-4S] clusters on IscU, their release is catalyzed with the help of the HscBA co/chaperones in the ISC system. This step in particular still needs to be clarified in the SUF machinery. After formation, the clusters are transferred by the A-type carrier proteins IscA and SufA to target proteins. Both systems are tightly regulated by IscR at the level of Fe-S cluster availability. Under normal conditions, the IscR regulator exists in an [2Fe-2S] cluster bound form and represses its own expression to control Fe-S cluster formation and the Fe-S state of the cell. Under iron limitation, the IscR regulator is accumulated and converted to its apo-form that activates the *suf* operon. The model shows a simplified version of the regulation of both systems, not depicting the regulation by small RNAs or oxygen.

For the release of Fe-S clusters from IscU, HscA, and HscB are involved in an ATP-dependent manner, two members of the DnaK/DnaJ chaperones/co-chaperone family [32]. HscA recognizes a specific motif on IscU, and their interaction is additionally regulated by the co-chaperone, HscB. The mechanism by which the chaperone facilitates cluster release from IscU has been proposed to involve two conformational states of IscU with different affinities to the bound Fe-S clusters [33]. The chaperones thereby favor the low-affinity IscU state and facilitate the release of the Fe-S cluster from IscU.

In comparison, the SUF system has also been well characterized from studies in bacteria [4,24,34–37]. Here, SufS forms a complex with SufE, which together mobilize the sulfur for cluster assembly. In the SUF machinery, SufB is the Fe-S scaffold protein that acts in conjunction with SufC (and in some cases additionally with SufD) [38]. After Fe-S cluster formation, SufA then transfers the Fe-S clusters to target apo-proteins [39] (Figure 3).

Numerous Fe-S carriers have been identified in both prokaryotes and eukaryotes [4]. These include as main carriers the so-called A-type carriers (ATC) IscA, SufA, and ErpA (Figure 4) [40,41]. Other carriers include the highly-conserved NFU-type proteins [42], the monothiol glutaredoxins (Grx 5 in yeast and GrxD in *E. coli*) [43,44], or the P-loop NTPases, (Ind1 in mitochondria, ApbC in Salmonella) [45,46].

Previous phylogenetic studies had classified ErpA and IscA into two different families; while ErpA belongs to family ATC-I, IscA was grouped into family ATC-II [40]. While ATC-I family members interact with the apo-target proteins, the ATC-II family members are predicted to interact with the scaffold proteins instead. However, the ATC proteins were also shown to replace each other in their roles.

The expression of the SUF and ISC system has been revealed to be tightly regulated in *E. coli*. While the ISC system is the house-keeping Fe-S cluster assembly system, the SUF system instead is mainly synthesized under iron-limiting conditions [24]. One of the main regulators that regulate the expression of either the ISC system or the SUF system is the IscR protein [4]. IscR is a transcriptional regulator that exists in the apo-form and in a [2Fe-2S] cluster bound form in the cell [47]. In its [2Fe-2S] cluster bound form, IscR represses its own expression in addition to that of *iscRSUA-hscBA-fdx-iscX*. In contrast, in its apo-form, IscR activates the expression of the SUF system (Figure 3) [4]. This mechanism allows IscR to fine-tune Fe-S cluster synthesis in response to the presence of synthesized Fe-S clusters and iron availability in the cell.

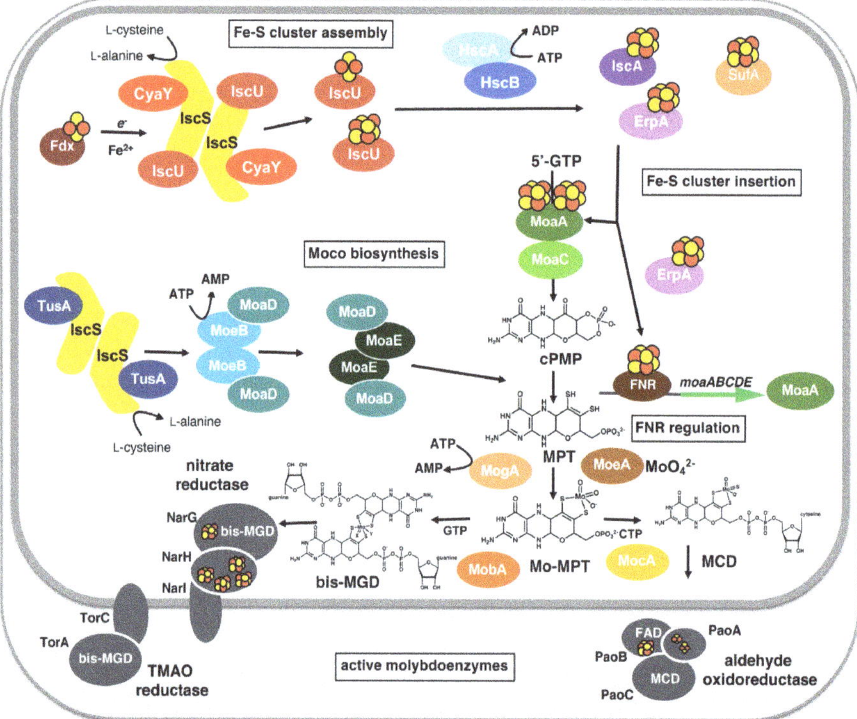

Figure 4. Moco biosynthesis and the link to Fe-S cluster assembly in *E. coli*. In Moco biosynthesis, Mo-MPT is formed from 5′GTP with cPMP and MPT as stable intermediates. For enzymes of the DMSO reductase family like TorCA or NarGHI, bis-MGD is formed by the addition of GMP to each MPT unit in the bis-Mo-MPT intermediate. For enzymes like PaoABC, Mo-MPT is further modified by the addition of CMP to form the MCD form of the cofactor. Molybdoenzymes, in general, are complex enzymes, containing additional cofactors like FAD, Fe-S clusters, or hemes. The sulfur for the synthesis of the dithiolene group in Moco is mobilized by IscS with the additional involvement of the TusA protein. CyaY interacts with the IscS-IscU complex and forms a central heterotetramer. Fe-S clusters assemble on the scaffold protein IscU, which receives the sulfur from the L-cysteine desulfurase IscS. Assembly and release of the clusters is catalyzed by the chaperones HscAB. Ferredoxin is delivering the electrons. The carrier proteins IscA and ErpA deliver the matured Fe-S clusters to recipient protein among which is MoaA in Moco biosynthesis.

3. A General Scheme for the Biosynthesis of the Molybdenum Cofactor

The biosynthesis of the molybdenum cofactor (Moco) is highly conserved among all kingdoms of life. The chemical nature of Moco was first determined by Rajagopalan and coworkers in 1982 [48]. The pathway for the biosynthesis can generally be divided into three steps, with the characteristic of each step being the formation of a stable intermediate (Figure 4) [49–52]. The first step represents the synthesis of cyclic pyranopterin monophosphate (cPMP) from 5′-GTP [53], the second step the introduction of two sulfur atoms into cPMP forming molybdopterin (MPT) by [54], and the third step is the insertion of molybdate into MPT, resulting in the formation of the so-called Mo-MPT cofactor [55]. In prokaryotes, Mo-MPT is further modified by the attachment of GMP or CMP to the phosphate group of MPT, forming the two dinucleotide variants of Moco, MPT-guanine dinucleotide (MGD), [56] and MPT-cytosine dinucleotide (MCD) [57,58] (Figure 4). The characteristics of different forms of Moco are represented further by different ligands at the molybdenum atom. This resulted in the categorization of molybdoenzymes into three families (Figure 5): the xanthine oxidase (XO) family, the sulfite oxidase (SO) family, and the dimethyl sulfoxide (DMSO) reductase family. Enzymes of the DMSO reductase family generally contain two MPT moieties ligated to the molybdenum atom and are present only in prokaryotes [50]. Here, we will briefly summarize the biosynthesis of Moco.

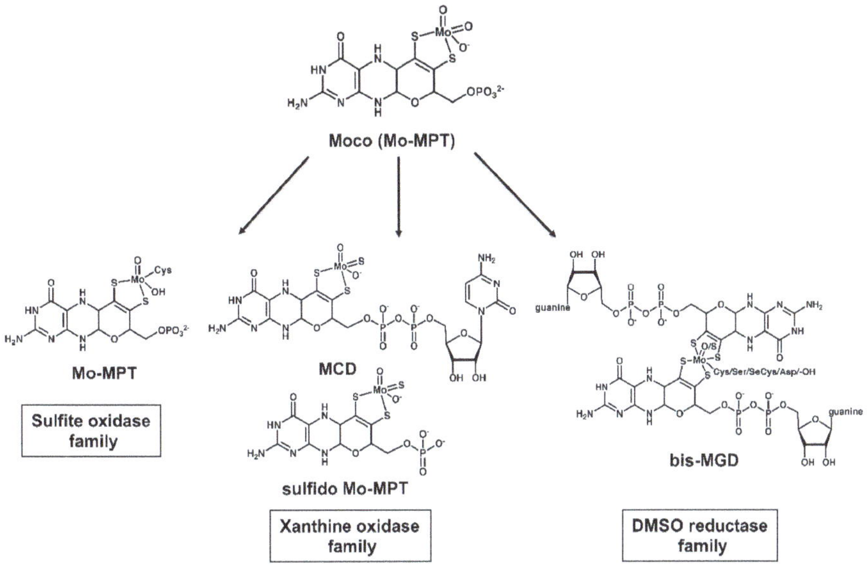

Moco (Mo-MPT)

Mo-MPT

Sulfite oxidase family

MCD

sulfido Mo-MPT

Xanthine oxidase family

bis-MGD

DMSO reductase family

Figure 5. The three families of molybdoenzymes in pro- and eukaryotes. Moco exists in different variants and is divided into three enzyme families. The basic form of Moco is Mo-MPT that coordinates the molybdenum atom in a tri-oxo form. The SO family is present both in pro-and eukaryotes and is characterized by a molybdenum ligation with one oxo-, one hydroxide- and one cysteine ligand from the protein backbone. The XO family is also present in both pro- and eukaryotes and is characterized by an equatorial sulfido-ligand, an apical oxo ligand, and one hydroxide or O^- ligand. The XO family contains the sulfurated form of Mo-MPT. In *E. coli*, an additional nucleotide is present in this family, forming the molybdopterin cytosine dinucleotide cofactor (MCD). The DMSO reductase family contains two MPTs (bis-Mo-MPT) or two MGDs (bis-MGD) ligated to one molybdenum atom with additional ligands being an O/S and a sixth ligand which can be a serine, a cysteine, a selenocysteine, an aspartate or a hydroxide ligand.

In the first step of Moco biosynthesis, 5′-GTP is converted to cPMP (Figure 4) [53]. cPMP is a 6-alkyl pterin with a cyclic phosphate group at the C2′ and C4′ atoms [59]. This reaction is catalyzed by two enzymes, MoaA and MoaC [53,60] (Figure 4). While the individual catalytic functions of MoaA and MoaC have long been unknown, recent studies showed that MoaA catalyzes the conversion of GTP to (8S)-3′,8-cyclo-7,8-dihydroguanosine 5′triphosphate (3′,8-cH_2GTP), and MoaC catalyzes the conversion of 3′,8-cH_2GTP to cPMP [15]. Details of the recent updates and the details of MoaA mechanism, where [4Fe-4S] clusters play central roles, will be discussed in the next chapter.

In the second step of Moco biosynthesis, two sulfur atoms are inserted into cPMP, and MPT is formed as a stable intermediate [54,61–65]. This reaction is catalyzed by MPT synthase, composed of two MoaD and two MoaE subunits (Figure 4) [66]. The sulfur atoms required for this reaction are present at the C-terminus of MoaD in form of a thiocarboxylate group [67,68]. Studies on the reaction mechanism showed that the first sulfur is added at the C2′ position of cPMP in a reaction that is coupled to the hydrolysis of the cPMP cyclic phosphate group [65]. The second sulfur is then transferred to the C1′ of the hemisulfurated intermediate and MPT is formed as the product.

In the third step of Moco biosynthesis, molybdate is inserted to the dithiolene sulfurs of MPT and Mo-MPT is formed, one of the forms of Moco. The specific insertion of molybdenum into MPT is catalyzed by the joined action of MoeA and MogA (Figure 4) [69,70]. MogA thereby hydrolyzes ATP and forms the activated MPT-AMP intermediate [71]. This intermediate is then transferred to MoeA, which mediates molybdenum ligation at low concentrations of MoO_4^{2-} [70]. Mo-MPT can be further modified by nucleotide addition in the next step of Moco biosynthesis, which is present only in prokaryotes (Figure 5) [72]. Alternatively, the Mo-MPT cofactor can be directly inserted into enzymes of the SO family. In this family, the molybdenum atom of Mo-MPT is coordinated by a cysteine from the polypeptide backbone of the protein, representing the cofactor in its MPT-$Mo^{VI}O_2$ form [50].

The proteins of the DMSO reductase family in bacteria contain the bis-MGD cofactor (Figure 5) [50]. The synthesis of the bis-MGD was shown to occur in a two-step reaction catalysed by MobA using Mo-MPT and Mg-GTP as substrates [72]. In the first reaction, the bis-Mo-MPT intermediate is formed on MobA from two Mo-MPT molecules [73]. In the second reaction, two GMP moieties from GTP are added to the C4′ phosphate of bis-Mo-MPT, forming the bis-MGD cofactor [74,75]. After the formation of bis-MGD, the cofactor can be released from MobA and inserted into target enzymes. However, it is expected that bis-MGD does not exist in a free form in the cell. Formed bis-MGD is rather expected to be immediately recruited by Moco-binding chaperones that protect the cofactor from oxidation and specifically interact with their target apo-molybdoenzymes for bis-MGD insertion. After insertion, bis-MGD is generally ligated by a serine, a cysteine, a selenocysteine, or an aspartate from the protein backbone. The other ligand in enzymes of the DMSO reductase family that bind the bis-MGD cofactor is an oxo- or sulfido atom.

The addition of CMP results in the formation of the MPT-cytosine dinucleotide (MCD) cofactor [58], a cofactor that is present in the xanthine oxidase family of molybdoenzymes in bacteria (Figure 5). The formation of MCD is catalyzed by MocA (Moco cytidylyltransferase). After the formation of MCD and the bis-MGD cofactor, a further modification at the molybdenum atom can occur by the addition of a terminal sulfido-ligand [76]. The sulfido-ligand at the equatorial position of the molybdenum atom is a characteristic feature of enzymes in the xanthine oxidase family. In eukaryotes, enzymes of this family do not harbor the additional CMP modification of the cofactor (Figure 5).

4. Linking Moco Biosynthesis and Fe-S Cluster Assembly in Bacteria

More than 60 different Moco-containing molybdoenzymes have been identified to date [77]. In recent years it has become evident that the biosynthesis of Moco and the assembly of Fe-S clusters are directly connected to each other. Moco biosynthesis directly depends on the presence of Fe-S clusters or components of the Fe-S cluster assembly machinery on several levels (Figure 4). Many molybdoenzymes bind Fe-S clusters as additional cofactors that are involved in intramolecular electron transfer reactions. In Moco biosynthesis, the MoaA protein harbors two [4Fe-4S] clusters and thus directly depends on Fe-S cluster assembly. In addition, the L-cysteine desulfurase IscS is shared between Fe-S cluster assembly and Moco biosynthesis since it also mobilizes the sulfur for the synthesis of the dithiolene group present in Moco. Further, the expression of most molybdoenzymes and proteins involved in Moco biosynthesis in bacteria is regulated by the transcriptional regulator for fumarate and nitrate reduction, FNR [78,79]. The activity of FNR itself is directly dependent on the availability of Fe-S clusters under anaerobic conditions; consequently, Moco is not synthesized and molybdoenzymes are not expressed when Fe-S clusters are not assembled (Figure 4).

4.1. The Involvement of Radical SAM Chemistry for Moco Biosynthesis

The first step of Moco biosynthesis, the conversion of GTP into cPMP, directly depends on the assembly of [4Fe-4S] clusters, which proceeds through a complex rearrangement reaction, where C8 of guanine is being inserted between C2′ and C3′ of ribose [60]. During this reaction, MoaA plays a key role. MoaA is the only protein that binds Fe-S clusters in the pathway of Moco biosynthesis and was grouped into the superfamily of radical/SAM enzymes. In general, the mechanism of radical SAM chemistry involves the $[4Fe\text{-}4S]^{2+}$ cluster bound to the C-X_3-C-X_2-C motif located at the N-terminus of radical SAM superfamily enzymes (Figure 6).

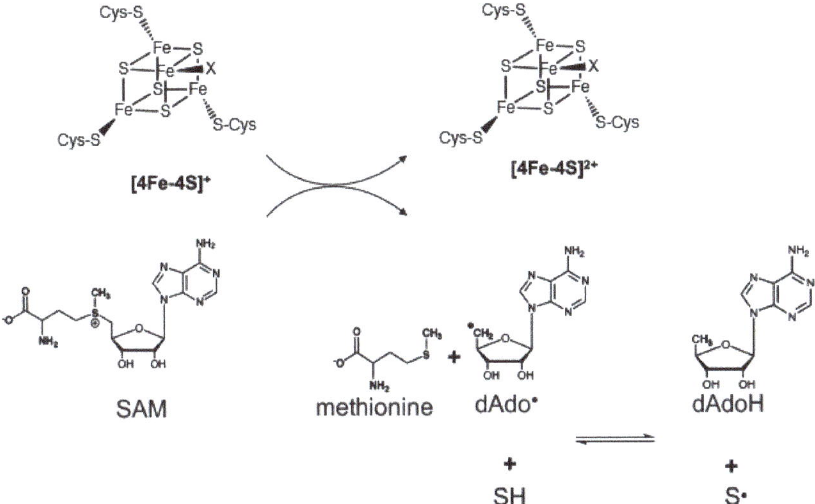

Figure 6. A general scheme for the mechanism of radical-SAM enzymes. The [4Fe-4S] that is bound to the conserved C-X_3-C-X_2-C motif provides the electron required for the reductive cleavage of SAM. In this reaction, methionine and the adenosyl radical (dAdo·) is generated. The formed dAdo· then abstracts a hydrogen atom from the substrate (SH), and a substrate radical (S·) and dAdoH are formed.

The cluster is reduced by one electron to the catalytically-active state and coordinates the substrate 5′S-adenosyl-methionine (SAM). In the reduced $[4Fe\text{-}4S]^{+}$ state, the cluster transfers an electron to the sulfonium sulfur of SAM, thereby promoting homolytic S–C bond cleavage to generate the 5′-deoxyadenosyl radical intermediate (dAdo·) and methionine. The dAdo· intermediate then abstracts a hydrogen atom from the substrate. The 5′-deoxyadenosine (dAdoH) is formed, and the resulting product radical intermediate may either be the end product or can undergo further transformations (Figure 6).

In the first step of Moco biosynthesis, MoaA and MoaC catalyze the conversion of 5′GTP into cyclic pyranopterin monophosphate (cPMP). MoaA is a member of the radical SAM superfamily and binds the characteristic [4Fe-4S] cluster at the N-terminus and an additional [4Fe-4S] at the C-terminus. For Fe-S cluster insertion into MoaA, recent studies showed that both ErpA and IscA can provide Fe-S clusters to MoaA. Since Δ*erpA*/Δ*iscA* double mutant strains were revealed to be completely devoid of Moco, it was concluded that SufA is unable to substitute the roles of both A-type carrier proteins for Moco biosynthesis (unpublished results).

The mechanism of cPMP formation was first investigated using isotope labeling experiments, which indicated that the C-8 atom of the guanine base of GTP is inserted between the C2′ and C3′ atoms of the ribose moiety.

Recent studies showed that MoaA catalyzes the conversion of GTP into a 3′,8-cyclo-7,8-dihydroguanosine 5′-triphosphate (3′,8-cH_2GTP) intermediate (Figure 7). The conversion of GTP into 3′,8-cH_2GTP by MoaA proceeds through a radical formation at C3′ by the abstraction of the H-3′ atom of GTP by 5′-dAdo· [15,80]. The free radical generated at the C3′ of the ribose has been revealed by isotope labeling studies [15,80], where a deuterium atom at the 3′-position was shown to be transferred to dAdoH. The resulting C3′ centered radical attacks C8 to form the aminyl radical intermediate, which is then reduced by transfer of an electron and a proton to form 3′,8-cH_2GTP. Therefore, MoaA catalyzes the C–C bond formation between the GTP C3′ and the C8 of the guanine, resulting in the

3′,8-cH_2GTP intermediate as the end product of the MoaA catalyzed reaction. The conversion of 3′,8-cH_2GTP to cPMP is catalyzed by MoaC in the next step of the reaction [15]. It has been proposed that two loops in MoaC provide conformational flexibility of the enzyme that facilitates a general acid/base-catalyzed mechanism for the formation of the pyranopterin structure that is coupled to the cyclic phosphate ring formation (Figure 7). This step catalyzed by MoaC involves the irreversible cleavage of the pyrophosphate group that has been proposed to provide the thermodynamic driving force for the overall reaction.

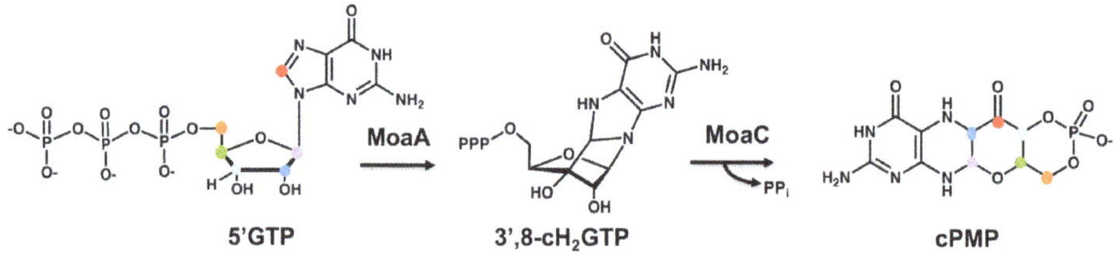

Figure 7. The conversion of 5′GTP to cPMP involving the functions of MoaA and MoaC. The colors on 5′GTP and cPMP indicate the source of the carbon and nitrogen atoms in cPMP as determined by isotope labeling studies. The reaction involves the formation of 3′,8-cH2GTP as the product of MoaA and the substrate for MoaC.

4.2. *Sulfur Mobilization Involves Sharing of Protein Functions in Prokaryotes*

In the next step of Moco biosynthesis, cPMP is converted to MPT in a reaction catalyzed by MPT synthase [54,61–64]. In this reaction, two sulfur atoms are inserted into the C2′ position of cPMP and the C1′ position of the formed hemisulfurated intermediate [65]. In bacteria, MPT synthase is composed as a $(\alpha\beta)_2$ heterotetramer of two central MoaE subunits, each binding one MoaD subunit (Figure 4) [66]. The sulfur atoms required for this reaction are bound as thiocarboxylate groups at the C-terminus of each MoaD subunit [67,68]. After the sulfur transfer reaction, the thiocarboxylate group is regenerated on each MoaD subunit. This reaction is catalyzed by MoeB under ATP consumption [81,82]. In the course of the reaction, MoaD dissociates from MoaE and reassociates with MoeB in a $(\alpha\beta)_2$ heterotetramer. In the first step, an acyl-adenylate group is formed at the C-terminus of MoaD [81,83,84]. In a second step, sulfur is directly transferred from a sulfur transferase to the activated MoaD-AMP C-terminus, and the MoaD thiocarboxylate (MoaD-COSH) is rebuilt. MoaD-COSH then dissociates from the $(\text{MoeB-MoaD})_2$ complex and MoaD reassociates with MoaE [84,85]. The proteins IscS and TusA are proposed to be involved in the sulfur transfer reaction for the formation of the MoaD thiocarboxylate group in *E. coli* (Figure 4) [86,87]. It has been proposed that IscS first forms a persulfide group on TusA that is then reductively cleaved and transferred to MoaD by attacking the MoaD–AMP bond [88,89].

After MPT formation, the dithiolene group of MPT serves as the backbone for molybdenum ligation. This step is catalyzed by MogA and MoeA under ATP consumption by the formation of an MPT-AMP intermediate catalyzed by MogA. After molybdate insertion into MPT-AMP in a MoeA-dependent reaction, the resulting tri-oxo Mo-MPT is either inserted into enzymes of the SO family or is further modified by nucleotide (GMP or CMP) addition, forming the MCD or bis-MGD cofactors, respectively.

4.3. *The Insertion of Different Cofactors into Molybdoenzymes in Bacteria*

Molybdoenzymes are generally composed of different subunits harboring additional prosthetic groups, such as cytochromes, the Fe-S cluster, or FAD/FMN that are involved in intramolecular electron transfer reactions (Figure 4) [50]. The molybdenum atom thereby exists in the oxidation states VI, V, or IV under physiological conditions and acts as a transducer between 2 electron transfer and 1 electron transfer processes often coupled to proton transfer.

In general, Moco is deeply buried within the enzyme that is accessible via a substrate-binding funnel leading to the molybdenum atom [90]. It has been suggested that Moco insertion is accomplished by molecular chaperones that induce the final folding of the enzymes after Moco insertion [91]. The insertion of the different forms of Moco has been best studied in bacteria. In *E. coli*, many molybdoenzymes are located in the periplasm and require the Tat system for their translocation. The transport of enzymes occurs in the folded state after the insertion of the Moco and Fe-S clusters or other cofactors. This translocation presents a "quality control" step accomplished by the chaperones as Moco insertases and ensures that only the matured enzymes that contain Moco are translocated. After subunit assembly, Moco is inserted, final folding of the enzyme is accomplished, the enzyme is directed to the Tat-translocon and finally exported to the periplasm.

For Fe-S cluster insertion into enzymes, several A-type carriers have been identified that facilitate this process [40]. These proteins, which are SufA, IscA, or ErpA in *E. coli*, carry and insert the Fe-S clusters to designated target proteins. Previous reports suggested that ErpA is essential for the formation of an active formate-nitrate reductase complex in *E. coli* [92]. It was shown that *E. coli erpA* mutant strains were devoid of formate dehydrogenase and nitrate reductase activities. In these studies, IscA was able to partially complement the *erpA* mutant, showing that these proteins might have overlapping roles, but ErpA seems to be the more specific enzyme for nitrate reductase and formate reductase maturation [92]. However, the overlapping effects of Fe-S cluster assembly and biosynthesis and insertion of Moco were not accounted for in this study. Since MoaA contains Fe-S clusters, the effect of the *erpA* mutant might have already been a decrease in MoaA activity, leading to a lack of Moco in the cell. Recent studies revealed that NarG is not expressed in *erpA* mutant strains and that inactive FNR precludes the expression of the *narGHJI* operon (unpublished results). Additionally, it was shown that ErpA and IscA are involved in Fe-S cluster insertion into MoaA (unpublished results), so that the inactivity of nitrate reductase and formate dehydrogenase in *erpA* mutant strains is likely rather based on the lack of Moco. It still remains possible that, in addition, ErpA and IscA are involved in Fe-S cluster insertion into these molybdoenzymes. However, based on the lack of expression of the FNR-regulated operons, the involvement of these proteins cannot be analyzed. Thus, it is difficult to dissect the combined effects of Moco biosynthesis and Fe-S cluster insertion for molybdoenzyme maturation.

5. Compartmentalization of Fe-S Cluster and Moco Biosynthesis in Eukaryotes: The Role of Mitochondria

5.1. Mitochondrial Fe-S Cluster Biosynthesis in Eukaryotes

In eukaryotes, the main Fe-S cluster assembly is localized in the mitochondria, which were derived from bacteria by endosymbiosis. The key component for Fe-S cluster biosynthesis in mitochondria is the L-cysteine desulfurase NFS1, which forms the central dimer (Figure 8).

The enzyme serves as the general sulfur donor for cellular Fe-S cluster synthesis, also for cytosolic Fe-S clusters. At the NFS1 dimer interface, a dimer of ISD11 is bound that stabilizes NFS1 [93–95]. While ISD11 is conserved in eukaryotes [96], a prokaryotic homolog of ISD11 is not known [93]. ISD11 belongs to the large family of LYRM proteins that fold into a triple-helical bundle [97]. ISD11 contributes to the interaction with the acyl carrier protein ACP1 [93]. While ACP1 is not needed for efficient synthesis of the [2Fe-2S] cluster on ISCU2, the ISD11-ACP1 sub-complex was proposed to regulate Fe-S cluster assembly by linking the energy load of the cells to the Fe-S cluster assembly complex [98]. Each monomer of the NFS1 dimer binds frataxin (FXN), ISCU2, and FDX2 at the two opposite ends. For the conversion of the persulfide sulfur (S^0) to sulfide (S^{2-}), electrons are required, which are provided by FDX2 [99–102]. First, a [2Fe-2S] cluster is formed on ISCU2, by combining Fe^{2+} entry from a still unresolved iron donor and S^{2-} provided by NFS1. The role of FXN in this process is still under debate.

The next step in mitochondrial Fe-S protein biogenesis involves the dissociation of the [2Fe-2S] cluster from ISCU2, its trafficking to the monothiol glutaredoxin GLRX5 as an intermediate Fe-S cluster binding partner, and the specific insertion into target proteins (Figure 8) [103–105]. This reaction requires the roles of the dedicated chaperone system of the Hsp40-Hsp70 (DnaJ-DnaK) class [106,107]. For cluster transfer, the DnaJ co-chaperone HSC20 binds to a hydrophobic motif in ISCU2 [97,108,109]. The HSC20-ISCU2 complex then recruits the Hsp70 chaperone HSPA9. In the next step, the cluster is transferred to GLRX5. For [2Fe-2S] cluster transfer to target proteins, no further proteins are required [110]. However, since GLRX5 is not essential, the protein function can be bypassed by direct transfer of [2Fe-2S] from ISCU2.

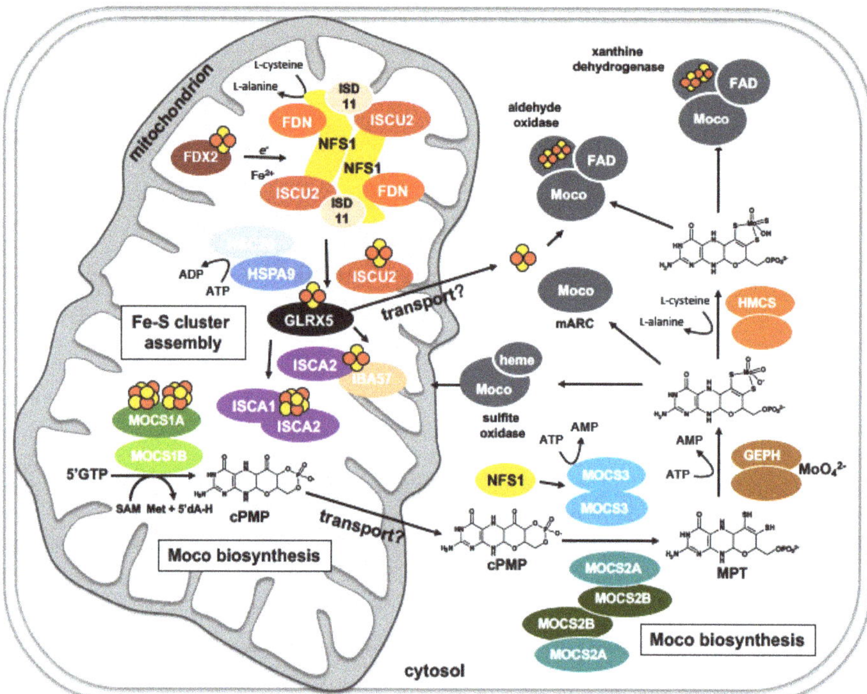

Figure 8. The different compartments for the biosynthesis of Moco in humans and the link to Fe-S cluster assembly in mitochondria. The first step of Moco biosynthesis, the conversion of 5′GTP to cPMP catalyzed by MOCS1A and MOCS1B, is localized in mitochondria. This is also the main compartment for Fe-S cluster biosynthesis in eukaryotes. Fe-S clusters assemble on the scaffold protein ISCU2, which receives the sulfur from the L-cysteine desulfurase complex NFS1/ISD11. Frataxin (FDN) interacts with the NFS1/Isd11-ISCU2 complex to form the quaternary complex. Ferredoxin (FDX2) delivers the electrons for the process. Assembly and release of the clusters from ISCU2 are facilitated by the chaperones HSC20/HSPA9. The carrier proteins GLRX5, ISCA2, and/or IBA57 deliver the Fe-S clusters to target proteins, like MOCS1A in Moco biosynthesis. Synthesized cPMP by MOCS1A/B needs to be transferred to the cytosol, where all further modification steps of Moco are catalyzed. These steps involve the conversion of cPMP to MPT by MOCS2A/MOCS2B (which are activated by MOCS3), the insertion of molybdate by GEPHYRIN (GEPH), and the insertion of Mo-MPT either to the mARC protein (localized at the outer mitochondrial membrane) or to sulfite oxidase (before its translocation to the mitochondrial intermembrane space). For aldehyde oxidase and xanthine dehydrogenase, the formation of the equatorial sulfido-ligand is transferred by HMCS. Dual localization of NFS1 both in mitochondria and the cytosol is predicted. Cytosolic NFS1 acts as a sulfur donor for MOCS3.

The assembly and insertion of [4Fe-4S] clusters additionally require the A-type ISC proteins ISCA1-ISCA2 and IBA57 (Figure 8) [111–114]. How these proteins mechanistically assist the fusion of the [2Fe-2S] into a [4Fe-4S] cluster is a complex process involving the transfer of [2Fe-2S] clusters from

GLRX5 to ISCA1-ISCA2 and the assembly of a [4Fe-4S] cluster on the ISCA1-ISCA2 complex [115–117]. IBA57 is not required in this step. IBA57 was shown to form a [2Fe-2S] cluster-mediated complex specifically with ISCA2, involving GLRX5 [111,118,119]. Mitochondrial [4Fe-4S] cluster biosynthesis is a dynamic system, also involving Nfu1, Ind1 and BolA3 in some cases (recently reviewed in Reference [120]). The Fe-S cluster insertion for the human MoaA homologue MOCS1A (see below) has not been investigated so far.

5.2. Cytosolic Fe-S Cluster Assembly: The CIA Machinery

For cytosolic Fe-S cluster assembly, the mitochondrial ISC system has been proposed to be essential [121,122]. Studies by several groups have shown that the mitochondrial ISC machinery generates a sulfur-containing factor "X-S" that is exported to the cytosol via the mitochondrial ABC transporter ABCB7 and is used for cytosolic Fe-S cluster assembly by the CIA machinery (Figure 9) [12,123–125]. However, models also exist that propose that cytosolic versions of NFS1, ISCU, and FDN are involved in cytosolic Fe-S cluster formation [120]. The CIA machinery is composed of up to 13 known proteins that assemble both cytosolic and nuclear Fe-S proteins. Initially, a [4Fe-4S] cluster is assembled on the CIA scaffold complex formed between CFD1-NBP35 (Figure 9) [125,126]. The initial cluster synthesis on CFD1-NBP35 further requires the electron transfer chain composed of the flavin-dependent oxidoreductase NDOR1 and CIAPIN1, however, the precise role of the electron-transfer for cluster synthesis is not completely understood yet [127,128]. The insertion of the two Fe-S clusters of CIAPIN1 additionally requires the cytosolic monothiol glutaredoxin reductase GLRX3 [129–131]. GLRX3 binds a bridging [2Fe-2S] cluster with BOLA2 which is further transferred to CIAPIN1 (Figure 9) [130,132]. CIAPIN1 can coordinate a pair of [2Fe-2S] clusters or a [2Fe-2S] cluster and a [4Fe-4S] cluster [115,128,133,134]. While the BOLA2-GLRX3 complex is able to transfer [2Fe-2S] clusters to CIAPIN1, the formation of the [4Fe-4S] cluster in CIAPIN1 is still fully undefined.

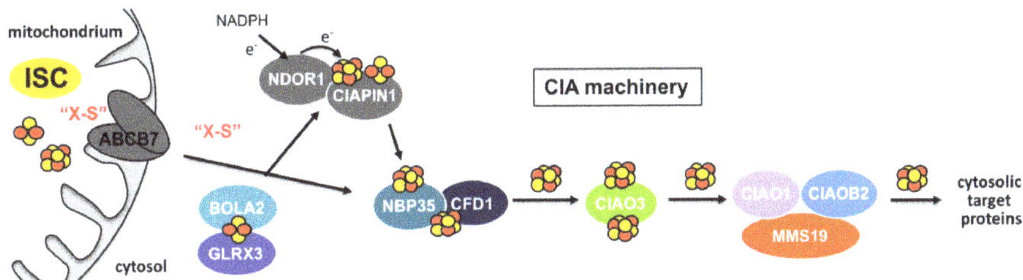

Figure 9. Model of the cytosolic Fe-S cluster assembly (CIA) in humans. Assembly of cytosolic Fe-S clusters starts in mitochondria with components of the early ISC machinery synthesizing a sulfur-containing precursor (X-S) that is subsequently exported to the cytosol by the ABC transporter ABCB7. The CIA machinery involves [4Fe-4S] cluster assembly by the CFD1-NBP35 complex serving as a cluster scaffold. This reaction requires electron input from the flavin-oxidoreductase NDOR1 and the Fe-S protein CIAPIN1. The complex of GLRX3-BOLA2 facilitates [2Fe-2S] cluster insertion into CIAPIN1. In the next step, the CFD1-NBP35-bound [4Fe-4S] cluster is released and transferred to the majority of target Fe-S apoproteins with the help of CIAO3. Fe-S cluster targeting to individual apoproteins is achieved by CIAO1, CIAO2B, and MMS19.

The next step of cytosolic Fe-S assembly involves the trafficking of the [4Fe-4S] cluster from the CFD1-NBP35 complex to CIAO3 and then to the CIA targeting complex (CTC) which is composed of CIAO1, CIAO2B, and MMS19 [135,136]. From this complex, which can be formed with different protein components, the cluster is transferred directly to target proteins (Figure 9). The specific interaction with the target proteins is mediated by the CTC proteins.

5.3. The Formation of cPMP Is Localized in Mitochondria

The conversion of 5′GTP to cPMP is localized in mitochondria in humans in a reaction that is catalyzed by MOCS1A and MOCS1B [137]. As explained above, this compartment is also the main compartment for the synthesis of Fe-S clusters (Figure 8). Since MOCS1A requires two [4Fe-4S] clusters for activity, it remains speculative that linking that first step of Moco biosynthesis to mitochondrial Fe-S cluster assembly provided a mechanistic advantage [138]. Since MOCS1A and MOCS1B are highly homologous to their bacterial counterparts, MoaA and MoaC, respectively, the conversion of 5′GTP to cPMP is catalyzed by the same mechanism and is therefore not described in detail again (Figure 7). After cPMP formation, the molecule has to be transported to the cytosol, where all further steps of Moco biosynthesis are catalyzed. For plants, it has been suggested that the export of cPMP involves the transporter protein Atm3 [139]. The human counterpart to Atm3 is ABCB7 (Figure 9). Surprisingly, Atm3, like ABCB7 has initially been suggested to transport the "X-S" species essential for cytosolic Fe-S cluster assembly. The precise role of Atm3 is consequently still unknown and the defects in molybdoenzyme activities in Atm3 mutants might also be explained by an overlapping defect in Fe-S cluster assembly. Future studies are necessary to further explore the export of cPMP from mitochondria to the cytosol.

5.4. Moco Is Formed in the Cytosol

After its transport to the cytosol, cPMP is converted to MPT in the reaction catalyzed by MPT synthase. In humans, MPT synthase is formed by MOCS2A and MOCS2B, the homologues of MoaD and MoeB, respectively (Figure 10). Both the human and bacterial MPT synthases catalyze a similar reaction that is explained in detail for the bacterial counterparts above (Figure 4) [68].

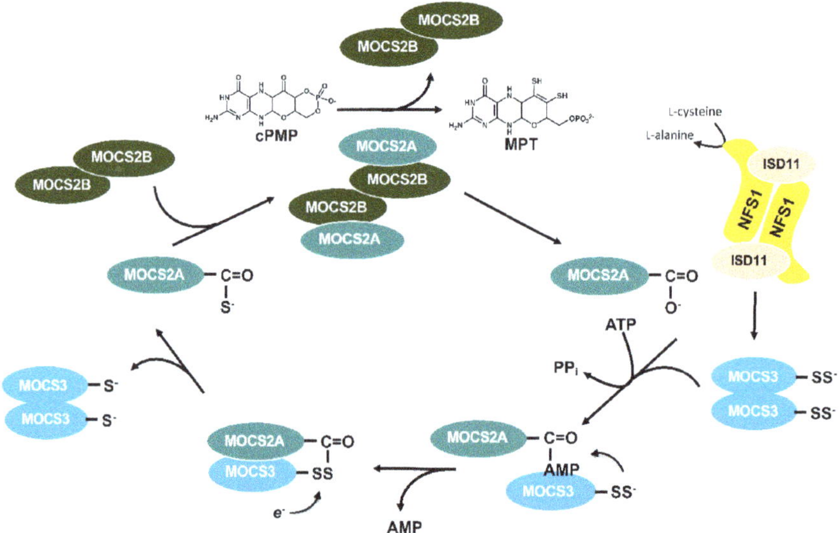

Figure 10. Regeneration of the thiocarboxylate group on MOCS2A. Conversion of cPMP to MPT requires the transfer of two sulfur, a reaction catalyzed by MPT synthase. The formation of the MOCS2A thiocarboxylate group is catalyzed after the formation of the (MOCS2A/MOCS3)$_2$ complex in humans. First, MOCS2A-AMP is formed under ATP consumption. MOCS3 receives the sulfur from NFS1, and a persulfide group is formed at the C-terminal domain of MOCS3. This sulfur is then transferred to MOCS2A with the formation of a perthiocarboxylate group as the intermediate. After reductive cleavage, the thiocarboxylate group on MOCS2A is formed and MOCS2A-SH reassociates with MOCS2B.

After the MPT synthase reaction, the C-terminal carboxylate group on MOCS2A is regenerated by MOCS3 (Figure 10) [140]. MOCS3 contains an N-terminal domain that is homologous to the *E. coli* MoeB protein and a C-terminal domain that shares homologies to sulfur transferases (rhodaneses). First, the N-terminal domain of MOCS3 activates the C-terminus of MOCS2A under ATP consumption (Figure 10). In the second part of the reaction, sulfur is transferred from the C-terminal rhodanese-like domain of MOSC3 to the formed MOCS2A-AMP [141,142]. This reaction differs from the reaction described for the *E. coli* proteins, since the MOCS3 protein itself catalyzes both the adenylation and the sulfur transfer reaction. The persulfide group on MOCS3 has been suggested to be formed by the L-cysteine desulfurase NFS1 in the cytosol [143,144]. It has been proposed that small amounts of NFS1 in the cytosol are sufficient for supplying the sulfur for MPT formation [143]. Since for the CIA machinery cytosolic NFS1 might not be required, it remains possible that the role of NFS1 in the cytosol is restricted to Moco biosynthesis. Further studies are necessary to confirm this.

After the MPT formation, molybdate ion is ligated to the dithiolene group of MPT. GEPHYRIN is the homolog of the bacterial MogA and MoeA proteins (Figure 8) [145]. GEPHYRIN is a two-domain protein with one domain being homologous to MogA and the other domain being homologous to MoeA [146].

The chemistry of molybdenum insertion has been studied in detail for the plant counterpart of GEPHYRIN, named CNX1. The AMP-part of MPT-AMP functions as an anchor on the E-domain while the dithiolene moiety of MPT-AMP points to a separate pocket on the E-domain where molybdate is bound and waits to be inserted. Finally, the pyrophosphate bond between AMP and MPT is hydrolyzed and the newly-formed Moco is released [147].

After the completion of Moco, the cofactor can be directly inserted into the molybdoenzymes sulfite oxidase or mARC which belong to the SO family of molybdoenzymes (Figure 8) [148]. For the two enzymes of the XO family, xanthine dehydrogenase and aldehyde oxidase, Moco is further modified by the formation of an equatorial sulfido-group [149,150]. This reaction is catalyzed by a Moco sulfurase, named HMCS (human molybdenum cofactor sulfurase) before the insertion of the cofactor into molybdoenzymes (Figure 8). HMCS is a homodimeric two-domain protein with an N-terminal domain sharing homologies to the bacterial L-cysteine desulfurases IscS [151] and a C-terminal domain that binds Moco. The Moco-sulfido group is directly formed on Mo-MPT bound to the C-terminal domain by transfer of a persulfide group from the N-terminal domain [152,153]. After sulfuration, Moco is then inserted into the target enzymes xanthine dehydrogenase and aldehyde oxidase (Figure 8) [152,153]. So far, it is not known which enzymes of the CIA machinery insert the two [2Fe-2S] clusters into xanthine dehydrogenase or aldehyde oxidase.

6. Conclusions

In this review, we highlighted the link between the biosynthesis and maturation of molybdoenzymes and the biosynthesis and distribution of Fe-S clusters. Several levels of this link were identified: (a) the synthesis of the first intermediate in Moco biosynthesis requires a radical/SAM-dependent protein; (b) the sulfurtransferase for the dithiolene group in Moco is shared with the synthesis of Fe-S clusters; (c) the modification of the active site with a sulfur atom additionally involves an L-cysteine desulfurase, and (d) most molybdoenzymes require Fe-S clusters as additional redox-active cofactors. While the general pathways of the biosynthesis/assembly of Moco and Fe-S clusters have been studied in detail, numerous open questions still remain. One of the most intriguing questions in Fe-S cluster assembly is the source of the iron atom in the cluster. In particular, iron has to be provided in the correct oxidation state for biological processes. Further, after the biosynthesis of the complex cofactors like Fe-S clusters and Moco, intricate mechanisms have to control distribution, trafficking, and insertion of these cofactors into their specific target proteins. The transfer mechanisms involve cofactor-binding chaperones, as most of these prosthetic groups are extremely fragile and oxygen-sensitive. The specificity of the cluster transfer process into target proteins, their recognition, and the hierarchy in which metal-center insertion for Fe-S clusters and Moco occurs, is still largely unresolved. Also, the compound "X-S" that is delivered by the mitochondrial Fe-S cluster machinery to the CIA machinery is still unknown.

For humans, the molybdoenzyme sulfite oxidase is most crucial for survival and it is located in the mitochondrial intermembrane space. A lack of sulfite oxidase activity usually results in death in early childhood. The dependency on Fe-S clusters for the biosynthesis of Moco directly links Fe-S cluster related diseases to the severe outcome of sulfite oxidase deficiency. The link of Fe-S cluster related diseases to Moco deficiency needs to be investigated in future studies.

Author Contributions: Conceptualization, R.R.M. and S.L.; writing—original draft preparation, R.R.M. and S.L.; writing—review and editing, R.R.M., T.W.H., A.Z., M.A.H. and S.L.; visualization, S.L.; supervision, R.R.M. and S.L.; project administration, R.R.M. and S.L.; funding acquisition, R.R.M. and S.L. All authors have read and agreed to the published version of the manuscript.

Funding: The research was supported by continuous individual grants from the Deutsche Forschungsgemeinschaft (DFG, German Research Foundation) (LE1171/15-2, LE1171/11-2 to S.L. and ME1266/31-1 to R.R.M.) including funding by the DFG Priority Programme SPP1927 'FeS for Life' and by the DFG Research Training Group GRK 2223/1 'PROCOMPAS'.

Acknowledgments: The authors thank all current and former members of their research groups in addition to collaboration partners who were involved in the work over the past years.

Conflicts of Interest: The authors declare no conflict of interest.

References

1. Andreini, C.; Bertini, I.; Cavallaro, G.; Holliday, G.L.; Thornton, J.M. Metal ions in biological catalysis: From enzyme databases to general principles. *JBIC J. Boil. Inorg. Chem.* **2008**, *13*, 1205–1218. [CrossRef] [PubMed]
2. Beinert, H.; Holm, R.H.; Munck, E. Iron-Sulfur Clusters: Nature's Modular, Multipurpose Structures. *Science* **1997**, *277*, 653–659. [CrossRef] [PubMed]
3. Stehling, O.; Wilbrecht, C.; Lill, R. Mitochondrial iron–sulfur protein biogenesis and human disease. *Biochimie* **2014**, *100*, 61–77. [CrossRef]
4. Roche, B.; Aussel, L.; Ezraty, B.; Mandin, P.; Py, B.; Barras, F. Iron/sulfur proteins biogenesis in prokaryotes: Formation, regulation and diversity. *Biochim. Biophys.* **2013**, *1827*, 455–469. [CrossRef]
5. Arnon, D.I.; Whatley, F.R.; Allen, M.B. Triphosphopyridine Nucleotide as a Catalyst of Photosynthetic Phosphorylation. *Nature* **1957**, *180*, 182–185. [CrossRef] [PubMed]
6. Mortenson, L.E.; Valentine, R.C.; Carnahan, J.E. An electron transport factor from Clostridium pasteurianum. *Biochem. Biophys. Res. Commun.* **1962**, *7*, 448–452. [CrossRef]
7. Beinert, H.; Sands, R.H. Studies on succinic and DPNH dehydrogenase preparations by paramagnetic resonance (EPR) spectroscopy. *Biochem. Biophys. Res. Commun.* **1960**, *3*, 41–46. [CrossRef]
8. Malkin, R.; Rabinowitz, J.C. The reconstitution of clostridial ferredoxin. *Biochem. Biophys. Res. Commun.* **1966**, *23*, 822–827. [CrossRef]
9. Wachtershäuser, G. Groundworks for an evolutionary biochemistry: The iron-sulphur world. *Prog. Biophys. Mol. Boil.* **1992**, *58*, 85–201. [CrossRef]
10. Lill, R.; Mühlenhoff, U. Iron-Sulfur Protein Biogenesis in Eukaryotes: Components and Mechanisms. *Annu. Rev. Cell Dev. Boil.* **2006**, *22*, 457–486. [CrossRef]
11. Zheng, L.; Cash, V.L.; Flint, D.H.; Dean, D.R. Assembly of iron-sulfur clusters. Identification of an iscSUA-hscBA-fdx gene cluster from Azotobacter vinelandii. *J. Biol. Chem.* **1998**, *273*, 13264–13272. [CrossRef]
12. Kispal, G.; Csere, P.; Prohl, C.; Lill, R. The mitochondrial proteins Atm1p and Nfs1p are essential for biogenesis of cytosolic Fe/S proteins. *EMBO J.* **1999**, *18*, 3981–3989. [CrossRef]
13. Sofia, H.J.; Chen, G.; Hetzler, B.G.; Reyes-Spindola, J.F.; Miller, N.E. Radical SAM, a novel protein superfamily linking unresolved steps in familiar biosynthetic pathways with radical mechanisms: Functional characterization using new analysis and information visualization methods. *Nucleic Acids Res.* **2001**, *29*, 1097–1106. [CrossRef]
14. Broderick, J.B.; Duffus, B.R.; Duschene, K.S.; Shepard, E.M. Radical *S*-Adenosylmethionine Enzymes. *Chem. Rev.* **2014**, *114*, 4229–4317. [CrossRef]

15. Hover, B.M.; Loksztejn, A.; Ribeiro, A.A.; Yokoyama, K. Identification of a Cyclic Nucleotide as a Cryptic Intermediate in Molybdenum Cofactor Biosynthesis. *J. Am. Chem. Soc.* **2013**, *135*, 7019–7032. [CrossRef]
16. Pang, H.; Yokoyama, K. Lessons from the Studies of a C–C Bond Forming Radical SAM Enzyme in Molybdenum Cofactor Biosynthesis. *Methods Enzym.* **2018**, *606*, 485–522.
17. Xu, X.M.; Moller, S. Iron–Sulfur Clusters: Biogenesis, Molecular Mechanisms, and Their Functional Significance. *Antioxidants Redox Signal.* **2011**, *15*, 271–307. [CrossRef]
18. Glasner, J.D.; Yang, C.-H.; Reverchon, S.; Hugouvieux-Cotte-Pattat, N.; Condemine, G.; Bohin, J.-P.; Van Gijsegem, F.; Yang, S.; Franza, T.; Expert, D.; et al. Genome Sequence of the Plant-Pathogenic Bacterium Dickeya dadantii 3937. *J. Bacteriol.* **2011**, *193*, 2076–2077. [CrossRef] [PubMed]
19. Tsaousis, A.D. On the Origin of Iron/Sulfur Cluster Biosynthesis in Eukaryotes. *Front. Microbiol.* **2019**, *10*, 2478. [CrossRef] [PubMed]
20. Beinert, H. Iron-sulfur proteins: Ancient structures, still full of surprises. *JBIC J. Boil. Inorg. Chem.* **2000**, *5*, 2–15. [CrossRef] [PubMed]
21. Blanc, B.; Gerez, C.; De Choudens, S.O. Assembly of Fe/S proteins in bacterial systems: Biochemistry of the bacterial ISC system. *Biochim. Biophys. Acta* **2015**, *1853*, 1436–1447. [CrossRef]
22. Bühning, M.; Valleriani, A.; Leimkühler, S. The Role of SufS Is Restricted to Fe–S Cluster Biosynthesis in *Escherichia coli*. *Biochemistry* **2017**, *56*, 1987–2000. [CrossRef] [PubMed]
23. Hidese, R.; Mihara, H.; Esaki, N. Bacterial cysteine desulfurases: Versatile key players in biosynthetic pathways of sulfur-containing biofactors. *Appl. Microbiol. Biotechnol.* **2011**, *91*, 47–61. [CrossRef]
24. Outten, F.W.; Djaman, O.; Storz, G. A suf operon requirement for Fe-S cluster assembly during iron starvation in *Escherichia coli*. *Mol. Microbiol.* **2004**, *52*, 861–872. [CrossRef]
25. Urbina, H.D.; Silberg, J.J.; Hoff, K.G.; Vickery, L.E. Transfer of Sulfur from IscS to IscU during Fe/S Cluster Assembly. *J. Boil. Chem.* **2001**, *276*, 44521–44526. [CrossRef]
26. Shi, R.; Proteau, A.; Villarroya, M.M.; Moukadiri, I.; Zhang, L.; Trempe, J.-F.; Matte, A.; Armengod, M.E.; Cygler, M. Structural Basis for Fe–S Cluster Assembly and tRNA Thiolation Mediated by IscS Protein–Protein Interactions. *PLoS Boil.* **2010**, *8*, e1000354. [CrossRef]
27. Roche, B.; Huguenot, A.; Barras, F.; Py, B. The iron-binding CyaY and IscX proteins assist the ISC-catalyzed Fe-S biogenesis in *Escherichia coli*. *Mol. Microbiol.* **2015**, *95*, 605–623. [CrossRef]
28. Cupp-Vickery, J.R.; Urbina, H.; Vickery, L.E. Crystal structure of IscS, a cysteine desulfurase from *Escherichia coli*. *J. Mol. Boil.* **2003**, *330*, 1049–1059. [CrossRef]
29. Chandramouli, K.; Unciuleac, M.C.; Naik, S.; Dean, D.R.; Huynh, B.H.; Johnson, M.K. Formation and Properties of [4Fe-4S] Clusters on the IscU Scaffold Protein†. *Biochemistry* **2007**, *46*, 6804–6811. [CrossRef] [PubMed]
30. Kim, J.H.; Frederick, R.O.; Reinen, N.M.; Troupis, A.T.; Markley, J.L. [2Fe-2S]-Ferredoxin Binds Directly to Cysteine Desulfurase and Supplies an Electron for Iron–Sulfur Cluster Assembly but Is Displaced by the Scaffold Protein or Bacterial Frataxin. *J. Am. Chem. Soc.* **2013**, *135*, 8117–8120. [CrossRef]
31. Yan, R.; Konarev, P.V.; Iannuzzi, C.; Adinolfi, S.; Roche, B.; Kelly, G.; Simon, L.; Martin, S.R.; Py, B.; Barras, F.; et al. Ferredoxin Competes with Bacterial Frataxin in Binding to the Desulfurase IscS. *J. Boil. Chem.* **2013**, *288*, 24777–24787. [CrossRef] [PubMed]
32. Kim, J.H.; Tonelli, M.; Frederick, R.O.; Chow, D.C.; Markley, J.L. Specialized Hsp70 Chaperone (HscA) Binds Preferentially to the Disordered Form, whereas J-protein (HscB) Binds Preferentially to the Structured Form of the Iron-Sulfur Cluster Scaffold Protein (IscU). *J. Boil. Chem.* **2012**, *287*, 31406–31413. [CrossRef] [PubMed]
33. Iametti, S.; Barbiroli, A.; Bonomi, F. Functional implications of the interaction between HscB and IscU in the biosynthesis of FeS clusters. *J. Boil. Inorg. Chem.* **2015**, *20*, 1039–1048. [CrossRef]
34. Loiseau, L.; Ollagnier-De-Choudens, S.; Nachin, L.; Fontecave, M.; Barras, F. Biogenesis of Fe-S Cluster by the Bacterial Suf System: SufS and SufE form a new type of cysteine desulfurase. *J. Boil. Chem.* **2003**, *278*, 38352–38359. [CrossRef]
35. Perard, J.; Ollagnier-De-Choudens, S. Correction to: Iron–sulfur clusters biogenesis by the SUF machinery: Close to the molecular mechanism understanding. *J. Boil. Inorg. Chem.* **2018**, *23*, 581–596. [CrossRef]
36. Zheng, C.; Dos Santos, P.C. Metallocluster transactions: Dynamic protein interactions guide the biosynthesis of Fe–S clusters in bacteria. *Biochem. Soc. Trans.* **2018**, *46*, 1593–1603. [CrossRef]
37. Selbach, B.P.; Pradhan, P.K.; Dos Santos, P.C. Protected Sulfur Transfer Reactions by the *Escherichia coli* Suf System. *Biochemistry* **2013**, *52*, 4089–4096. [CrossRef]

38. Black, K.A.; Dos Santos, P.C. Shared-intermediates in the biosynthesis of thio-cofactors: Mechanism and functions of cysteine desulfurases and sulfur acceptors. *Biochim. Biophys. Acta* **2015**, *1853*, 1470–1480. [CrossRef]
39. Tan, G.; Lu, J.; Bitoun, J.P.; Huang, H.; Ding, H. IscA/SufA paralogues are required for the [4Fe-4S] cluster assembly in enzymes of multiple physiological pathways in *Escherichia coli* under aerobic growth conditions. *Biochem. J.* **2009**, *420*, 463–472. [CrossRef]
40. Vinella, D.; Brochier-Armanet, C.; Loiseau, L.; Talla, E.; Barras, F. Iron-Sulfur (Fe/S) Protein Biogenesis: Phylogenomic and Genetic Studies of A-Type Carriers. *PLoS Genet.* **2009**, *5*, e1000497. [CrossRef]
41. Loiseau, L.; Gerez, C.; Bekker, M.; Choudens, S.O.-D.; Py, B.; Sanakis, Y.; De Mattos, J.T.; Fontecave, M.; Barras, F. ErpA, an iron–sulfur (Fe–S) protein of the A-type essential for respiratory metabolism in *Escherichia coli*. *Proc. Natl. Acad. Sci. USA* **2007**, *104*, 13626–13631. [CrossRef]
42. Angelini, S.; Gerez, C.; Ollagnier-de Choudens, S.; Sanakis, Y.; Fontecave, M.; Barras, F.; Py, B. NfuA, a new factor required for maturing Fe/S proteins in *Escherichia coli* under oxidative stress and iron starvation conditions. *J. Biol. Chem.* **2008**, *283*, 14084–14091. [CrossRef] [PubMed]
43. Picciocchi, A.; Saguez, C.; Boussac, A.; Cassier-Chauvat, C.; Chauvat, F.; Cassier-Chauvat, C. CGFS-Type Monothiol Glutaredoxins from the Cyanobacterium *Synechocystis* PCC6803 and Other Evolutionary Distant Model Organisms Possess a Glutathione-Ligated [2Fe-2S] Cluster. *Biochemistry* **2007**, *46*, 15018–15026. [CrossRef] [PubMed]
44. Boutigny, S.; Saini, A.; Baidoo, E.E.; Yeung, N.; Keasling, J.D.; Butland, G. Physical and Functional Interactions of a Monothiol Glutaredoxin and an Iron Sulfur Cluster Carrier Protein with the Sulfur-donating Radical S-Adenosyl-L-methionine Enzyme MiaB. *J. Boil. Chem.* **2013**, *288*, 14200–14211. [CrossRef]
45. Boyd, J.M.; Lewis, J.A.; Escalante-Semerena, J.C.; Downs, D.M. Salmonella enterica Requires ApbC Function for Growth on Tricarballylate: Evidence of Functional Redundancy between ApbC and IscU. *J. Bacteriol.* **2008**, *190*, 4596–4602. [CrossRef] [PubMed]
46. Boyd, J.M.; Sondelski, J.L.; Downs, D.M. Bacterial ApbC protein has two biochemical activities that are required for in vivo function. *J. Biol. Chem.* **2009**, *284*, 110–118. [CrossRef]
47. Schwartz, C.J.; Giel, J.L.; Patschkowski, T.; Luther, C.; Ruzicka, F.J.; Beinert, H.; Kiley, P.J. IscR, an Fe-S cluster-containing transcription factor, represses expression of *Escherichia coli* genes encoding Fe-S cluster assembly proteins. *Proc. Natl. Acad. Sci. USA* **2001**, *98*, 14895–14900. [CrossRef]
48. Rajagopalan, K.V.; Johnson, J.L. The pterin molybdenum cofactors. *J. Boil. Chem.* **1992**, *267*, 10199–10202.
49. Rajagopalan, K.V. Biosynthesis of the molybdenum cofactor. In *Escherichia coli and Salmonella. Cellular and Molecular Biology*; Neidhardt, F.C., Ed.; ASM Press: Washington, DC, USA, 1996; pp. 674–679.
50. Hille, R.; Hall, J.; Basu, P. The Mononuclear Molybdenum Enzymes. *Chem. Rev.* **2014**, *114*, 3963–4038. [CrossRef]
51. Leimkühler, S.; Wuebbens, M.M.; Rajagopalan, K. The history of the discovery of the molybdenum cofactor and novel aspects of its biosynthesis in bacteria. *Coord. Chem. Rev.* **2011**, *255*, 1129–1144. [CrossRef]
52. Mendel, R.; Leimkühler, S. The biosynthesis of the molybdenum cofactors. *J. Boil. Inorg. Chem.* **2015**, *20*, 337–347. [CrossRef] [PubMed]
53. Wuebbens, M.M.; Rajagopalan, K.V. Structural characterization of a molybdopterin precursor. *J. Boil. Chem.* **1993**, *268*, 13493–13498.
54. Pitterle, D.M.; Johnson, J.L.; Rajagopalan, K.V. In Vitro synthesis of molybdopterin from precursor Z using purified converting factor. Role of protein-bound sulfur in formation of the dithiolene. *J. Boil. Chem.* **1993**, *268*, 13506–13509.
55. Joshi, M.S.; Johnson, J.L.; Rajagopalan, K.V. Molybdenum cofactor biosynthesis in *Escherichia coli* mod and mog mutants. *J. Bacteriol.* **1996**, *178*, 4310–4312. [CrossRef]
56. Johnson, J.L.; Bastian, N.R.; Rajagopalan, K.V. Molybdopterin guanine dinucleotide: A modified form of molybdopterin identified in the molybdenum cofactor of dimethyl sulfoxide reductase from *Rhodobacter sphaeroides* forma specialis *denitrificans*. *Proc. Natl. Acad. Sci. USA* **1990**, *87*, 3190–3194. [CrossRef]
57. Meyer, O.; Rajagopalan, K.V. Molybdopterin in carbon monoxide oxidase from carboxydotrophic bacteria. *J. Bacteriol.* **1984**, *157*, 643–648. [CrossRef]
58. Neumann, M.; Mittelstadt, G.; Seduk, F.; Iobbi-Nivol, C.; Leimkuhler, S. MocA Is a Specific Cytidylyltransferase Involved in Molybdopterin Cytosine Dinucleotide Biosynthesis in *Escherichia coli*. *J. Boil. Chem.* **2009**, *284*, 21891–21898. [CrossRef]

59. Santamaria-Araujo, J.A.; Fischer, B.; Otte, T.; Nimtz, M.; Mendel, R.R.; Wray, V.; Schwarz, G. The Tetrahydropyranopterin Structure of the Sulfur-free and Metal-free Molybdenum Cofactor Precursor. *J. Boil. Chem.* **2004**, *279*, 15994–15999. [CrossRef]
60. Wuebbens, M.M.; Rajagopalan, K.V. Investigation of the Early Steps of Molybdopterin Biosynthesis in *Escherichia coli* through the Use of In Vivo Labeling Studies. *J. Boil. Chem.* **1995**, *270*, 1082–1087. [CrossRef]
61. Pitterle, D.M.; Johnson, J.L.; Rajagopalan, K.V. Molybdopterin formation by converting factor of *E. coli chlA1*. *FASEB J.* **1990**, *4*, A1957.
62. Pitterle, D.M.; Rajagopalan, K.V. Two proteins encoded at the *chlA* locus constitute the converting factor of *Escherichia coli chlA1*. *J. Bacteriol.* **1989**, *171*, 3373–3378. [CrossRef] [PubMed]
63. Pitterle, D.M.; Rajagopalan, K.V. Purification and characterization of the converting factor from *E. coli chlA1*. *FASEB J.* **1991**, *5*, A468.
64. Pitterle, D.M.; Rajagopalan, K.V. The biosynthesis of molybdopterin in *Escherichia coli*. Purification and characterization of the converting factor. *J. Boil. Chem.* **1993**, *268*, 13499–13505.
65. Daniels, J.N.; Wuebbens, M.M.; Rajagopalan, K.V.; Schindelin, H. Crystal Structure of a Molybdopterin Synthase–Precursor Z Complex: Insight into Its Sulfur Transfer Mechanism and Its Role in Molybdenum Cofactor Deficiency. *Biochemistry* **2008**, *47*, 615–626. [CrossRef] [PubMed]
66. Rudolph, M.J.; Wuebbens, M.M.; Rajagopalan, K.V.; Schindelin, H. Crystal structure of molybdopterin synthase and its evolutionary relationship to ubiquitin activation. *Nat. Genet.* **2001**, *8*, 42–46.
67. Gutzke, G.; Fischer, B.; Mendel, R.R.; Schwarz, G. Thiocarboxylation of Molybdopterin Synthase Provides Evidence for the Mechanism of Dithiolene Formation in Metal-binding Pterins. *J. Boil. Chem.* **2001**, *276*, 36268–36274. [CrossRef] [PubMed]
68. Leimkühler, S.; Freuer, A.; Araujo, J.A.S.; Rajagopalan, K.V.; Mendel, R.R. Mechanistic Studies of Human Molybdopterin Synthase Reaction and Characterization of Mutants Identified in Group B Patients of Molybdenum Cofactor Deficiency. *J. Boil. Chem.* **2003**, *278*, 26127–26134. [CrossRef]
69. Nichols, J.; Rajagopalan, K.V. *Escherichia coli* MoeA and MogA. Function in metal incorporation step of molybdenum cofactor biosynthesis. *J. Biol. Chem.* **2002**, *277*, 24995–25000. [CrossRef]
70. Nichols, J.D.; Rajagopalan, K.V. In Vitro Molybdenum Ligation to Molybdopterin Using Purified Components. *J. Boil. Chem.* **2005**, *280*, 7817–7822. [CrossRef]
71. Kuper, J.; Llamas, Á.; Hecht, H.-J.; Mendel, R.R.; Schwarz, G. Structure of the molybdopterin-bound Cnx1G domain links molybdenum and copper metabolism. *Nature* **2004**, *430*, 803–806. [CrossRef]
72. Reschke, S.; Sigfridsson, K.G.; Kaufmann, P.; Leidel, N.; Horn, S.; Gast, K.; Schulzke, C.; Haumann, M.; Leimkühler, S. Identification of a Bis-molybdopterin Intermediate in Molybdenum Cofactor Biosynthesis in *Escherichia coli*. *J. Boil. Chem.* **2013**, *288*, 29736–29745. [CrossRef] [PubMed]
73. Temple, C.A.; Rajagopalan, K.V. Mechanism of Assembly of the Bis(Molybdopterin Guanine Dinucleotide)Molybdenum Cofactor in *Rhodobacter sphaeroides* Dimethyl Sulfoxide Reductase. *J. Boil. Chem.* **2000**, *275*, 40202–40210. [CrossRef] [PubMed]
74. Palmer, T.; Santini, C.-L.; Iobbi-Nivol, C.; Eaves, D.J.; Boxer, D.H.; Giordano, G. Involvement of the narJ and mob gene products in distinct steps in the biosynthesis of the molybdoenzyme nitrate reductase in *Escherichia coli*. *Mol. Microbiol.* **1996**, *20*, 875–884. [CrossRef] [PubMed]
75. Lake, M.W.; Temple, C.A.; Rajagopalan, K.V.; Schindelin, H. The Crystal Structure of the *Escherichia coli* MobA Protein Provides Insight into Molybdopterin Guanine Dinucleotide Biosynthesis. *J. Boil. Chem.* **2000**, *275*, 40211–40217. [CrossRef] [PubMed]
76. Bohmer, N.; Hartmann, T.; Leimkühler, S. The chaperone FdsC for *Rhodobacter capsulatus* formate dehydrogenase binds the bis-molybdopterin guanine dinucleotide cofactor. *FEBS Lett.* **2014**, *588*, 531–537. [CrossRef]
77. Leimkühler, S.; Iobbi-Nivol, C. Bacterial Molybdoenzymes: Chaperones, Assembly and Insertion. *FEMS Microbiol. Rev.* **2016**, *40*. [CrossRef]
78. Zupok, A.; Iobbi-Nivol, C.; Mejean, V.; Leimkühler, S. The regulation of Moco biosynthesis and molybdoenzyme gene expression by molybdenum and iron in bacteria. *Metallomics* **2019**, *11*, 1602–1624. [CrossRef]
79. Zupok, A.; Gorka, M.; Siemiatkowska, B.; Skirycz, A.; Leimkühler, S. Iron-Dependent Regulation of Molybdenum Cofactor Biosynthesis Genes in *Escherichia coli*. *J. Bacteriol.* **2019**, *201*, e00382-19. [CrossRef]

80. Mehta, A.P.; Hanes, J.W.; Abdelwahed, S.H.; Hilmey, D.G.; Hänzelmann, P.; Begley, T.P. Catalysis of a New Ribose Carbon-Insertion Reaction by the Molybdenum Cofactor Biosynthetic Enzyme MoaA. *Biochemistry* **2013**, *52*, 1134–1136. [CrossRef] [PubMed]
81. Leimkühler, S.; Wuebbens, M.M.; Rajagopalan, K.V. Characterization of *Escherichia coli* MoeB and Its Involvement in the Activation of Molybdopterin Synthase for the Biosynthesis of the Molybdenum Cofactor. *J. Boil. Chem.* **2001**, *276*, 34695–34701. [CrossRef]
82. Schindelin, H. Evolutionary Origin of the Activation Step During Ubiquitin-dependent Protein Degradation. In *Protein Degradation: Ubiquitin and the Chemistry of Life*; Mayer, R.J., Ciechanover, A., Rechsteiner, M., Eds.; WILEY-VCH: Weinheim, Germany, 2005; pp. 21–43.
83. Lake, M.W.; Wuebbens, M.M.; Rajagopalan, K.V.; Schindelin, H. Mechanism of ubiquitin activation revealed by the structure of a bacterial MoeB–MoaD complex. *Nature* **2001**, *414*, 325–329. [CrossRef] [PubMed]
84. Schmitz, J.; Wuebbens, M.M.; Rajagopalan, K.V.; Leimkühler, S. Role of the C-Terminal Gly-Gly Motif of *Escherichia Coli* MoaD, a Molybdenum Cofactor Biosynthesis Protein with a Ubiquitin Fold. *Biochemistry* **2007**, *46*, 909–916. [CrossRef] [PubMed]
85. Tong, Y.; Wuebbens, M.M.; Rajagopalan, K.V.; Fitzgerald, M.C. Thermodynamic Analysis of Subunit Interactions in *Escherichia coli* Molybdopterin Synthase. *Biochemistry* **2005**, *44*, 2595–2601. [CrossRef] [PubMed]
86. Leimkühler, S.; Rajagopalan, K.V. An *Escherichia coli* NifS-like sulfurtransferase is required for the transfer of cysteine sulfur in the In Vitro synthesis of molybdopterin from precursor Z. *J. Biol. Chem.* **2001**, *276*, 22024–22031. [CrossRef]
87. Zhang, W.; Urban, A.; Mihara, H.; Leimkühler, S.; Kurihara, T.; Esaki, N. IscS Functions as a Primary Sulfur-donating Enzyme by Interacting Specifically with MoeB and MoaD in the Biosynthesis of Molybdopterin in *Escherichia coli*. *J. Boil. Chem.* **2010**, *285*, 2302–2308. [CrossRef]
88. Iobbi-Nivol, C.; Leimkühler, S. Molybdenum enzymes, their maturation and molybdenum cofactor biosynthesis in *Escherichia coli*. *Biochim. Biophys. Acta* **2013**, *1827*, 1086–1101. [CrossRef]
89. Dahl, J.U.; Radon, C.; Bühning, M.; Nimtz, M.; Leichert, L.I.; Denis, Y.; Jourlin-Castelli, C.; Iobbi-Nivol, C.; Méjean, V.; Leimkühler, S. The Sulfur Carrier Protein TusA Has a Pleiotropic Role in *Escherichia coli* That Also Affects Molybdenum Cofactor Biosynthesis. *J. Boil. Chem.* **2013**, *288*, 5426–5442. [CrossRef]
90. Kisker, C.; Schindelin, H.; Rees, D.C. Molybdenum-Cofactor–Containing Enzymes: Structure and Mechanism. *Annu. Rev. Biochem.* **1997**, *66*, 233–267. [CrossRef]
91. Schumann, S.; Saggu, M.; Möller, N.; Anker, S.D.; Lendzian, F.; Hildebrandt, P.; Leimkühler, S. The Mechanism of Assembly and Cofactor Insertion into *Rhodobacter capsulatus* xanthine dehydrogenase. *J. Boil. Chem.* **2008**, *283*, 16602–16611. [CrossRef]
92. Pinske, C.; Sawers, R.G. A-Type Carrier Protein ErpA Is Essential for Formation of an Active Formate-Nitrate Respiratory Pathway in *Escherichia coli* K-12. *J. Bacteriol.* **2011**, *194*, 346–353. [CrossRef]
93. Boniecki, M.T.; Freibert, S.A.; Mühlenhoff, U.; Lill, R.; Cygler, M. Structure and functional dynamics of the mitochondrial Fe/S cluster synthesis complex. *Nat. Commun.* **2017**, *8*, 1287. [CrossRef] [PubMed]
94. Wiedemann, N.; Urzica, E.; Guiard, B.; Müller, H.; Lohaus, C.; E Meyer, H.; Ryan, M.T.; Meisinger, C.; Mühlenhoff, U.; Lill, R.; et al. Essential role of Isd11 in mitochondrial iron–sulfur cluster synthesis on Isu scaffold proteins. *EMBO J.* **2006**, *25*, 184–195. [CrossRef] [PubMed]
95. Adam, A.C.; Bornhövd, C.; Prokisch, H.; Neupert, W.; Hell, K. The Nfs1 interacting protein Isd11 has an essential role in Fe/S cluster biogenesis in mitochondria. *EMBO J.* **2006**, *25*, 174–183. [CrossRef] [PubMed]
96. Lim, S.C.; Friemel, M.; Marum, J.E.; Tucker, E.J.; Bruno, D.L.; Riley, L.G.; Christodoulou, J.; Kirk, E.; Boneh, A.; DeGennaro, C.M.; et al. Mutations in LYRM4, encoding iron-sulfur cluster biogenesis factor ISD11, cause deficiency of multiple respiratory chain complexes. *Hum. Mol. Genet.* **2013**, *22*, 4460–4473. [CrossRef] [PubMed]
97. Maio, N.; Singh, A.; Uhrigshardt, H.; Saxena, N.; Tong, W.H.; Rouault, T.A. Cochaperone binding to LYR motifs confers specificity of iron sulfur cluster delivery. *Cell Metab.* **2014**, *19*, 445–457. [CrossRef]
98. Lill, R.; Freibert, S.A. Mechanisms of Mitochondrial Iron-Sulfur Protein Biogenesis. *Annu. Rev. Biochem.* **2020**, *89*, 471–499. [CrossRef]

99. Gervason, S.; Larkem, D.; Ben Mansour, A.; Botzanowski, T.; Müller, C.S.; Pecqueur, L.; Le Pavec, G.; Delaunay-Moisan, A.; Brun, O.; Agramunt, J.; et al. Physiologically relevant reconstitution of iron-sulfur cluster biosynthesis uncovers persulfide-processing functions of ferredoxin-2 and frataxin. *Nat. Commun.* **2019**, *10*, 3566. [CrossRef] [PubMed]
100. Lange, H.; Kaut, A.; Kispal, G.; Lill, R. A mitochondrial ferredoxin is essential for biogenesis of cellular iron-sulfur proteins. *Proc. Natl. Acad. Sci. USA* **2000**, *97*, 1050–1055. [CrossRef]
101. Sheftel, A.D.; Stehling, O.; Pierik, A.J.; Elsässer, H.P.; Mühlenhoff, U.; Webert, H.; Hobler, A.; Hannemann, F.; Bernhardt, R.; Lill, R. Humans possess two mitochondrial ferredoxins, Fdx1 and Fdx2, with distinct roles in steroidogenesis, heme, and Fe/S cluster biosynthesis. *Proc. Natl. Acad. Sci. USA* **2010**, *107*, 11775–11780. [CrossRef]
102. Webert, H.; Freibert, S.A.; Gallo, A.; Heidenreich, T.; Linne, U.; Amlacher, S.; Hurt, E.; Mühlenhoff, U.; Banci, L.; Lill, R. Functional reconstitution of mitochondrial Fe/S cluster synthesis on Isu1 reveals the involvement of ferredoxin. *Nat. Commun.* **2014**, *5*, 5013. [CrossRef]
103. Herrero, E.; De La Torre-Ruiz, M.A. Monothiol glutaredoxins: A common domain for multiple functions. *Cell. Mol. Life Sci.* **2007**, *64*, 1518–1530. [CrossRef] [PubMed]
104. Rodriguez-Manzaneque, M.T.; Tamarit, J.; Belli, G.; Ros, J.; Herrero, E. Grx5 Is a Mitochondrial Glutaredoxin Required for the Activity of Iron/Sulfur Enzymes. *Mol. Boil. Cell* **2002**, *13*, 1109–1121. [CrossRef] [PubMed]
105. Rouhier, N.; Couturier, J.; Johnson, M.K.; Jacquot, J.P. Glutaredoxins: Roles in iron homeostasis. *Trends Biochem. Sci.* **2010**, *35*, 43–52. [CrossRef] [PubMed]
106. Dutkiewicz, R.; Nowak, M. Molecular chaperones involved in mitochondrial iron-sulfur protein biogenesis. *J. Boil. Inorg. Chem.* **2018**, *23*, 569–579. [CrossRef]
107. Kampinga, H.H.; Craig, E.A. The HSP70 chaperone machinery: J proteins as drivers of functional specificity. *Nat. Rev. Mol. Cell Boil.* **2010**, *11*, 579–592. [CrossRef]
108. Andrew, A.J.; Dutkiewicz, R.; Knieszner, H.; Craig, E.A.; Marszalek, J. Characterization of the Interaction between the J-protein Jac1p and the Scaffold for Fe-S Cluster Biogenesis, Isu1p. *J. Boil. Chem.* **2006**, *281*, 14580–14587. [CrossRef]
109. Ciesielski, S.J.; Schilke, B.A.; Osipiuk, J.; Bigelow, L.; Mulligan, R.; Majewska, J.; Joachimiak, A.; Marszalek, J.; Craig, E.A.; Dutkiewicz, R. Interaction of J-Protein Co-Chaperone Jac1 with Fe–S Scaffold Isu Is Indispensable In Vivo and Conserved in Evolution. *J. Mol. Boil.* **2012**, *417*. [CrossRef]
110. Lill, R. From the discovery to molecular understanding of cellular iron-sulfur protein biogenesis. *Boil. Chem.* **2020**. [CrossRef]
111. Beilschmidt, L.K.; De Choudens, S.O.; Fournier, M.; Sanakis, I.; Hograindleur, M.-A.; Clémancey, M.; Blondin, G.; Schmucker, S.; Eisenmann, A.; Weiss, A.; et al. ISCA1 is essential for mitochondrial Fe_4S_4 biogenesis In Vivo. *Nat. Commun.* **2017**, *8*, 15124. [CrossRef]
112. Gelling, C.; Dawes, I.W.; Richhardt, N.; Lill, R.; MühlenhoffU. Mitochondrial Iba57p Is Required for Fe/S Cluster Formation on Aconitase and Activation of Radical SAM Enzymes. *Mol. Cell. Boil.* **2007**, *28*, 1851–1861. [CrossRef]
113. Muhlenhoff, U.; Richter, N.; Pines, O.; Pierik, A.J.; Lill, R. Specialized Function of Yeast Isa1 and Isa2 Proteins in the Maturation of Mitochondrial [4Fe-4S] Proteins. *J. Boil. Chem.* **2011**, *286*, 41205–41216. [CrossRef]
114. Sheftel, A.D.; Wilbrecht, C.; Stehling, O.; Niggemeyer, B.; Elsasser, H.P.; Muhlenhoff, U.; Lill, R. The human mitochondrial ISCA1, ISCA2, and IBA57 proteins are required for [4Fe-4S] protein maturation. *Mol. Biol. Cell.* **2012**, *23*, 1157–1166. [CrossRef] [PubMed]
115. Banci, L.; Brancaccio, D.; Ciofi-Baffoni, S.; Del Conte, R.; Gadepalli, R.; Mikolajczyk, M.; Neri, S.; Piccioli, M.; Winkelmann, J. [2Fe-2S] cluster transfer in iron-sulfur protein biogenesis. *Proc. Natl. Acad Sci. USA* **2014**, *111*, 6203–6208. [CrossRef] [PubMed]
116. Brancaccio, D.; Gallo, A.; Mikolajczyk, M.; Zovo, K.; Palumaa, P.; Novellino, E.; Piccioli, M.; Ciofi-Baffoni, S.; Banci, L. Formation of [4Fe-4S] Clusters in the Mitochondrial Iron–Sulfur Cluster Assembly Machinery. *J. Am. Chem. Soc.* **2014**, *136*, 16240–16250. [CrossRef] [PubMed]
117. Brancaccio, D.; Gallo, A.; Piccioli, M.; Novellino, E.; Ciofi-Baffoni, S.; Banci, L. [4Fe-4S] Cluster Assembly in Mitochondria and Its Impairment by Copper. *J. Am. Chem. Soc.* **2017**, *139*, 719–730. [CrossRef] [PubMed]
118. Gourdoupis, S.; Nasta, V.; Calderone, V.; Ciofi-Baffoni, S.; Banci, L. IBA57 Recruits ISCA2 to Form a [2Fe-2S] Cluster-Mediated Complex. *J. Am. Chem. Soc.* **2018**, *140*, 14401–14412. [CrossRef]

119. Nasta, V.; Da Vela, S.; Gourdoupis, S.; Ciofi-Baffoni, S.; Svergun, D.I.; Banci, L. Structural properties of [2Fe-2S] ISCA2-IBA57: A complex of the mitochondrial iron-sulfur cluster assembly machinery. *Sci. Rep.* **2019**, *9*, 1–12. [CrossRef]
120. Maio, N.; Rouault, T.A. Outlining the Complex Pathway of Mammalian Fe-S Cluster Biogenesis. *Trends Biochem. Sci.* **2020**, *45*, 411–426. [CrossRef]
121. Netz, D.J.; Mascarenhas, J.; Stehling, O.; Pierik, A.J.; Lill, R. Maturation of cytosolic and nuclear iron–sulfur proteins. *Trends Cell Boil.* **2014**, *24*, 303–312. [CrossRef]
122. Paul, V.D.; Lill, R. Biogenesis of cytosolic and nuclear iron–sulfur proteins and their role in genome stability. *Biochim. Biophys. Acta* **2015**, *1853*, 1528–1539. [CrossRef]
123. Gerber, S.; Comellas-Bigler, M.; Goetz, B.A.; Locher, K.P. Structural Basis of Trans-Inhibition in a Molybdate/Tungstate ABC Transporter. *Science* **2008**, *321*, 246–250. [CrossRef]
124. Biederbick, A.; Stehling, O.; Rosser, R.; Niggemeyer, B.; Nakai, Y.; Elsasser, H.-P.; Lill, R. Role of Human Mitochondrial Nfs1 in Cytosolic Iron-Sulfur Protein Biogenesis and Iron Regulation. *Mol. Cell. Boil.* **2006**, *26*, 5675–5687. [CrossRef] [PubMed]
125. Stehling, O.; Netz, D.J.; Niggemeyer, B.; Rosser, R.; Eisenstein, R.S.; Puccio, H.; Pierik, A.J.; Lill, R. Human Nbp35 Is Essential for both Cytosolic Iron-Sulfur Protein Assembly and Iron Homeostasis. *Mol. Cell. Boil.* **2008**, *28*, 5517–5528. [CrossRef] [PubMed]
126. Stehling, O.; Jeoung, J.-H.; Freibert, S.A.; Paul, V.D.; Bänfer, S.; Niggemeyer, B.; Rösser, R.; Dobbek, H.; Lill, R. Function and crystal structure of the dimeric P-loop ATPase CFD1 coordinating an exposed [4Fe-4S] cluster for transfer to apoproteins. *Proc. Natl. Acad. Sci. USA* **2018**, *115*, E9085–E9094. [CrossRef] [PubMed]
127. Netz, D.J.; Stümpfig, M.; Doré, C.; Mühlenhoff, U.; Pierik, A.J.; Lill, R. Tah18 transfers electrons to Dre2 in cytosolic iron-sulfur protein biogenesis. *Nat. Methods* **2010**, *6*, 758–765. [CrossRef] [PubMed]
128. Netz, D.J.; Genau, H.M.; Weiler, B.D.; Bill, E.; Pierik, A.J.; Lill, R. The conserved protein Dre2 uses essential [2Fe–2S] and [4Fe–4S] clusters for its function in cytosolic iron–sulfur protein assembly. *Biochem. J.* **2016**, *473*, 2073–2085. [CrossRef]
129. Banci, L.; Ciofi-Baffoni, S.; Gajda, K.; Muzzioli, R.; Peruzzini, R.; Winkelmann, J. N-terminal domains mediate [2Fe-2S] cluster transfer from glutaredoxin-3 to anamorsin. *Nat. Methods* **2015**, *11*, 772–778. [CrossRef]
130. Frey, A.G.; Palenchar, D.J.; Wildemann, J.D.; Philpott, C.C. A Glutaredoxin·BolA Complex Serves as an Iron-Sulfur Cluster Chaperone for the Cytosolic Cluster Assembly Machinery. *J. Boil. Chem.* **2016**, *291*, 22344–22356. [CrossRef]
131. Haunhorst, P.; Hanschmann, E.-M.; Bräutigam, L.; Stehling, O.; Hoffmann, B.; Mühlenhoff, U.; Lill, R.; Berndt, C.; Lillig, C.H. Crucial function of vertebrate glutaredoxin 3 (PICOT) in iron homeostasis and hemoglobin maturation. *Mol. Boil. Cell* **2013**, *24*, 1895–1903. [CrossRef]
132. Banci, L.; Camponeschi, F.; Ciofi-Baffoni, S.; Muzzioli, R. Elucidating the Molecular Function of Human BOLA2 in GRX3-Dependent Anamorsin Maturation Pathway. *J. Am. Chem. Soc.* **2015**, *137*, 16133–16143. [CrossRef]
133. Banci, L.; Ciofi-Baffoni, S.; Mikolajczyk, M.; Winkelmann, J.; Bill, E.; Pandelia, M.-E. Human anamorsin binds [2Fe–2S] clusters with unique electronic properties. *JBIC J. Boil. Inorg. Chem.* **2013**, *18*, 883–893. [CrossRef] [PubMed]
134. Zhang, Y.; Lyver, E.R.; Nakamaru-Ogiso, E.; Yoon, H.; Amutha, B.; Lee, D.-W.; Bi, E.; Ohnishi, T.; Daldal, F.; Pain, D.; et al. Dre2, a Conserved Eukaryotic Fe/S Cluster Protein, Functions in Cytosolic Fe/S Protein Biogenesis. *Mol. Cell. Boil.* **2008**, *28*, 5569–5582. [CrossRef] [PubMed]
135. Gari, K.; Ortiz, A.M.L.; Borel, V.; Flynn, H.R.; Skehel, J.M.; Boulton, S.J. MMS19 Links Cytoplasmic Iron-Sulfur Cluster Assembly to DNA Metabolism. *Science* **2012**, *337*, 243–245. [CrossRef] [PubMed]
136. Srinivasan, V.; Netz, D.J.; Webert, H.; Mascarenhas, J.; Pierik, A.J.; Michel, H.; Lill, R. Structure of the Yeast WD40 Domain Protein Cia1, a Component Acting Late in Iron-Sulfur Protein Biogenesis. *Struct.* **2007**, *15*, 1246–1257. [CrossRef] [PubMed]
137. Hänzelmann, P.; Hernandez, H.L.; Menzel, C.; García-Serres, R.; Huynh, B.H.; Johnson, M.K.; Mendel, R.R.; Schindelin, H. Characterization of MOCS1A, an Oxygen-sensitive Iron-Sulfur Protein Involved in Human Molybdenum Cofactor Biosynthesis. *J. Boil. Chem.* **2004**, *279*, 34721–34732. [CrossRef] [PubMed]
138. Reiss, J.; Christensen, E.; Kurlemann, G.; Zabot, M.-T.; Dorche, C. Genomic structure and mutational spectrum of the bicistronic MOCS1 gene defective in molybdenum cofactor deficiency type A. *Hum. Genet.* **1998**, *103*, 639–644. [CrossRef] [PubMed]

139. Teschner, J.; Lachmann, N.; Schulze, J.; Geisler, M.; Selbach, K.; Santamaria-Araujo, J.; Balk, J.; Mendel, R.; Bittner, F. A novel role for Arabidopsis mitochondrial ABC transporter ATM3 in molybdenum cofactor biosynthesis. *Plant Cell* **2010**, *22*, 468–480. [CrossRef]
140. Stallmeyer, B.; Coyne, K.E.; Wuebbens, M.M.; Johnson, J.L.; Rajagopalan, K.V.; Mendel, R.R. *The cDNA Sequence of MOCS3, Human Molybdopterin Synthase Sulfurylase*; GenBank: National Center for Biotechnology Information: Bethesda, MD, USA, 1998; Accession Number AF102544.
141. Mendel, R.; Schwarz, G. Molybdenum cofactor biosynthesis in plants and humans. *Coord. Chem. Rev.* **2011**, *255*, 1145–1158. [CrossRef]
142. Matthies, A.; Nimtz, A.M.; Leimkühler, S. Molybdenum Cofactor Biosynthesis in Humans: Identification of a Persulfide Group in the Rhodanese-like Domain of MOCS3 by Mass Spectrometry. *Biochemistry* **2005**, *44*, 7912–7920. [CrossRef]
143. Marelja, Z.; Chowdhury, M.; Dosche, C.; Hille, C.; Baumann, O.; Löhmannsröben, H.-G.; Leimkühler, S. The L-Cysteine Desulfurase NFS1 Is Localized in the Cytosol where it Provides the Sulfur for Molybdenum Cofactor Biosynthesis in Humans. *PLoS ONE* **2013**, *8*, e60869. [CrossRef]
144. Marelja, Z.; Stöcklein, W.; Nimtz, M.; Leimkühler, S. A novel role for human Nfs1 in the cytoplasm: Nfs1 acts as a sulfur donor for MOCS3, a protein involved in molybdenum cofactor biosynthesis. *J. Boil. Chem.* **2008**, *283*, 25178–25185. [CrossRef] [PubMed]
145. Stallmeyer, B.; Schwarz, G.; Schulze, J.; Nerlich, A.; Reiss, J.; Kirsch, J.; Mendel, R. The neurotransmitter receptor-anchoring protein gephyrin reconstitutes molybdenum cofactor biosynthesis in bacteria, plants, and mammalian cells. *Proc. Natl. Acad. Sci. USA* **1999**, *96*, 1333–1338. [CrossRef] [PubMed]
146. Belaidi, A.A.; Schwarz, G. Metal insertion into the molybdenum cofactor: Product–substrate channelling demonstrates the functional origin of domain fusion in gephyrin. *Biochem. J.* **2013**, *450*, 149–157. [CrossRef] [PubMed]
147. Krausze, J.; Hercher, T.W.; Zwerschke, D.; Kirk, M.L.; Blankenfeldt, W.; Mendel, R.R.; Kruse, T. The functional principle of eukaryotic molybdenum insertases. *Biochem. J.* **2018**, *475*, 1739–1753. [CrossRef] [PubMed]
148. Hille, R.; Nishino, T.; Bittner, F. Molybdenum enzymes in higher organisms. *Coord. Chem. Rev.* **2011**, *255*, 1179–1205. [CrossRef] [PubMed]
149. Wahl, R.C.; Rajagopalan, K.V. Evidence for the inorganic nature of the cyanolyzable sulfur of molybdenum hydroxylases. *J. Boil. Chem.* **1982**, *257*, 1354–1359.
150. Hille, R. The Mononuclear Molybdenum Enzymes. *Chem. Rev.* **1996**, *96*, 2757–2816. [CrossRef]
151. Bittner, F.; Oreb, M.; Mendel, R.R. ABA3 Is a Molybdenum Cofactor Sulfurase Required for Activation of Aldehyde Oxidase and Xanthine Dehydrogenase in *Arabidopsis thaliana*. *J. Boil. Chem.* **2001**, *276*, 40381–40384. [CrossRef] [PubMed]
152. Wollers, S.; Heidenreich, T.; Zarepour, M.; Zachmann, D.; Kraft, C.; Zhao, Y.; Mendel, R.R.; Bittner, F. Binding of Sulfurated Molybdenum Cofactor to the C-terminal Domain of ABA3 fromArabidopsis thalianaProvides Insight into the Mechanism of Molybdenum Cofactor Sulfuration. *J. Boil. Chem.* **2008**, *283*, 9642–9650. [CrossRef]
153. Lehrke, M.; Rump, S.; Heidenreich, T.; Wissing, J.; Mendel, R.R.; Bittner, F. Identification of persulfide-binding and disulfide-forming cysteine residues in the NifS-like domain of the molybdenum cofactor sulfurase ABA3 by cysteine-scanning mutagenesis. *Biochem. J.* **2012**, *441*, 823–839. [CrossRef]

Review

Tungstoenzymes: Occurrence, Catalytic Diversity and Cofactor Synthesis

Carola S. Seelmann [1], Max Willistein [1], Johann Heider [2] and Matthias Boll [1,*]

[1] Faculty of Biology-Microbiology, Albert-Ludwigs-Universität Freiburg, 79104 Freiburg, Germany; carola.seelmann@biologie.uni-freiburg.de (C.S.S.); max.willistein@biologie.uni-freiburg.de (M.W.)

[2] Faculty of Biology, Philipps-Universität Marburg, 35037 Marburg, Germany; heider@staff.uni-marburg.de

* Correspondence: matthias.boll@biologie.uni-freiburg.de

Received: 30 June 2020; Accepted: 28 July 2020; Published: 31 July 2020

Abstract: Tungsten is the heaviest element used in biological systems. It occurs in the active sites of several bacterial or archaeal enzymes and is ligated to an organic cofactor (metallopterin or metal binding pterin; MPT) which is referred to as tungsten cofactor (Wco). Wco-containing enzymes are found in the dimethyl sulfoxide reductase (DMSOR) and the aldehyde:ferredoxin oxidoreductase (AOR) families of MPT-containing enzymes. Some depend on Wco, such as aldehyde oxidoreductases (AORs), class II benzoyl-CoA reductases (BCRs) and acetylene hydratases (AHs), whereas others may incorporate either Wco or molybdenum cofactor (Moco), such as formate dehydrogenases, formylmethanofuran dehydrogenases or nitrate reductases. The obligately tungsten-dependent enzymes catalyze rather unusual reactions such as ones with extremely low-potential electron transfers (AOR, BCR) or an unusual hydration reaction (AH). In recent years, insights into the structure and function of many tungstoenzymes have been obtained. Though specific and unspecific ABC transporter uptake systems have been described for tungstate and molybdate, only little is known about further discriminative steps in Moco and Wco biosynthesis. In bacteria producing Moco- and Wco-containing enzymes simultaneously, paralogous isoforms of the metal insertase MoeA may be specifically involved in the molybdenum- and tungsten-insertion into MPT, and in targeting Moco or Wco to their respective apo-enzymes. Wco-containing enzymes are of emerging biotechnological interest for a number of applications such as the biocatalytic reduction of CO_2, carboxylic acids and aromatic compounds, or the conversion of acetylene to acetaldehyde.

Keywords: tungsten enzymes; tungsten cofactor; aldehyde:ferredoxin oxidoreductase; benzoyl-CoA reductase; acetylene hydratase; formate dehydrogenase

---

1. Introduction

Tungsten and molybdenum are transition metals of the sixth group and occur in nature predominantly in form of their oxyanions—tungstate (WO_4^{2-}) and molybdate (MoO_4^{2-}). While their average abundance in the earth's crust is highly similar, their bioavailability may differ in various aqueous environments. Both metals are present in biological systems in ligation to the so-called metallopterin (or metal binding pterin, MPT, initially introduced as molybdopterin), which occurs as a three-ring pyranopterin compound in most known molybdo- or tungstoenzymes. The entire metal cofactors are referred to as Moco or Wco, respectively [1–6]. There are many variants of MPT-derived cofactors with regard to the metal inserted, the presence of additional ligands of the metal, the number of pyranopterin molecules bound per metal (one or two), and the attachment of additional nucleotides to the MPT core. According to this structural diversity and the underlying amino acid sequence similarities, MPT-containing enzymes can be divided into four unrelated families: (i) xanthine oxidase (XO), (ii) sulfite oxidase (SO), (iii) dimethyl sulfoxide reductase (DMSOR), and (iv) aldehyde oxidoreductase (AOR) families. Wco-containing enzymes belong to the DMSOR and AOR families, whereas Moco is predominantly found in members of the XO, SO and DMSOR families. In the AOR family, tungsten is always bound by two MPTs, referred to as W-bis-MPT; in DMSOR family members, tungsten or molybdenum is coordinated by two MPT guanine dinucleotide (MGD) moieties, referred to as W-/Mo-bis-MGD cofactor (Figure 1). All tungstoenzymes contain at least one Fe-S cluster next to Wco, and especially the multi-subunit enzymes harbor a number of additional redox-active cofactors such as Fe-S clusters, flavins or hemes.

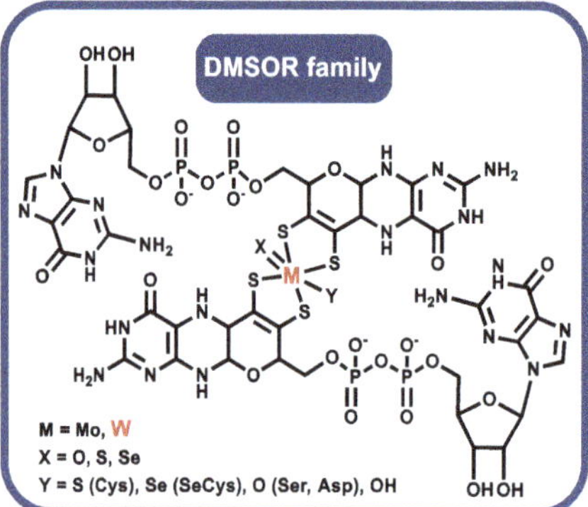

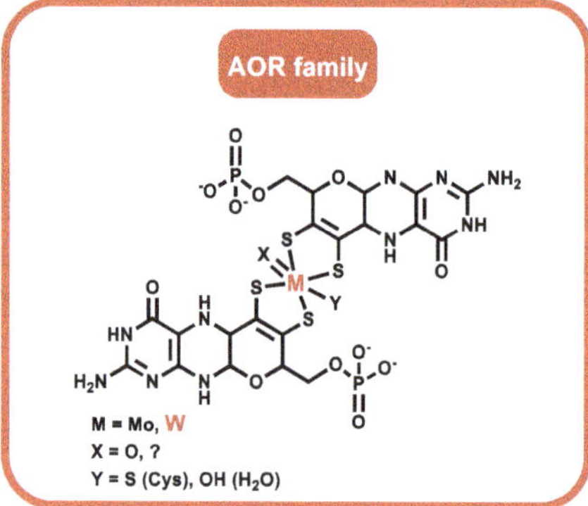

Figure 1. Tungsten cofactor (Wco) found in members of the dimethyl sulfoxide reductase (DMSOR) and aldehyde oxidoreductase (AOR) enzyme families. The former contains either Mo-bis-MPT guanine dinucleotide (MGD) or W-bis-MGD, the latter in most cases use W-bis-metallopterin (MPT) as the active site cofactor.

The majority of Moco- or Wco-containing enzymes catalyze hydroxy- or oxo-transfer reactions such as water-dependent hydroxylations or hydrations via water-, hydroxyl-, or oxo-intermediates bound to the metal in the catalytic course. There are also examples of MPT-dependent enzymes catalyzing hydride or hydrogen atom transfers (e.g., formate dehydrogenase or class II benzoyl-CoA reductase) or sulfur atom transfer reactions (polysulfide reductase). With the exception of acetylene hydratase, all Wco- and Moco-containing enzymes catalyze redox reactions (Figure 2). It is evident that most naturally occurring tungstoenzymes are involved in low-potential ($E°' < -400$ mV) electron transfer reactions. This finding can be rationalized by the generally lower redox potential of the biologically relevant redox transitions of W(IV/V/VI) vs. Mo(IV/V/VI). Though acetylene hydratase does not catalyze a redox reaction, the enzyme depends on a low-potential redox activation (see below).

Based on the highly similar physicochemical properties of molybdate and tungstate, the latter was initially considered to act as a general inhibitor of Moco-dependent enzymes, which has been observed in many cases if the replacement of molybdenum by tungsten is enforced by high tungstate concentrations (for an excellent review on this aspect, see [7]). As an example, a recent study used tungstate to inhibit Moco-dependent enzymes from facultatively anaerobic bacteria involved in gut disorders [8]. In facultatively tungsten-dependent enzymes that occur with either of the two metals, the bioavailability of tungstate or molybdate may govern which of the two metals is incorporated into bis-MPT-containing enzymes.

Today, it has become evident that Wco is an essential cofactor of many archaeal and bacterial enzymes that have been discussed in previous reviews with various focuses [7,9–12]. Here, we first aim to provide a state-of-the-art overview of the occurrence and function of obligately and facultatively tungsten-containing enzymes. We then discuss processes that may be involved in discriminating between molybdenum and tungsten during uptake, Mo/W-MPT synthesis and the incorporation into their respective apo-enzymes. This aspect is of particular importance for the emerging number of organisms that simultaneously produce Wco- and Moco-dependent enzymes.

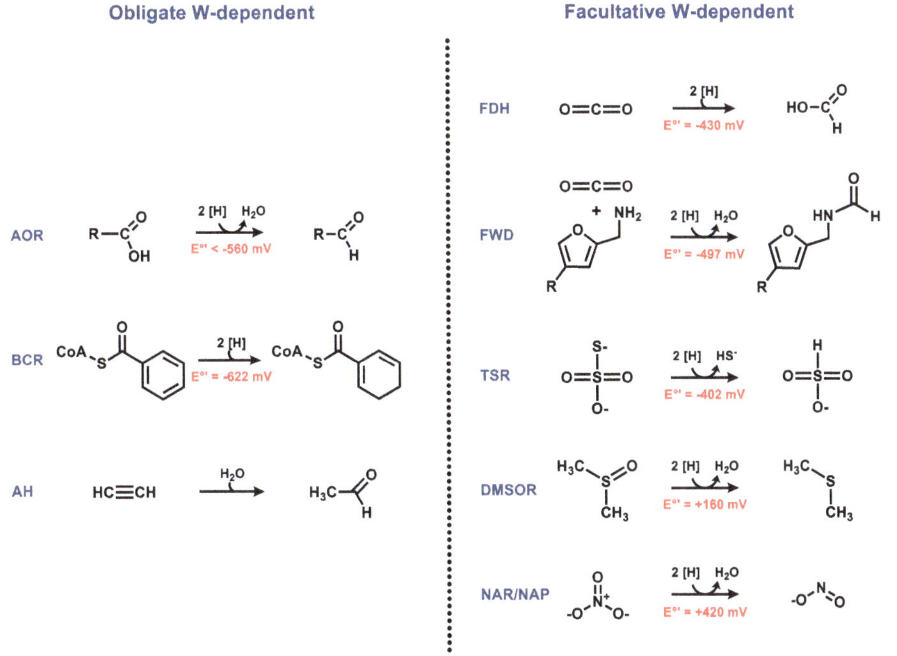

Figure 2. Reactions catalyzed by obligately and facultatively tungsten-dependent enzymes. Abbreviations: AOR: aldehyde oxidoreductase; BCR: class II benzoyl-CoA reductase; AH: acetylene hydratase; FDH: formate dehydrogenase; FWD: tungsten-containing formylmethanofuran dehydrogenase; TSR: thiosulfate reductase; DMSOR: dimethyl sulfoxide reductase; NAR: membrane-bound nitrate reductase; NAP: periplasmic nitrate reductase.

2. Affiliation of Tungsten Enzymes in the AOR and DMSOR Enzyme Families

The presence of Wco is restricted to members of a few clades of the DMSOR and AOR enzyme families. The biochemically characterized tungstoenzymes of the AOR family are affiliated to five clades of AORs (AOR, FOR, GAPOR, GOR, and WOR5) and one clade of the active site subunit BamB of class II benzoyl-CoA reductases (Figure 3A). In addition, the enzyme family contains one clade of archaeal enzymes and two of bacterial enzymes, whose biological functions are yet unknown (WOR4, YdhV and AOR1). Enzymes of the YdhV clade are unique in containing Moco instead of Wco, whereas WOR4 has been identified as a Wco-containing protein. The metal content of the enzymes of the AOR1 clade is unknown, since these have been only identified from genome sequences and not on the protein level. AORs of the XO family [13,14] are solely Moco-dependent and therefore not further discussed in this review.

The DMSOR family is subdivided into three subfamilies and several additional clades, and the tungstoenzymes of this family are affiliated to at least eight clades, which mostly contain either Moco- or Wco-dependent members (in case of formate or formylmethanofuran dehydrogenases, and nitrate-, DMSO-, or thiosulfate reductases [15]). Only the acetylene hydratase clade seems to be predominantly W-dependent, but this is based on only a single biochemically characterized enzyme, and it is unknown whether all these enzymes or closely related clades share the same metal preference (Figure 3B).

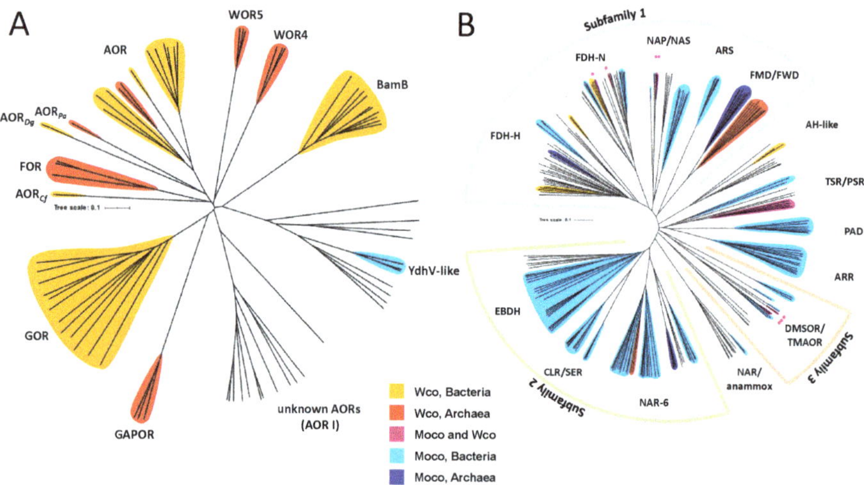

Figure 3. Phylogenetic trees of the large subunits of selected enzymes of the AOR family (**A**) and the DMSOR family (**B**). Clades or subclades containing tungsten enzymes are labeled in light-orange (bacteria), dark-orange (archaea) or violet (enzymes working with either tungsten or molybdenum, for better visualization additionally marked with a violet dot), and those containing molybdenum enzymes are labeled in light- or dark-blue (bacterial and archaeal enzymes, respectively). Enzyme clades of unknown metal content are not colored. Members of the DMSOR subfamilies 1–3 are highlighted by the outer circle as indicated. Species epithets for unaffiliated AORs: *Pa: Pyrobaculum aerophilum; Dg: Desulfovibrio gigas; Cf: Clostridium formicoaceticum.* Abbreviations: AOR, FOR, GAPOR, WOR5: archaeal aldehyde oxidoreductases of various specificities (for details see text); BamB: active site subunit of class II benzoyl-CoA reductases; WOR4, YdhV, AOR1: archaeal or bacterial Mo- or W-containing enzymes of unknown function; GOR: glyceraldehyde-3-phosphate oxidoreductases from thermophilic bacteria; FDH: formate dehydrogenases; NAP/NAS: periplasmic and assimilatory nitrate reductases; FMD/FWD: molybdenum- and tungsten-dependent formylmethanofuran dehydrogenases; AH-like: acetylene hydratases and similar hypothetical enzymes; TSR/PSR: thiosulfate/polysulfide reductases; PAD: phenylacetyl-CoA dehydrogenases; ARR, ARS: arsenate reductases/arsenite oxidases; DMSOR: dimethyl sulfoxide reductases; TMAOR: trimethylaminoxide reductases; NAR/anammox: chemolithotrophic nitrite oxidoreductases and putative nitrate reductases from anammox bacteria; NAR-6: nitrate reductases and similar clades; CLR/SER: chlorate/selenate reductases and dimethyl sulfide dehydrogenases; EBDH: ethylbenzene dehydrogenases and related enzymes.

3. Obligately W-Containing Enzymes

3.1. Aldehyde Oxidoreductases

Most known clades of obligately tungsten-dependent enzymes show activities as aldehyde oxidoreductases (Figure 2). Four of the five enzymes of the AOR family encoded in the genome of the archaeon *Pyrococcus furiosus* have been characterized as hyperthermophilic and extremely O_2-sensitive aldehyde oxidoreductases of various specificities. Together with the tungsten-dependent oxidoreductase WOR4, which was shown to contain Wco and a [3Fe-4S] cluster, but did not show any detectable activity [16], these enzymes define five separate phylogenetic clades of the AOR enzyme family (Figure 3A), namely the subfamilies of archaeal AOR, formaldehyde oxidoreductase (FOR), glyceraldehyde-3-phosphate oxidoreductases (GAPOR), and tungsten-dependent oxidoreductases (WOR4 and WOR5). Attempts to replace tungsten in some of these proteins with molybdenum or vanadium led either to proteins lacking any metal or to molybdenum-containing, inactive variants, proving the necessity of the Wco for their activities [17,18]. Additional members of the enzyme family emerged recently from obligately or facultatively anaerobic bacteria, namely bacterial AOR and GAPOR (GOR), which represent sister clades of the archaeal AOR and GAPOR, respectively (Figure 3A).

Archaeal and bacterial AORs (*sensu stricto*) represent the most abundant and best-studied clades exhibiting aldehyde oxidoreductase activity. Both contain a subunit of a similar size harboring the W-bis-MPT cofactor with an additional Mg^{2+} ion bridging the phosphate groups of the two cofactors, as well as an additional [4Fe-4S] cluster [19–21] (Figure 4). The archaeal AORs, as well as some AORs from strictly anaerobic bacteria, occur as homodimers with a bridging iron atom [20], and utilize ferredoxin as a physiological electron acceptor. In contrast, AORs from *Aromatoleum aromaticum* and other facultatively anaerobic denitrifying bacteria, as well as from the strictly anaerobic *Moorella thermoacetica*, consist of three different subunits in an $\alpha_2\beta_2\gamma_2$ composition. The additional electron-transferring small subunit with four [4Fe-4S] clusters and the medium-size FAD-containing subunit allows for the use of a broader electron acceptor range, including NAD^+ [21]. All characterized enzymes of the archaeal or bacterial AOR clades oxidize a variety of aldehydes to their respective carboxylic acids. Their main physiological function was proposed to be the detoxification of aldehydes that accumulate in different metabolic pathways. Though known for decades, there is still little knowledge of the reaction mechanisms of AORs and the internal electron transfer events involved, largely because of the difficult genetic accessibility of the respective host organisms. Remarkably, AORs also catalyze the reverse reaction, namely the reduction of non-activated acids to the corresponding aldehydes at $E^{\circ\prime} \approx -560$ mV, albeit at a rate lower than 1% compared to those of aldehyde oxidation. This property is reflected in the alternative name "carbonic acid reductase" (CAR) for the enzyme from *M. thermoacetica*. By utilizing this property, the metabolism of fermentative archaea has been artificially shifted from the production of acetic acid to ethanol. In spite of the very low observed reverse in vitro activities of AORs, the reaction seems to work well in whole-cell systems [22]. A similar potential of AORs in the biotechnologically interesting utilization of syngas as a substrate for acetogenic bacteria has been suggested [22,23].

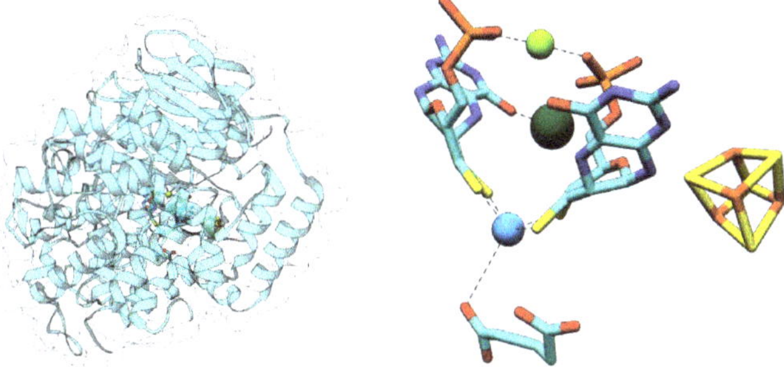

Figure 4. Structure of formaldehyde:ferredoxin oxidoreductase from *P. furiosus*, in complex with glutarate (pdb 1B4N) at 2.4 Å solution. Left panel: overall structure; right panel W-bis-MPT and [4Fe-4S] cofactor. The tungsten atom is depicted in blue, Mg^{2+} is shown in light green, and Ca^{2+} in dark green.

FORs are known from various hyperthermophilic archaea and account together with AOR and GAPOR for the most abundant tungstoenzymes in these organisms. Unlike other AORs, FORs show only low specific activities with short chain aldehydes. They turn over formaldehyde with reasonable rates but show very high K_m-values for almost all substrates. A possible exception is glutardialdehyde with the lowest K_m-value of the tested substrates, but the data available do not allow to propose a rational physiological function of FOR [24]. The X-ray structure of the FOR for *P. furiosus* confirms its composition as a homotetramer without a bridging iron atom, as in AORs (Figure 4). Instead,

FOR contains an additional Ca^{2+} ion ligated to one of the MPT cofactors, which is attributed to a structural function. The binding of W-bis-MPT and the [4Fe-4S] cluster is highly similar in AOR and FOR, and the FOR structure with a bound glutarate as a substrate mimic led to the speculation that diacid semialdehydes may serve as potential physiological substrates [25].

Two further clades comprise GAPOR from archaea and GOR from bacteria, which are involved in glucose metabolism via a modified glycolytic pathway. They specifically couple the oxidation of glyceraldehyde-3-phosphate to 3-phosphoglycerate without 1,3-bisphosphoglycerate as an intermediate to the reduction of ferredoxin. GAPORs are monomeric enzymes containing Zn^{2+} as an additional metal [26,27].

Finally, two further clades of tungsten-containing AOR family members of unknown physiological relevance are represented by WOR4 and WOR5 from *P. furiosus*. The former showed no aldehyde-oxidizing activity with any substrate. Furthermore, it is the only enzyme of the family containing a [3Fe-4S] cluster, consistent with the loss of one of the cysteine ligands [16], the latter was found to oxidize a similar aldehyde as AOR [28]. Orthologous genes putatively encoding AOR-like enzymes are found in many hyperthermophilic archaea and anaerobic bacteria, and some of them have been biochemically characterized, such as AORs and FORs from *Thermococcus litoralis*, AORs from *T. paralvinellae*, *Clostridium formicoaceticum*, *Eubacterium acidaminophilum*, *Desulfovibrio gigas*, *Methanobacterium thermoautotrophicum*, and GAPOR from *Pyrobaculum aerophilum* [7,29–31].

3.2. Class II Benzoyl-CoA Reductases

Benzoyl-CoA-reductases (BCRs) are key enzymes for aromatic compound degradation in anaerobic prokaryotes [32–35]. They catalyze the reduction of the central intermediate benzoyl-CoA to cyclohexa-1,5-diene-1-carboxyl-CoA (dienoyl-CoA) at a redox potential of $E'° = -622$ mV [36]. Two phylogenetically unrelated classes of BCRs have been identified that both catalyze the same reaction. Class I BCRs are present in facultative anaerobes such as the denitrifying bacterium *Thauera aromatica* [37]. They employ [4Fe-4S] clusters as the only cofactors and use reduced ferredoxin as an electron donor. Class I BCRs couple the thermodynamically unfavorable reduction of benzoyl-CoA to the stoichiometric hydrolysis of ATP. Obligately anaerobic bacteria employ an ATP-independent, Wco-containing one megadalton class II BCR complex [38]. Class II BCRs have been purified from the Fe(III)-respiring *Geobacter metallireducens* [39] and the sulfate-respiring *Desulfosarcina cetonica* [40], both are composed of eight subunits ($Bam[(BC)_2DEFGHI]_2$, Bam = benzoic acid metabolism). With four W-bis-MPTs, four Zn, >50 Fe-S clusters, four selenocysteines, and six FADs, class II BCRs represent one of the most complex metalloenzyme machineries known (Figure 5). BamB harbors the Wco-binding active site subunit and can be purified along with the electron transferring BamC subunit [41]. The remaining BamDEFGHI subunits are proposed to be involved in the endergonic electron transfer to Wco, driven by a flavin-based electron bifurcation [39]. In this process, the endergonic electron transfer from a donor to a low-potential acceptor, here benzoyl-CoA, is driven by the exergonic reduction of a second high-potential acceptor using the same donor [42]. The anticipated high-potential acceptor(s) of class II BCRs are still unclear, but recent studies suggest that class II BCRs transfer electrons to menaquinone, as evident from the membrane-association of the complex from *G. metallireducens* [39].

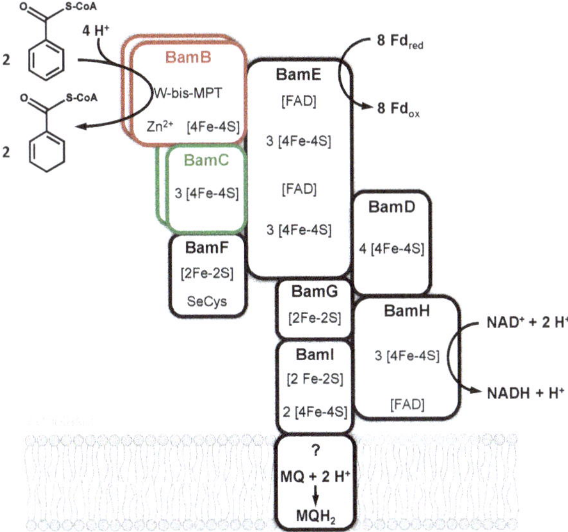

Figure 5. Scheme of the class II BCR-complex from *G. metallireducens*. The endergonic electron transfer from reduced ferredoxin to benzoyl-CoA may be driven by an exergonic reduction of NAD^+ and/or menaquinone (MQ).

The BamB subunit belongs to the AOR enzyme family and contains Wco, Zn^{2+} and a [4Fe-4S] cluster as cofactors. Bioinformatic analyses suggest that all strictly anaerobic bacteria employ a tungsten-containing BamB for benzoyl-CoA reductions during growth with aromatic substrates [34]. Initial studies with *Desulfococcus multivorans* [43] and *G. metallireducens* [38] indicated that growth with aromatics depends on molybdenum and selenium, however, the isolated $Bam(BC)_2$ contained stoichiometric amounts of tungsten, whereas molybdenum was virtually absent [41]. This finding suggests that traces of tungstate in the molybdate stock solution were sufficient to produce an active, Wco-containing class II BCR. A TupABC transporter is induced during growth with aromatics that guarantees selective tungstate uptake from the medium [44].

The 1.9 Å resolution X-ray structure of $Bam(BC)_2$ crystals revealed that Wco is accommodated in a highly hydrophobic pocket where it is coordinated by four dithiolene sulfurs of W-bis-MPT, a thiolate of a conserved cysteine (Cys322) and a sixth non-proteinogenic ligand [45] (Figure 6). Its identity could so far not be resolved unambiguously by spectroscopic methods, whereas computational studies favor a water ligand [46,47]. Neither the substrate nor the product are directly bound to tungsten. The spatial separation of the potential proton donor His260 and the W-atom with the aromatic ring positioned in-between provides the molecular basis for the anticipated biological Birch reduction of aromatic rings [48]. Continuum electrostatic and quantum-mechanical/molecular mechanics calculations suggest that benzoyl-CoA reductions are initiated by a hydrogen atom transfer from a W(IV) species via bound water yielding a W(V)-(OH^-) species and a substrate radical intermediate [46,47]. These studies also suggested that the second electron derives from the pyranopterin cofactor, rather than from W(V). Proton transfers from an invariant histidine (His260) likely assist this step. A BamB catalysis is one of the rare examples of an MPT-dependent enzyme that does not involve an oxo- or hydroxyl-transfer. The unusual properties of class II BCRs among MPT-enzymes may be explained by the extremely low redox potential of the substrate/product couple that affords a radical biochemistry.

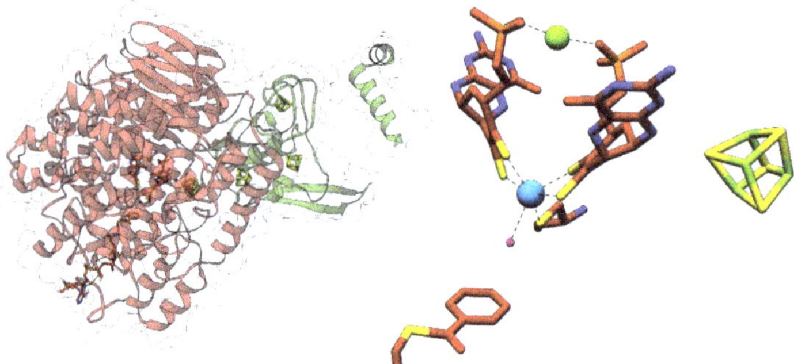

Figure 6. Structure of the BamBC components of class II benzoyl-CoA reductase from *G. metallireducens* (pdb 4Z3W-Z). Left panel: overall structure; the catalytic BamB subunit depicted in red contains the W-bis-MPT cofactor, BamC (light green) contains 3 [4Fe-4S] clusters. Right panel: W-bis-MPT, [4Fe-4S] cluster and the substrate benzoyl-CoA. Tungsten (blue) is coordinated by five thiols (four from the dithiolenes and one from Cys322), and by an inorganic ligand (magenta). A Mg^{2+} bridging the phosphate groups of the two MPTs is shown in green.

The Birch reduction of aromatic rings is widely used in synthetic chemistry for many applications [49]. Due to its negative attributes, such as its dependence on alkali metals, and ammonia and cryogenic reaction conditions, alternative procedures are being studied such as the use of photo- [50] and electrocatalytic [51] methodologies. In this light, biocatalytic BCRs may be attractive for future Birch reduction applications under mild and environmentally friendly conditions.

3.3. Acetylene Hydratases

Anaerobic acetylene degradation by *Pelobacter acetylenicus* [52] is initiated by the hydration and tautomerization to acetaldehyde, catalyzed by acetylene hydratase (AH). So far, only the enzyme from *P. acetylenicus* has been isolated and studied, and a high resolution crystal structure is available (Figure 7) [53–56]. The oxygen-sensitive enzyme belongs to the DMSOR enzyme family and contains a W-bis-MGD and a [4Fe-4S] cluster [53]. When AH was isolated from *P. acetylenicus* grown under tungsten-depletion (2 nm) and a 1000-fold excess of molybdate, a molybdenum-containing variant of AH was identified (45–50% metal occupation) that converted acetylene at 10% of the rate of the Wco-containing enzyme [57]. Because of these highly unphysiological expression conditions, we consider AH here as an obligately tungsten-containing enzyme, but further research on the viability of Mo-containing AH variants and their occurrence is required. Almost all studies were carried out with Wco-containing AH.

AH is unique among all Moco/Wco enzymes in terms of catalyzing a non-redox reaction, and it is the only known enzyme that acts on acetylene as a physiological substrate. Notably, acetylene inhibits many key enzymes of anaerobic metabolism, with the exception of nitrogenase, which reduces acetylene to ethylene [58,59]. AH itself is inhibited by acetylene analogues such as cyanide, carbon monoxide, nitrous oxide or substituted acetylenes [53,54]. Though the reaction catalyzed by AH is redox neutral, AH requires a strong reductant like titanium(III)-citrate or sodium dithionite to be catalytically active *in vitro* [53,56]. The mandatory reductive activation of AH is assigned to the reduction of Wco to the W(IV) state, though the exact redox potential of the Wco is unknown [53,55]. An intact [4Fe-4S] cluster appears to be dispensable for catalysis [56] as supported by the observation that degradation to a [3Fe-4S] cluster did not affect AH activity [53].

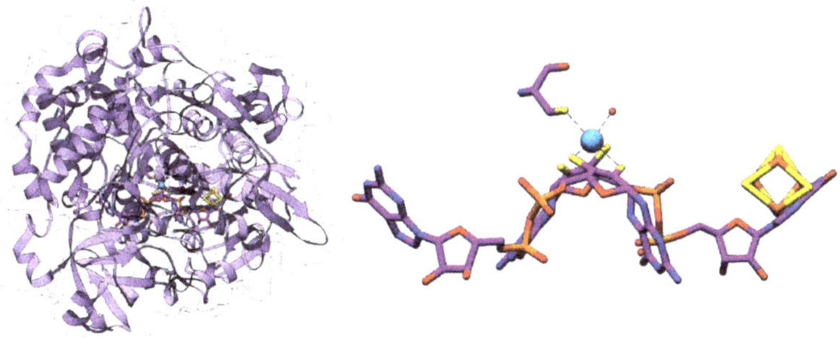

Figure 7. Structure of acetylene hydratase from *P. acetylenicus* (pdb 2E7Z) at 1.26 Å resolution. Left panel: overall structure. Right panel: W-bis-MGD and [4Fe-4S] cofactors. Tungsten is coordinated by five thiols (four from the dithiolenes and one from Cys141), and by a water molecule. The tungsten atom is depicted in blue.

The mechanism of AH has been subject to many discussions, and two types of mechanistic scenarios were suggested with different roles for the catalytically active W(IV): (i) a first-shell mechanism, where tungsten coordinates acetylene and primes the substrate for a nucleophilic attack by water (or hydroxide), and (ii) a second-shell mechanism that relies on the polarization of tungsten-bound water, followed by an electrophilic attack on the C≡C bond of acetylene [55]. The second-shell mechanism was favored by density-functional theory calculations based on a high resolution crystal structure, but no spectroscopic evidence was obtained to corroborate this mechanism [55]. Based on calculations for the active site mimics of AH, another study concluded that the direct coordination of acetylene by tungsten was favored over the binding of water [60]. It was reported that an inorganic model complex $(NEt_4)_2[W^{IV}O(mnt)_2]$, where mnt = malonitrile, was able to catalyze the same reaction as AH, whereas the W(VI) analogue failed [61]. While these results could not be reproduced, recent developments and thorough characterizations of other biomimetic tungsten complexes strongly support a mechanism of acetylene binding directly to tungsten (first-shell mechanism) [62,63]. Such a metal-acetylene coordination is known from other organometallic molybdenum-/tungsten-complexes [64].

Acetaldehyde is an important building block for many chemicals and pharmaceuticals. Its industrial production from acetylene is typically achieved in the presence of a mercury catalyst-containing mercuric chloride that exhibits adverse effects on human health and the environment [65]. Thus, enzymatic solutions for this industrial process are of potential interest.

4. Enzymes Containing Either Tungsten or Molybdenum

A number of isofunctional enzymes of the DMSOR family exist in nature that may contain either Moco or Wco. This apparent promiscuity may be explained by an adaptation to molybdate or tungstate enriched/depleted environments rather than to substantial differences in enzyme function. In some organisms, sets of genes exist that specifically code either for Moco- and/or for Wco-dependent variants of MPT-dependent enzymes. The induction of the individual gene clusters in response to varying molybdate/tungstate concentrations has been described in some cases. The molybdenum- or tungsten-containing variants may differ in their activity or affinity to the individual substrates. Owing to the lower redox potential of the biologically relevant W(IV/V/VI) vs. Mo(IV/V/VI) states, Wco-containing variants appear to be catalytically more efficient for reactions occurring at low redox potentials and Moco for those at more positive redox potentials.

4.1. Formate Dehydrogenases

The widely abundant tungsten- or molybdenum-dependent FDHs catalyze the reversible conversion of formate to CO_2 at $E°' = -430$ mV with various electron donors/acceptors [66–70]. Metal-dependent FDHs are involved in formate-dependent respiration, syntrophy and methanogenesis, acetogenesis, methylotrophy or in fermentations as components of formate hydrogen lyase complexes. Unlike most other Moco/Wco enzymes, an FDH catalysis does not involve an oxygen atom transfer reaction. Though the reaction mechanism of FDHs is still under debate, a hydride transfer from formate to a sulfido-ligand at Mo(VI)/W(VI) appears to be the most plausible as it circumvents the difficult deprotonation of the C_α proton of formate [68,71–73]. Here, we briefly summarize the occurrence and function of Wco-dependent FDHs and refer to recent in-depth reviews of the diversity, structure and function of metal-containing FDHs [66–68].

Many FDHs depend on molybdenum, and elevated tungstate concentrations in the medium yield inactive Mo-dependent variants as reported for FDHs from *E. coli* [74] or *Methanobacterium formicium* [75]. On the other side, a number of FDHs have been isolated with a clear preference for tungsten over molybdenum, whereas only a few FDHs appear to be active with either of the two metals. Wco-containing FDHs are affiliated to two enzyme clades, the membrane-bound FDH-N-like enzymes and the soluble or membrane-associated FDH-H-like enzymes, whose active sites are oriented towards the periplasm or the cytoplasm, respectively. Cytoplasmic Wco-containing FDHs have initially been described in a number of acetogenic clostridia such as *Moorella thermoaceticum* [76], *Clostridium formicoaceticum* [77] or *C. carboxidivorans* [78], where they catalyze the NAD(P)H-dependent reduction of CO_2. A tungsten-containing FDH of *Peptoclostridium* (formerly *Eubacterium*) *acidaminophilum* serves either as an electron donor system for amino acid fermentations or as component of a formate hydrogen lyase complex [79]. A tungsten-dependent FDH of this clade is also involved in the oxidation of reduced C1 compounds to CO_2 with dioxygen as an electron acceptor in the aerobic methylotrophic *Methylobacterium extorquens* [80]. This finding indicates that the use of Wco is not restricted to anaerobic organisms. Two tungsten-specific FDHs, an FDH-H and a FDH-N type enzyme, were reported from *Syntrophobacter fumaroxidans* during syntrophic propionate fermentations, where they play a role in electron transfers to the methanogenic partner [81]. Further Wco-containing FDHs of the FDH-N clade are known from *Desulfovibrio* species and other sulfate reducing bacteria [82–85]. One of these species, *D. alaskensis*, is known to incorporate either Moco or Wco into the same FDH, depending on the tungstate and molybdate concentrations in the medium [83,85]. In contrast, *D. vulgaris* Hildenborough produces a specific Moco-dependent FDH during growth with molybdate and an FDH active with either metal during growth with tungstate [84]. A similar case of an active FDH-N type enzyme with either Moco or Wco in differently grown cells was reported for *Campylobacter jejuni* [86], which contains only one set of FDH encoding genes.

Biocatalysts that sequester CO_2 are of increasing general interest. In this light, FDHs that reduce CO_2 to formate as a viable energy source are of a potential biotechnological use. A tungsten-containing FDH has been adsorbed to an electrode and successfully used for electrocatalytical CO_2 reduction [87]. Furthermore, an enzyme complex of FDH and hydrogenase (formate hydrogen lyase) reversibly interconverts $H_2 + CO_2$ to formate demonstrating that formate may be used as an energy storage/transport compound [88,89]. Finally, the solar-driven CO_2 reduction by FDHs in combination with artificial and/or biological photosystems has been demonstrated [90,91].

4.2. Formylmethanofuran Dehydrogenases

Formylmethanofuran dehydrogenase catalyzes the first step of methanogenesis that is related to the reaction catalyzed by FDHs: the reduction of CO_2 yielding formate is bound in an activated form at the methanofuran cofactor [69]. Methanogens have been found to contain isoenzymes being either specific for tungsten or molybdenum [92,93]. The genome of *Methanososarcina acetivorans* encodes four putative formylmethanofuran dehydrogenases. Two of these are proposed to be specific for tungsten or molybdenum, respectively, with one of the tungsten-dependent isoenzymes being specifically required for growth with carbon monoxide [94]. The crystal structure of the tungsten-containing formylmethanofuran dehydrogenase from *Methanothermobacter wolfei* revealed that the enzyme harbors two active sites being separated by a 43 Å tunnel which allows the directed diffusion of the formate from the first active site (the Wco center) to the second one, the formylmethanofuran forming center [95].

4.3. Respiratory Nitrate Reductases

Two versions of respiratory nitrate reductases belong to the DMSOR enzyme family: the membrane bound enzymes (NarGHI) and the periplasmic ones (NAP). The active site subunit of the former, NarG, is largely conserved among all nitrate-respiring bacteria and archaea and usually contains Moco in the active site subunit. Tungstate usually acts as an inhibitor of the assembly of Moco into NarG, NAP or other respiratory Moco-containing enzymes [8,74]. However, *Paracoccus pantotrophus* is able to grow under denitrifying conditions in the presence of 100 μm tungstate, and indirect evidence was obtained that tungsten was incorporated into its periplasmic nitrate reductase NAP [96]. Furthermore, the nitrate-respiring, hyperthermophilic archaea *Pyrobaculum aerophilum* and *Aeropyrum pernix* grow at temperatures and environments where tungstate is enriched and molybdate is depleted. Under these special growth conditions, a tungsten-containing NarGHI-type enzyme was isolated from of *P. aerophilum* with a two-fold lower turnover number than the Mo-containing enzyme [97].

4.4. Dimethyl Sulfoxide and Trimethylamine N-Oxide Reductases

DMSOR and trimethylaminoxide reductase (TMAOR) are both required to use DMSO or TMAO as terminal electron acceptors in oxygen-independent respiratory chains. In the DMSOR from *Rhodobacter capsulatus* and *R. sphaeroides*, the Mo bound to the bis-Mo-MGD cofactor was substituted with tungsten, and the resulting Wco-containing enzyme reduced DMSO even at a higher rate, but was inactive in terms of catalyzing the reverse dimethyl sulfide oxidation [15,98]. This finding is rationalized by the generally lower redox potential of Wco in comparison to Moco [99]. Likewise, TMAOR from *E. coli* was also active when molybdenum was substituted by tungsten even with a slightly increased catalytic efficiency [100].

4.5. Thiosulfate Reductases

A recent report shows that a thiosulfate reductase (TSR) of the DMSOR family is involved in thiosulfate respiration in the archaeon *P. aerophilum*. The enzyme was recombinantly produced in *P. furiosus* and contained either Mo or W, dependent on the growth conditions [101]. The molybdenum-containing enzyme showed a ten-fold higher conversion rate and a twice as high affinity to thiosulfate than the tungsten-containing one. This study represents the first case of producing an active Moco- or Wco-containing enzyme in *P. furiosus*.

5. Tungsten Uptake and Assembly of the Wco

5.1. Tungsten Uptake

Tungsten is taken up by prokaryotes in form of the tungstate oxyanion (WO_4^{2-}) by three high-affinity ATP-binding cassette (ABC)-type systems: the highly tungstate-specific TupABC system, and the ModABC/WtpABC systems that transport either tungstate or molybdate [102]. They all consist of a periplasmic binding protein (component A), a transmembrane channel (component B), and an ATPase subunit (component C). The ModABC system with similar affinities to molybdate and tungstate in the low to high nM range is highly abundant in bacteria and archaea and allows for the specific transport of molybdate and tungstate even in the presence of high concentrations of similar other oxyanions (e.g., 28 mm sulfate in marine environments). The WtpABC transporter is present in archaea that lack TupABC or ModABC systems, such as *P. furiosus*, and is rather distantly related to these uptake systems. The binding protein WtpA shows by far the highest affinity for tungstate with K_d = 19 pm [103]. The TupABC system has originally been studied in *P. acidaminophilum* [104,105], and later in C. *jejuni* [106,107] and *D. alaskensis* [108,109]. The encoding TupABC genes are widely abundant in bacterial and archaeal genomes [110]. The structural basis by which TupA discriminates between tungstate and molybdate, considering their highly similar atomic radii, is still unknown.

5.2. Metallopterin Cofactor Synthesis

Most knowledge of MPT biosynthesis derives from research on Moco (for initial and recent reviews on Moco synthesis see [4,5,111–113]). As almost all genes required for Moco synthesis are present in the genomes of prokaryotes that synthesize Wco-dependent enzymes, it is generally accepted that MPT synthesis is catalyzed by similar enzymes and identical intermediates [114]. In contrast, very little is known about the steps that discriminate between molybdenum and tungsten during the insertion of the metals into MPT. Here, we first briefly summarize knowledge of Moco synthesis in the prokaryotic model organism *E. coli*, followed by the presentation of initial insights and hypotheses about Wco synthesis. There, we will address the special challenges of simultaneously synthesizing Moco- and Wco-dependent enzymes.

Moco biosynthesis proceeds via a four-step process: (i) conversion of guanosine-5′-triphosphate (GTP) to the four-ringed cyclic intermediate pyranopterin monophosphate (cPMP), (ii) introduction of the two thiols of the dithiolene group, (iii) adenylation to MPT-adenosine monophosphate (AMP), and (iv) metal insertion and release of AMP; the latter two steps are often merged as a single one (Figure 8) [4,5,111,113]. The formation of cPMP (or "precursor Z") from GTP occurs in two steps involving complex molecule rearrangements. The radical S-adenosylmethionine (SAM) enzyme MoaA catalyzes an initial remodeling of the carbon backbone of GTP and afterwards, MoaC completes the synthesis of cPMP with the concomitant release of the pyrophosphate moiety of the original GTP [115,116].

In the second step, two sulfur atoms are inserted into cPMP at the C1′ and C2′ position, resulting in MPT formation. This step is catalyzed by MoaD and MoaE, which form a heterotetrameric complex, also referred to as MPT synthase. Two MoaEs bind one cPMP each, while MoaD transfers one sulfur atom at a time from a thiocarboxylate group on its conserved C-terminal glycine residue, producing a hemisulfurated intermediate in the process [117–120]. MoaD itself receives its thiogylcyl modification by the sulfurylase MoeB. MoaD forms a complex with MoeB, which catalyses the Mg^{2+}-and ATP-dependent adenylation of the C-terminal glycine residue of MoaD. This activated MoaD is then sulfurated by the cysteine desulfurase IscS [121,122].

In the third step, MPT is adenylated in an ATP-dependent reaction by MogA (in bacteria) or MoaB (in archaea), yielding MPT-AMP. This activated intermediate is transferred to MoeA, where molybdate is bound to the dithiolene sulfurs of MPT, accompanied by hydrolysis and the release of the AMP moiety [114,123,124]. The product is referred to as Mo-MPT, which represents the Moco found in SO family members, and which serves as a building block for the cofactors of the three other families of MPT-containing enzymes. To generate the cofactor for the XO-family enzymes, MocA catalyzes the cytidylation of MPT with CTP while releasing pyrophosphate. The generated MPT cytosine dinucleotide cofactor is then further modified by the addition of an equatorial sulfido-ligand in the active site of the enzymes [13,125]. Ligation of a molybdenum and a second MPT yields Mo-bis-MPT, which is found in the AOR family member YdhV [126]. The Mo-bis-MGD cofactor of the DMSOR family enzymes is synthesized by the attachment of two GMP moieties from GTP [127,128].

Wco synthesis is considered to involve similar or even the same enzymes for the steps up to metal insertion. So far, the adenylation of MPT by MoaB in the course of Wco synthesis was shown in the hyperthermophilic *P. furiosus* [114]. No other enzyme specifically involved in Wco synthesis has been studied so far.

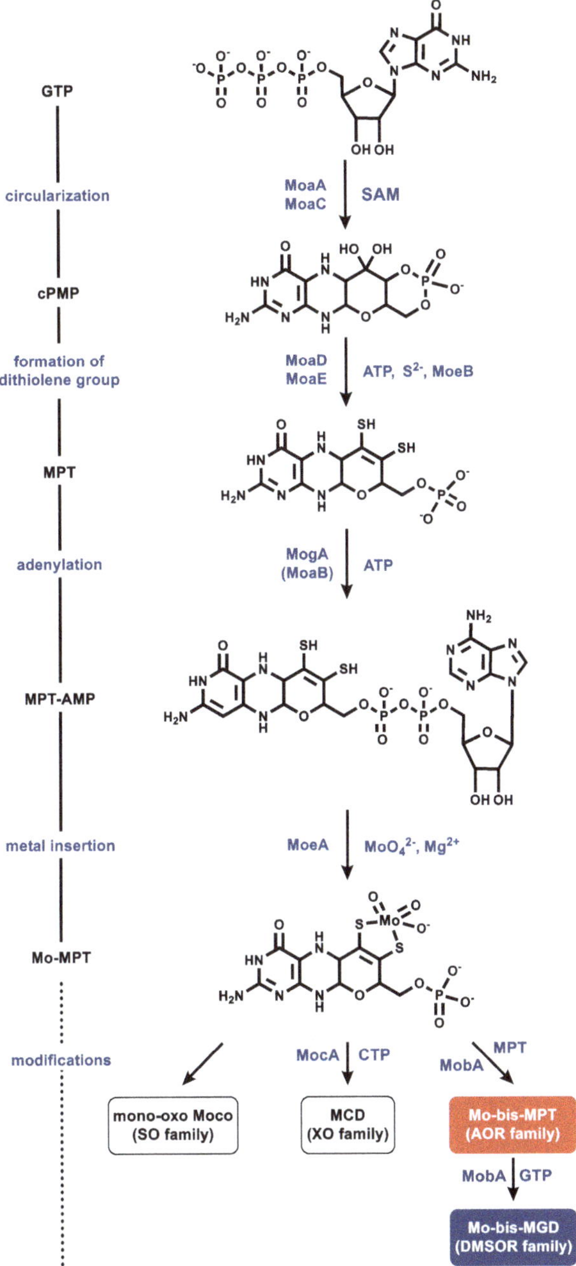

Figure 8. Moco biosynthesis in *E. coli*. In prokaryotes, the steps from GTP to MPT-adenosine monophosphate (AMP) are catalyzed by similar enzymes for Moco and Wco synthesis. YdhV from *E. coli* is the only characterized molybdenum-containing AOR family member [126].

5.3. Strategies for the Selective Insertion of Molybdenum/Tungsten into Target Proteins

The highly similar physicochemical properties of molybdate and tungstate immediately lead to the question regarding which mechanism guarantees the correct transfer of both oxyanions into their target proteins. In particular, organisms producing enzymes that are only active with either of the two metals depend on a highly selective metal insertion system. We distinguish here between three different scenarios: (i) organisms that exclusively produce either Wco- or Moco-dependent enzymes, (ii) organisms that contain genes for both type of enzymes, but only produce either Moco- or Wco-enzymes under certain conditions, and (iii) organisms that produce Moco- and Wco-dependent enzymes simultaneously. In principle, selectivity may be achieved during molybdate/tungstate uptake, Mo-MPT/W-MPT formation, or the insertion of the mature W-bis-MPT/bis-Mo-MPT into the respective target proteins.

In the first scenario, the selective uptake of either molybdate or tungstate would already be sufficient to guarantee selectivity. Specific tungstate uptake may be achieved by the periplasmic binding protein TupA from the TupABC transporter [105], or WtpA from *P. furiosus*, which show much higher affinities for tungstate vs. molybdate [103]. However, no transporter system exhibiting a comparable selectivity for molybdate over tungstate has been described so far, which explains the frequently observed antagonistic effect of tungstate for Moco-dependent enzymatic processes.

In the second scenario, the selective uptake may also play a role but would require at least two selective uptake systems for either of the two metal oxyanions, which may be induced under certain conditions. In addition, MoeAs have been proposed earlier as selectivity generating enzymes that may be specific for either molybdate or tungstate during metal insertion into MPT-AMP [12,85,114]. This proposal was based on the finding that at least two versions of MoeA encoding genes were identified in genomes of archaea [12] or in *D. alaskensis* [85]. The latter produces either Moco or Wco containing versions of the same FDH, depending on the molybdate/tungsten concentrations in the medium. In *D. vulgaris* Hildenborough, the production of Wco- or Moco-containing FDH isoenzymes is regulated by the molybdate/tungstate availability at the transcriptional level [84]. These results are in line with a metal-dependent induction of either tungsten- or molybdenum-specific pairs of MoeA/apo-FDH paralogues.

Such a mechanism is not feasible when Moco- and Wco-depending enzymes have to be produced in parallel (scenario iii). In recent years, a number of bacteria were identified that simultaneously produce obligately Wco- and Moco-dependent enzymes, and they all contain at least two *moeA* copies (Figure 9). *C. jejuni* produces a tungsten-containing FDH under certain growth conditions that may be coupled with molybdenum-dependent nitrate reductase in a menaquinone-dependent respiratory chain [86]. In *A. aromaticum*, a tungsten-dependent AOR and two molybdenum-dependent enzymes, phenylacetyl-CoA dehydrogenase and nitrate reductase, are produced during growth with phenylalanine under denitrifying conditions [21]. When *G. metallireducens* grows with *p*-cresol or 4-hydroxybenzoate under ammonifying conditions, it produces one strictly Wco-dependent class II benzoyl-CoA reductase [38,41], and two Moco-containing enzymes from different families: the 4-hydroxybenzoyl-CoA reductase of the XO family [129,130] and nitrate reductase of the DMSOR family [44]. While *C. jejuni* contains two paralogous *moeA* genes [106], even three paralogues are found in the genomes of *A. aromatoleum* and *G. metallireducens*. In the latter bacterium, the genetic context of the three *moeA* genes indicates the specificity for molybdenum or tungsten (Figure 9). In particular, *moeA-1* is located adjacent to the genes encoding the tungstate-dependent TupABC transporter, and *moeA-3* is adjacent to paralogues of the *bamBC* genes, which putatively code for an isoenzyme of tungsten-dependent class II benzoyl-CoA reductase. Thus, it is tempting to speculate that *moeA-1* and *moeA-3* encode a tungstate-specific metal insertase. In contrast, *moeA-2* is located in a gene cluster together with genes encoding molybdenum-dependent NarGHI-type nitrate reductase. Thus, *moeA-2* likely encodes a molybdenum-specific enzyme. These findings support the idea that the simultaneous production of Moco- and Wco-dependent enzymes is accomplished by MoeA isoenzymes that are selective for either molybdate or tungstate. Though this assumption appears plausible, any experimental evidence for it is lacking.

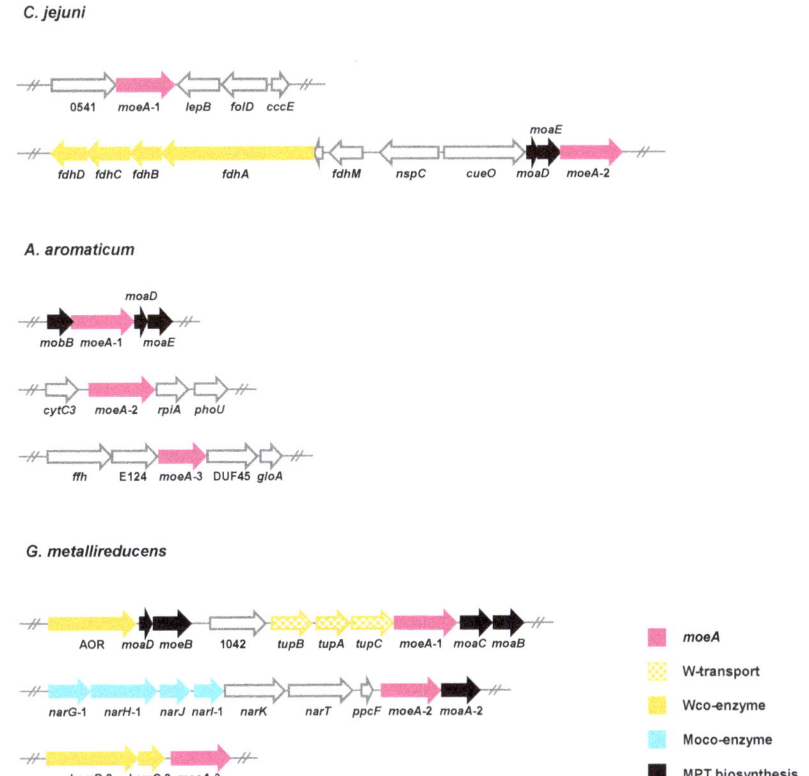

Figure 9. Genomic location of *moeA* paralogues in bacteria that produce Wco- and Moco-containing enzymes simultaneously.

Finally, the question rises as to how the Mo-MPT/W-MPT cofactors formed by specific MoeAs are specifically processed and targeted into the individual apo-enzymes. Specific targeting excludes that Mo-MPT/W-MPT formed are released from individual MoeAs because the modifying enzymes that bind the Mo-MPT/W-MPT in the next step will hardly distinguish between the molybdenum- or tungsten-containing versions. In tungstoenzymes, the metal is always bound to two pyranopterins (bis-MPT), which affords the conversion of MoeA-bound W-MPT to W-bis-MPT in a similar manner as the MobA-catalyzed process known for molybdenum enzymes of the DMSOR family [128]. Thus, the reported complex formation between MoeA and MobA [131] appears to be essential for maintaining the selectivity of the individual bis-MPT intermediates formed. It is conceivable that individual MoeA/MobA complexes specifically interact with the respective apo-enzymes either before or after the GTP-dependent W-bis-MGD formation, including further modifying enzymes such as FdhD. The latter is involved in sulfuration of the Mo-/W-bis-MPT of FDHs and in its insertion into apo-FDHs [132–134]. MoeAs, specific for either tungsten or molybdenum, may enable or disable such additional interactions. Finally, the formation of supercomplexes between all the components involved in the conversion of the last common MPT-AMP intermediate to the mature cofactors inserted in individual target proteins could substantially facilitate selective Moco/Wco targeting.

6. Conclusions

Recent research has revealed a higher abundance and catalytic diversity of tungstoenzymes than previously anticipated. While tungsten was originally mostly associated with AORs from hyperthermophilic archaea and some clostridial FDHs, a number of tungstoenzymes with novel functions have been discovered in a large variety of microorganisms comprising fermenting, sulfate-reducing, metal oxide-reducing, denitrifying, aerobic and pathogenic bacteria. Tungstoenzymes are of emerging biotechnological interest due to their involvement in the low-potential reduction processes of CO_2, carboxylic acids or aromatic rings, or in the industrially important conversion of acetylene to acetaldehyde. Thus, tungsten appears to be the bio-metal of choice for a number of challenging enzymatic reactions. While knowledge of the structure and function of tungstoenzymes has continuously been increasing over the past two decades, there is still a high research demand concerning the biosynthesis and incorporation of Wco. In particular, the mechanisms that discriminate between tungstate- and molybdate-insertion, and the selective targeting of Mo- or W-bis-MPTs to the individual apo-proteins, are still unknown. Considering the high similarities of molybdate and tungstate, the observed high selectivity for either of the two metals affords sophisticated molecular solutions. Insights into the molecular basis of this selectivity will help to explain how enzymes that strictly rely on tungsten are specifically loaded with this cofactor even in the presence of high molybdate concentrations in the medium. They will also provide a rationale as to why whole-cells may be inhibited by tungstate due to the antagonistic effects of tungstate and molybdate. Bacteria that simultaneously produce tungsto- and molybdoenzymes will serve as valuable model organisms to address the question of selectivity for tungstate vs. molybdate in the future.

Author Contributions: Phylogenetic analyses and visualization: J.H., C.S.S.; structural analysis: M.W.; visualization of reaction schemes and genetic organization: C.S.S., M.B.; project supervision: M.B.; the manuscript was written and revised through contributions of all authors. All authors have read and agreed to the published version of the manuscript.

Funding: This work was funded by the German Research Foundation (DFG), grants BO 1565 15-1 and HE 2190 11-1, the collaborative research center SFB 1381, project ID 403222702 (M.B.) and the Synmikro center Marburg (J.H.).

Conflicts of Interest: The authors declare no conflict of interest.

References

1. Hille, R. The molybdenum oxotransferases and related enzymes. *Dalt. Trans.* **2013**, *42*, 3029–3042. [CrossRef] [PubMed]
2. Hille, R.; Hall, J.; Basu, P. The mononuclear molybdenum enzymes. *Chem. Rev.* **2014**, *114*, 3963–4038. [CrossRef] [PubMed]
3. Leimkühler, S.; Iobbi-Nivol, C. Bacterial molybdoenzymes: Old enzymes for new purposes. *FEMS Microbiol. Rev.* **2015**, *40*, 1–18. [CrossRef]
4. Mendel, R.R.; Leimkühler, S. The biosynthesis of the molybdenum cofactors. *J. Biol. Inorg. Chem.* **2015**, *20*, 337–347. [CrossRef]
5. Leimkühler, S. The biosynthesis of the molybdenum cofactors in *Escherichia coli*. *Environ. Microbiol.* **2020**, *22*, 2007–2026. [CrossRef]
6. Hille, R.; Schulzke, C.; Kirk, M.L. *Molybdenum and Tungsten Enzymes: Spectroscopic and Theoretical Investigations*; Metallobiology; Royal Society of Chemistry: Cambridge, UK, 2016; ISBN 978-1-78262-878-1.
7. Andreesen, J.R.; Makdessi, K. Tungsten, the surprisingly positively acting heavy metal element for prokaryotes. *Ann. N. Y. Acad. Sci.* **2008**, *1125*, 215–229. [CrossRef]
8. Zhu, W.; Winter, M.G.; Byndloss, M.X.; Spiga, L.; Duerkop, B.A.; Hughes, E.R.; Büttner, L.; de Lima Romão, E.; Behrendt, C.L.; Lopez, C.A.; et al. Precision editing of the gut microbiota ameliorates colitis. *Nature* **2018**, *553*, 208–211. [CrossRef]
9. Hille, R. Molybdenum and tungsten in biology. *Trends Biochem. Sci.* **2002**, *27*, 360–367. [CrossRef]
10. Pushie, M.J.; Cotelesage, J.J.; George, G.N. Molybdenum and tungsten oxygen transferases-structural and functional diversity within a common active site motif. *Metallomics* **2014**, *6*, 15–24. [CrossRef]

11. Romão, M.J. Molybdenum and tungsten enzymes: A crystallographic and mechanistic overview. *Dalt. Trans.* **2009**, 4053–4068. [CrossRef]
12. Bevers, L.E.; Hagedoorn, P.L.; Hagen, W.R. The bioinorganic chemistry of tungsten. *Coord. Chem. Rev.* **2009**, *253*, 269–290. [CrossRef]
13. Neumann, M.; Mittelstädt, G.; Iobbi-Nivol, C.; Saggu, M.; Lendzian, F.; Hildebrandt, P.; Leimkühler, S. A periplasmic aldehyde oxidoreductase represents the first molybdopterin cytosine dinucleotide cofactor containing molybdo-flavoenzyme from *Escherichia coli*. *FEBS J.* **2009**, *276*, 2762–2774. [CrossRef] [PubMed]
14. Romão, M.J.; Archer, M.; Moura, I.; Moura, J.J.; LeGall, J.; Engh, R.; Schneider, M.; Hof, P.; Huber, R. Crystal structure of the xanthine oxidase-related aldehyde oxido-reductase from *D. Gigas*. *Science* **1995**, *270*, 1170–1176. [CrossRef] [PubMed]
15. Pacheco, J.; Niks, D.; Hille, R. Kinetic and spectroscopic characterization of tungsten-substituted DMSO reductase from *Rhodobacter sphaeroides*. *J. Biol. Inorg. Chem.* **2018**, *23*, 295–301. [CrossRef] [PubMed]
16. Roy, R.; Adams, M.W.W. Characterization of a fourth tungsten-containing enzyme from the hyperthermophilic archaeon *Pyrococcus furiosus*. *J. Bacteriol.* **2002**, *184*, 6952–6956. [CrossRef]
17. Mukund, S.; Adams, M.W. Molybdenum and vanadium do not replace tungsten in the catalytically active forms of the three tungstoenzymes in the hyperthermophilic archaeon *Pyrococcus furiosus*. *J. Bacteriol.* **1996**, *178*, 163–167. [CrossRef]
18. Sevcenco, A.-M.; Bevers, L.E.; Pinkse, M.W.H.; Krijger, G.C.; Wolterbeek, H.T.; Verhaert, P.D.E.M.; Hagen, W.R.; Hagedoorn, P.-L. Molybdenum incorporation in tungsten aldehyde oxidoreductase enzymes from *Pyrococcus furiosus*. *J. Bacteriol.* **2010**, *192*, 4143–4152. [CrossRef]
19. Mukund, S.; Adams, M.W. Characterization of a novel tungsten-containing formaldehyde ferredoxin oxidoreductase from the hyperthermophilic archaeon, *Thermococcus litoralis*. A role for tungsten in peptide catabolism. *J. Biol. Chem.* **1993**, *268*, 13592–13600.
20. Chan, M.K.; Mukund, S.; Kletzin, A.; Adams, M.W.; Rees, D.C. Structure of a hyperthermophilic tungstopterin enzyme, aldehyde ferredoxin oxidoreductase. *Science* **1995**, *267*, 1463–1469. [CrossRef]
21. Arndt, F.; Schmitt, G.; Winiarska, A.; Saft, M.; Seubert, A.; Kahnt, J.; Heider, J. Characterization of an aldehyde oxidoreductase from the mesophilic bacterium *Aromatoleum aromaticum* Ebn1, a member of a new subfamily of tungsten-containing enzymes. *Front. Microbiol.* **2019**, *10*, 71. [CrossRef]
22. Basen, M.; Schut, G.J.; Nguyen, D.M.; Lipscomb, G.L.; Benn, R.A.; Prybol, C.J.; Vaccaro, B.J.; Poole, F.L.; Kelly, R.M.; Adams, M.W.W. Single gene insertion drives bioalcohol production by a thermophilic archaeon. *Proc. Natl. Acad. Sci. USA* **2014**, *111*, 17618–17623. [CrossRef]
23. Keller, M.W.; Lipscomb, G.L.; Nguyen, D.M.; Crowley, A.T.; Schut, G.J.; Scott, I.; Kelly, R.M.; Adams, M.W.W. Ethanol production by the hyperthermophilic archaeon *Pyrococcus furiosus* by expression of bacterial bifunctional alcohol dehydrogenases. *Microb. Biotechnol.* **2017**, *10*, 1535–1545. [CrossRef] [PubMed]
24. Roy, R.; Mukund, S.; Schut, G.J.; Dunn, D.M.; Weiss, R.; Adams, M.W. Purification and molecular characterization of the tungsten-containing formaldehyde ferredoxin oxidoreductase from the hyperthermophilic archaeon *Pyrococcus furiosus*: The third of a putative five-member tungstoenzyme family. *J. Bacteriol.* **1999**, *181*, 1171–1180. [CrossRef]
25. Hu, Y.; Faham, S.; Roy, R.; Adams, M.W.; Rees, D.C. Formaldehyde ferredoxin oxidoreductase from *Pyrococcus furiosus*: The 1.85 Å resolution crystal structure and its mechanistic implications. *J. Mol. Biol.* **1999**, *286*, 899–914. [CrossRef] [PubMed]
26. Mukund, S.; Adams, M.W.W. Glyceraldehyde-3-phosphate ferredoxin oxidoreductase, a novel tungsten-containing enzyme with a potential glycolytic role in the hyperthermophilic archaeon *Pyrococcus furiosus*. *J. Biol. Chem.* **1995**, *270*, 8389–8392. [CrossRef] [PubMed]
27. Scott, I.M.; Rubinstein, G.M.; Poole, F.L.; Lipscomb, G.L.; Schut, G.J.; Williams-Rhaesa, A.M.; Stevenson, D.M.; Amador-Noguez, D.; Kelly, R.M.; Adams, M.W.W. The thermophilic biomass-degrading bacterium *Caldicellulosiruptor bescii* utilizes two enzymes to oxidize glyceraldehyde 3-phosphate during glycolysis. *J. Biol. Chem.* **2019**, *294*, 9995–10005. [CrossRef]
28. Bevers, L.E.; Bol, E.; Hagedoorn, P.-L.; Hagen, W.R. WOR5, a novel tungsten-containing aldehyde oxidoreductase from *Pyrococcus furiosus* with a broad substrate specificity. *J. Bacteriol.* **2005**, *187*, 7056–7061. [CrossRef]
29. Kletzin, A.; Adams, M.W. Tungsten in biological systems. *FEMS Microbiol. Rev.* **1996**, *18*, 5–63. [CrossRef]

30. Hagedoorn, P.L.; Chen, T.; Schröder, I.; Piersma, S.R.; De Vries, S.; Hagen, W.R. Purification and characterization of the tungsten enzyme aldehyde:ferredoxin oxidoreductase from the hyperthermophilic denitrifier *Pyrobaculum aerophilum*. *J. Biol. Inorg. Chem.* **2005**, *10*, 259–269. [CrossRef]
31. Reher, M.; Gebhard, S.; Schönheit, P. Glyceraldehyde-3-phosphate ferredoxin oxidoreductase (GAPOR) and nonphosphorylating glyceraldehyde-3-phosphate dehydrogenase (GAPN), key enzymes of the respective modified Embden-Meyerhof pathways in the hyperthermophilic crenarchaeota *Pyrobaculum aerophilum*. *FEMS Microbiol. Lett.* **2007**, *273*, 196–205.
32. Heider, J.; Boll, M.; Breese, K.; Breinig, S.; Ebenau-Jehle, C.; Feil, U.; Gad'on, N.; Laempe, D.; Leuthner, B.; Mohamed, M.E.S.; et al. Differential induction of enzymes involved in anaerobic metabolism of aromatic compounds in the denitrifying bacterium *Thauera aromatica*. *Arch. Microbiol.* **1998**, *170*, 120–131. [CrossRef] [PubMed]
33. Fuchs, G.; Boll, M.; Heider, J. Microbial degradation of aromatic compounds-From one strategy to four. *Nat. Rev. Microbiol.* **2011**, *9*, 803–816. [CrossRef] [PubMed]
34. Löffler, C.; Kuntze, K.; Vazquez, J.R.; Rugor, A.; Kung, J.W.; Böttcher, A.; Boll, M. Occurrence, genes and expression of the W/Se-containing class II benzoyl-coenzyme A reductases in anaerobic bacteria. *Environ. Microbiol.* **2011**, *13*, 696–709. [CrossRef] [PubMed]
35. Boll, M.; Löffler, C.; Morris, B.E.L.; Kung, J.W. Anaerobic degradation of homocyclic aromatic compounds via arylcarboxyl-coenzyme A esters: Organisms, strategies and key enzymes. *Environ. Microbiol.* **2014**, *16*, 612–627. [CrossRef]
36. Kung, J.W.; Baumann, S.; von Bergen, M.; Müller, M.; Hagedoorn, P.-L.; Hagen, W.R.; Boll, M. Reversible biological birch reduction at an extremely low redox potential. *J. Am. Chem. Soc.* **2010**, *132*, 9850–9856. [CrossRef]
37. Boll, M.; Fuchs, G. Benzoyl-coenzyme A reductase (dearomatizing), a key enzyme of anaerobic aromatic metabolism. *Eur. J. Biochem.* **1995**, *234*, 921–933. [CrossRef]
38. Wischgoll, S.; Heintz, D.; Peters, F.; Erxleben, A.; Sarnighausen, E.; Reski, R.; Van Dorsselaer, A.; Boll, M. Gene clusters involved in anaerobic benzoate degradation of *Geobacter metallireducens*. *Mol. Microbiol.* **2005**, *58*, 1238–1252. [CrossRef]
39. Huwiler, S.G.; Löffler, C.; Anselmann, S.E.L.; Stärk, H.-J.; von Bergen, M.; Flechsler, J.; Rachel, R.; Boll, M. One-megadalton metalloenzyme complex in *Geobacter metallireducens* involved in benzene ring reduction beyond the biological redox window. *Proc. Natl. Acad. Sci. USA* **2019**, *116*, 2259–2264. [CrossRef]
40. Anselmann, S.E.L.; Löffler, C.; Stärk, H.; Jehmlich, N.; Bergen, M.; Brüls, T.; Boll, M. The class II benzoyl-coenzyme A reductase complex from the sulfate-reducing *Desulfosarcina cetonica*. *Environ. Microbiol.* **2019**, *21*, 4241–4252. [CrossRef]
41. Kung, J.W.; Löffler, C.; Dorner, K.; Heintz, D.; Gallien, S.; Van Dorsselaer, A.; Friedrich, T.; Boll, M. Identification and characterization of the tungsten-containing class of benzoyl-coenzyme A reductases. *Proc. Natl. Acad. Sci. USA* **2009**, *106*, 17687–17692. [CrossRef]
42. Buckel, W.; Thauer, R.K. Flavin-based electron bifurcation, a new mechanism of biological energy coupling. *Chem. Rev.* **2018**, *118*, 3862–3886. [CrossRef] [PubMed]
43. Peters, F.; Rother, M.; Boll, M. Selenocysteine-containing proteins in anaerobic benzoate metabolism of *Desulfococcus multivorans*. *J. Bacteriol.* **2004**, *186*, 2156–2163. [CrossRef] [PubMed]
44. Heintz, D.; Gallien, S.; Wischgoll, S.; Ullmann, A.K.; Schaeffer, C.; Kretzschmar, A.K.; Van Dorsselaer, A.; Boll, M. Differential membrane proteome analysis reveals novel proteins involved in the degradation of aromatic compounds in *Geobacter metallireducens*. *Mol. Cel Proteom.* **2009**, *8*, 2159–2169. [CrossRef]
45. Weinert, T.; Huwiler, S.G.; Kung, J.W.; Weidenweber, S.; Hellwig, P.; Stärk, H.-J.; Biskup, T.; Weber, S.; Cotelesage, J.J.H.; George, G.N.; et al. Structural basis of enzymatic benzene ring reduction. *Nat. Chem. Biol.* **2015**, *11*, 586–591. [CrossRef] [PubMed]
46. Culka, M.; Huwiler, S.G.; Boll, M.; Ullmann, G.M. Breaking benzene aromaticity-computational insights into the mechanism of the tungsten-containing benzoyl-CoA reductase. *J. Am. Chem. Soc.* **2017**, *139*, 14488–14500. [CrossRef]
47. Qian, H.X.; Liao, R.Z. QM/MM Study of tungsten-dependent benzoyl-coenzyme a reductase: Rationalization of regioselectivity and predication of W vs Mo selectivity. *Inorg. Chem.* **2018**, *57*, 10667–10678. [CrossRef]
48. Birch, A.J. Reduction by dissolving metals. *Nature* **1946**, *158*. [CrossRef]
49. Birch, A.J. The Birch reduction in organic synthesis. *Pure Appl. Chem.* **1996**, *68*, 553–556. [CrossRef]

50. Chatterjee, A.; König, B. Birch-type photoreduction of arenes and heteroarenes by sensitized electron transfer. *Angew. Chemie. Int. Ed.* **2019**, *58*, 14289–14294. [CrossRef]
51. Peters, B.K.; Rodriguez, K.X.; Reisberg, S.H.; Beil, S.B.; Hickey, D.P.; Kawamata, Y.; Collins, M.; Starr, J.; Chen, L.; Udyavara, S.; et al. Scalable and safe synthetic organic electroreduction inspired by Li-ion battery chemistry. *Science* **2019**, *363*, 838–845. [CrossRef]
52. Schink, B. Fermentation of acetylene by an obligate anaerobe, *Pelobacter acetylenicus* sp. nov. *Arch. Microbiol.* **1985**, *142*, 295–301. [CrossRef]
53. Meckenstock, R.U.; Krieger, R.; Ensign, S.; Kroneck, P.M.H.; Schink, B. Acetylene hydratase of *Pelobacter acetylenicus*. Molecular and spectroscopic properties of the tungsten iron-sulfur enzyme. *Eur. J. Biochem.* **1999**, *264*, 176–182. [CrossRef]
54. Rosner, B.M.; Schink, B. Purification and characterization of acetylene hydratase of *Pelobacter acetylenicus*, a tungsten iron-sulfur protein. *J. Bacteriol.* **1995**, *177*, 5767–5772. [CrossRef]
55. Seiffert, G.B.; Ullmann, G.M.; Messerschmidt, A.; Schink, B.; Kroneck, P.M.H.; Einsle, O. Structure of the non-redox-active tungsten/[4Fe:4S] enzyme acetylene hydratase. *Proc. Natl. Acad. Sci. USA* **2007**, *104*, 3073–3077. [CrossRef] [PubMed]
56. tenBrink, F.; Schink, B.; Kroneck, P.M.H. Exploring the active site of the tungsten, iron-sulfur enzyme acetylene hydratase. *J. Bacteriol.* **2011**, *193*, 1229–1236. [CrossRef] [PubMed]
57. Boll, M.; Einsle, O.; Ermler, U.; Kroneck, P.M.H.P.M.H.; Ullmann, G.M.M. Structure and function of the unusual tungsten enzymes acetylene hydratase and class ii benzoyl-coenzyme a reductase. *J. Mol. Microbiol. Biotechnol.* **2016**, *26*, 119–137. [CrossRef]
58. Hyman, M.R.; Daniel, A. Acetylene inhibition of metalloenzymes. *Anal. Biochem.* **1988**, *173*, 207–220. [CrossRef]
59. Stewart, W.D.; Fitzgerald, G.P.; Burris, R.H. In situ studies on N_2 fixation using the acetylene reduction technique. *Proc. Natl. Acad. Sci. USA* **1967**, *58*, 2071–2078. [CrossRef]
60. Antony, S.; Bayse, C.A. Theoretical studies of models of the active site of the tungstoenzyme acetylene hydratase. *Organometallics* **2009**, *28*, 4938–4944. [CrossRef]
61. Yadav, J.; Das, S.K.; Sarkar, S. A functional mimic of the new class of tungstoenzyme, acetylene hydratase. *J. Am. Chem. Soc.* **1997**, *119*, 4315–4316. [CrossRef]
62. Schreyer, M.; Hintermann, L. Is the tungsten(IV) complex $(NEt_4)_2[WO(mnt)_2]$ a functional analogue of acetylene hydratase? *Beilstein J. Org. Chem.* **2017**, *13*, 2332–2339. [CrossRef] [PubMed]
63. Vidovič, C.; Peschel, L.M.; Buchsteiner, M.; Belaj, F.; Mösch-Zanetti, N.C. Structural mimics of acetylene hydratase: Tungsten complexes capable of intramolecular nucleophilic attack on acetylene. *Chem. Eur. J.* **2019**, *25*, 14267–14272. [CrossRef] [PubMed]
64. Templeton, J.L.; Ward, B.C.; Chen, G.J.J.; Mcdonald, J.W.; Newton, W.E. Oxotungsten(IV)-acetylene complexes: Synthesis via intermetal oxygen atom transfer and nuclear magnetic resonance studies. *Inorg. Chem.* **1981**, *20*, 1248–1253. [CrossRef]
65. Schobert, H. Production of acetylene and acetylene-based chemicals from coal. *Chem. Rev.* **2014**, *114*, 1743–1760. [CrossRef] [PubMed]
66. Hartmann, T.; Schwanhold, N.; Leimkühler, S. Assembly and catalysis of molybdenum or tungsten-containing formate dehydrogenases from bacteria. *Biochim. Biophys. Acta Proteins Proteom.* **2015**, *1854*, 1090–1100. [CrossRef]
67. Maia, L.B.; Moura, J.J.G.; Moura, I. Molybdenum and tungsten-dependent formate dehydrogenases. *J. Biol. Inorg. Chem.* **2015**, *20*, 287–309. [CrossRef]
68. Niks, D.; Hille, R. Molybdenum- and tungsten-containing formate dehydrogenases and formylmethanofuran dehydrogenases: Structure, mechanism, and cofactor insertion. *Protein Sci.* **2019**, *28*, 111–122. [CrossRef]
69. Vorholt, J.A.; Thauer, R.K. Molybdenum and tungsten enzymes in C1 metabolism. *Met. Ions Biol. Syst.* **2002**, *39*, 571–619.
70. Moura, J.J.G.; Brondino, C.D.; Trincão, J.; Romão, M.J. Mo and W bis-MGD enzymes: Nitrate reductases and formate dehydrogenases. *J. Biol. Inorg. Chem.* **2004**, *9*, 791–799. [CrossRef]
71. Hartmann, T.; Schrapers, P.; Utesch, T.; Nimtz, M.; Rippers, Y.; Dau, H.; Mroginski, M.A.; Haumann, M.; Leimkühler, S. The Molybdenum Active Site of Formate Dehydrogenase Is Capable of Catalyzing C-H Bond Cleavage and Oxygen Atom Transfer Reactions. *Biochemistry* **2016**, *55*, 2381–2389.

72. Niks, D.; Duvvuru, J.; Escalona, M.; Hille, R. Spectroscopic and kinetic properties of the molybdenum-containing, NAD^+-dependent formate dehydrogenase from *Ralstonia eutropha*. *J. Biol. Chem.* **2016**, *291*, 1162–1174. [CrossRef] [PubMed]
73. Maia, L.B.; Fonseca, L.; Moura, I.; Moura, J.J.G. Reduction of carbon dioxide by a molybdenum-containing formate dehydrogenase: A kinetic and mechanistic study. *J. Am. Chem. Soc.* **2016**, *138*, 8834–8846. [CrossRef]
74. Enoch, H.G.; Lester, R.L. Effects of molybdate, tungstate, and selenium compounds on formate dehydrogenase and other enzyme systems in *Escherichia coli*. *J. Bacteriol.* **1972**, *110*, 1032–1040. [CrossRef] [PubMed]
75. May, H.D.; Patel, P.S.; Ferry, J.G. Effect of molybdenum and tungsten on synthesis and composition of formate dehydrogenase in *Methanobacterium formicicum*. *J. Bacteriol.* **1988**, *170*, 3384–3389. [CrossRef]
76. Ljungdahl, L.G.; Andreesen, J.R. Formate Dehydrogenase, a selenium-tungsten enzyme from *Clostridium thermoaceticum*. *Methods Enzymol.* **1978**, *53*, 360–372.
77. Strobl, G.; Feicht, R.; White, H.; Lottspeich, F.; Simon, H. The tungsten-containing aldehyde oxidoreductase from *Clostridium thermoaceticum* and its complex with a viologen-accepting NADPH oxidoreductase. *Biol. Chem. Hoppe. Seyler* **1992**, *373*, 123–132. [CrossRef] [PubMed]
78. Alissandratos, A.; Kim, H.K.; Matthews, H.; Hennessy, J.E.; Philbrook, A.; Easton, C.J. *Clostridium carboxidivorans* strain P7T recombinant formate dehydrogenase catalyzes reduction of CO_2 to formate. *Appl. Environ. Microbiol.* **2013**, *79*, 741–744. [CrossRef]
79. Graentzdoerffer, A.; Rauh, D.; Pich, A.; Andreesen, J.R. Molecular and biochemical characterization of two tungsten-and selenium-containing formate dehydrogenases from *Eubacterium acidaminophilum* that are associated with components of an iron-only hydrogenase. *Arch. Microbiol.* **2003**, *179*, 116–130. [CrossRef]
80. Laukel, M.; Chistoserdova, L.; Lidstrom, M.E.; Vorholt, J.A. The tungsten-containing formate dehydrogenase from *Methylobacterium extorquens* AM1: Purification and properties. *Eur. J. Biochem.* **2003**, *270*, 325–333. [CrossRef]
81. De Bok, F.A.M.; Hagedoorn, P.L.; Silva, P.J.; Hagen, W.R.; Schiltz, E.; Fritsche, K.; Stams, A.J.M. Two W-containing formate dehydrogenases (CO_2-reductases) involved in syntrophic propionate oxidation by *Syntrophobacter fumaroxidans*. *Eur. J. Biochem.* **2003**, *270*, 2476–2485. [CrossRef]
82. Almendra, M.J.; Brondino, C.D.; Gavel, O.; Pereira, A.S.; Tavares, P.; Bursakov, S.; Duarte, R.; Caldeira, J.; Moura, J.J.G.; Moura, I. Purification and characterization of a tungsten-containing formate dehydrogenase from *Desulfovibrio gigas*. *Biochemistry* **1999**, *38*, 16366–16372. [CrossRef] [PubMed]
83. Brondino, C.D.; Passeggi, M.C.G.; Caldeira, J.; Almendra, M.J.; Feio, M.J.; Moura, J.J.G.; Moura, I. Incorporation of either molybdenum or tungsten into formate dehydrogenase from *Desulfovibrio alaskensis* NCIMB 13491; EPR assignment of the proximal iron-sulfur cluster to the pterin cofactor in formate dehydrogenases from sulfate-reducing bacteria. *J. Biol. Inorg. Chem.* **2004**, *9*, 145–151. [CrossRef] [PubMed]
84. da Silva, S.M.; Pimentel, C.; Valente, F.M.A.; Rodrigues-Pousada, C.; Pereira, I.A.C. Tungsten and molybdenum regulation of formate dehydrogenase expression in *Desulfovibrio vulgaris* Hildenborough. *J. Bacteriol.* **2011**, *193*, 2909–2916. [CrossRef] [PubMed]
85. Mota, C.S.; Valette, O.; González, P.J.; Brondino, C.D.; Moura, J.J.G.; Moura, I.; Dolla, A.; Rivas, M.G. Effects of molybdate and tungstate on expression levels and biochemical characteristics of formate dehydrogenases produced by *Desulfovibrio alaskensis* NCIMB 13491. *J. Bacteriol.* **2011**, *193*, 2917–2923. [CrossRef]
86. Taylor, A.J.; Kelly, D.J. The function, biogenesis and regulation of the electron transport chains in *Campylobacter jejuni*: New insights into the bioenergetics of a major food-borne pathogen. *Adv. Microb. Physiol.* **2019**, *74*, 239–329.
87. Reda, T.; Plugge, C.M.; Abram, N.J.; Hirst, J. Reversible interconversion of carbon dioxide and formate by an electroactive enzyme. *Proc. Natl. Acad. Sci. USA* **2008**, *105*, 10654–10658. [CrossRef]
88. Schuchmann, K.; Müller, V. Direct and reversible hydrogenation of CO_2 to formate by a bacterial carbon dioxide reductase. *Science* **2013**, *342*, 1382–1385. [CrossRef]
89. Sokol, K.P.; Robinson, W.E.; Oliveira, A.R.; Zacarias, S.; Lee, C.Y.; Madden, C.; Bassegoda, A.; Hirst, J.; Pereira, I.A.C.; Reisner, E. Reversible and selective interconversion of hydrogen and carbon dioxide into formate by a semiartificial formate hydrogenlyase mimic. *J. Am. Chem. Soc.* **2019**, *141*, 17498–17502. [CrossRef]
90. Sokol, K.P.; Robinson, W.E.; Oliveira, A.R.; Warnan, J.; Nowaczyk, M.M.; Ruff, A.; Pereira, I.A.C.; Reisner, E. Photoreduction of CO_2 with a formate dehydrogenase driven by photosystem ii using a semi-artificial z-scheme architecture. *J. Am. Chem. Soc.* **2018**, *140*, 16418–16422. [CrossRef]

91. Miller, M.; Robinson, W.E.; Oliveira, A.R.; Heidary, N.; Kornienko, N.; Warnan, J.; Pereira, I.A.C.; Reisner, E. Interfacing formate dehydrogenase with metal oxides for the reversible electrocatalysis and solar-driven reduction of carbon dioxide. *Angew. Chemie. Int. Ed.* **2019**, *58*, 4601–4605. [CrossRef]
92. Hochheimer, A.; Hedderich, R.; Thauer, R.K. The formylmethanofuran dehydrogenase isoenzymes in *Methanobacterium wolfei* and *Methanobacterium thermoautotrophicum*: Induction of the molybdenum isoenzyme by molybdate and constitutive synthesis of the tungsten isoenzyme. *Arch. Microbiol.* **1998**, *170*, 389–393. [CrossRef] [PubMed]
93. Bertram, P.A.; Schmitz, R.A.; Linder, D.; Thauer, R.K. Tungstate can substitute for molybdate in sustaining growth of *Methanobacterium thermoautotrophicum*—Identification and characterization of a tungsten isoenzyme of formylmethanofuran dehydrogenase. *Arch. Microbiol.* **1994**, *161*, 220–228. [CrossRef] [PubMed]
94. Matschiavelli, N.; Rother, M. Role of a putative tungsten-dependent formylmethanofuran dehydrogenase in *Methanosarcina acetivorans*. *Arch. Microbiol.* **2015**, *197*, 379–388. [CrossRef] [PubMed]
95. Wagner, T.; Ermler, U.; Shima, S. The methanogenic CO_2 reducing-and-fixing enzyme is bifunctional and contains 46 [4Fe-4S] clusters. *Science* **2016**, *354*, 114–117. [CrossRef] [PubMed]
96. Gates, A.J.; Hughes, R.O.; Sharp, S.R.; Millington, P.D.; Nilavongse, A.; Cole, J.A.; Leach, E.R.; Jepson, B.; Richardson, D.J.; Butler, C.S. Properties of the periplasmic nitrate reductases from *Paracoccus pantotrophus* and *Escherichia coli* after growth in tungsten-supplemented media. *FEMS Microbiol. Lett.* **2003**, *220*, 261–269. [CrossRef]
97. De Vries, S.; Momcilovic, M.; Strampraad, M.J.F.; Whitelegge, J.P.; Baghai, A.; Schröder, I. Adaptation to a high-tungsten environment: *Pyrobaculum aerophilum* contains an active tungsten nitrate reductase. *Biochemistry* **2010**, *49*, 9911–9921.
98. Stewart, L.J.; Bailey, S.; Bennett, B.; Charnock, J.M.; Garner, C.D.; McAlpine, A.S. Dimethylsulfoxide reductase: An enzyme capable of catalysis with either molybdenum or tungsten at the active site. *J. Mol. Biol.* **2000**, *299*, 593–600. [CrossRef]
99. Johnson, M.K.; Rees, D.C.; Adams, M.W.W. Tungstoenzymes. *Chem. Rev.* **1996**, *96*, 2817–2839. [CrossRef]
100. Buc, J.; Santini, C.-L.; Giordani, R.; Czjzek, M.; Wu, L.-F.; Giordano, G. Enzymatic and physiological properties of the tungsten-substituted molybdenum TMAO reductase from *Escherichia coli*. *Mol. Microbiol.* **1999**, *32*, 159–168. [CrossRef]
101. Haja, D.K.; Wu, C.H.; Poole, F.L.; Sugar, J.; Williams, S.G.; Jones, A.K.; Adams, M.W.W. Characterization of thiosulfate reductase from *Pyrobaculum aerophilum* heterologously produced in *Pyrococcus furiosus*. *Extremophiles* **2020**, *24*, 53–62. [CrossRef]
102. Aguilar-Barajas, E.; Díaz-Pérez, C.; Ramírez-Díaz, M.I.; Riveros-Rosas, H.; Cervantes, C. Bacterial transport of sulfate, molybdate, and related oxyanions. *BioMetals* **2011**, *24*, 687–707. [CrossRef]
103. Bevers, L.E.; Hagedoorn, P.L.; Krijger, G.C.; Hagen, W.R. Tungsten transport protein A (WtpA) in *Pyrococcus furiosus*: The first member of a new class of tungstate and molybdate transporters. *J. Bacteriol.* **2006**, *188*, 6498–6505. [CrossRef]
104. Makdessi, K.; Fritsche, K.; Pich, A.; Andreesen, J.R. Identification and characterization of the cytoplasmic tungstate/molybdate-binding protein (Mop) from *Eubacterium acidaminophilum*. *Arch. Microbiol.* **2004**, *181*, 45–51. [CrossRef] [PubMed]
105. Makdessi, K.; Andreesen, J.R.; Pich, A. Tungstate uptake by a highly specific ABC transporter in *Eubacterium acidaminophilum*. *J. Biol. Chem.* **2001**, *276*, 24557–24564. [CrossRef] [PubMed]
106. Smart, J.P.; Cliff, M.J.; Kelly, D.J. A role for tungsten in the biology of *Campylobacter jejuni*: Tungstate stimulates formate dehydrogenase activity and is transported via an ultra-high affinity ABC system distinct from the molybdate transporter. *Mol. Microbiol.* **2009**, *74*, 742–757. [CrossRef] [PubMed]
107. Taveirne, M.E.; Sikes, M.L.; Olson, J.W. Molybdenum and tungsten in *Campylobacter jejuni*: Their physiological role and identification of separate transporters regulated by a single ModE-like protein. *Mol. Microbiol.* **2009**, *74*, 758–771. [CrossRef] [PubMed]
108. Otrelo-Cardoso, A.R.; Nair, R.R.; Correia, M.A.S.; Rivas, M.G.; Santos-Silva, T. TupA: A tungstate binding protein in the periplasm of *Desulfovibrio alaskensis* G20. *Int. J. Mol. Sci.* **2014**, *15*, 11783–11798. [CrossRef]
109. Otrelo-Cardoso, A.R.; Nair, R.R.; Correia, M.A.S.; Cordeiro, R.S.C.; Panjkovich, A.; Svergun, D.I.; Santos-Silva, T.; Rivas, M.G. Highly selective tungstate transporter protein TupA from *Desulfovibrio alaskensis* G20. *Sci. Rep.* **2017**, *7*, 1–12. [CrossRef] [PubMed]

110. Zhang, Y.; Gladyshev, V.N. Molybdoproteomes and evolution of molybdenum utilization. *J. Mol. Biol.* **2008**, *379*, 881–899. [CrossRef]
111. Rajagopalan, K.V. Biosynthesis and processing of the molybdenum cofactors. *Biochem. Soc. Trans.* **1997**, *25*, 757–761. [CrossRef]
112. Leimkühler, S. Shared function and moonlighting proteins in molybdenum cofactor biosynthesis. *Biol. Chem.* **2017**, *398*, 1009–1026. [CrossRef] [PubMed]
113. Schwarz, G.; Mendel, R.R.; Ribbe, M.W. Molybdenum cofactors, enzymes and pathways. *Nature* **2009**, *460*, 839–847. [CrossRef] [PubMed]
114. Bevers, L.E.; Hagedoorn, P.L.; Santamaria-Araujo, J.A.; Magalon, A.; Hagen, W.R.; Schwarz, G. Function of MoaB proteins in the biosynthesis of the molybdenum and tungsten cofactors. *Biochemistry* **2008**, *47*, 949–956. [CrossRef] [PubMed]
115. Wuebbens, M.M.; Rajagopalan, K.V. Structural characterization of a molybdopterin precursor. *J. Biol. Chem.* **1993**, *268*, 13493–13498. [PubMed]
116. Hänzelmann, P.; Schindelin, H. Binding of 5′-GTP to the C-terminal FeS cluster of the radical *S*-adenosylmethionine enzyme MoaA provides insights into its mechanism. *Proc. Natl. Acad. Sci. USA* **2006**, *103*, 6829–6834. [CrossRef]
117. Gutzke, G.; Fischer, B.; Mendel, R.R.; Schwarz, G. Thiocarboxylation of molybdopterin synthase provides evidence for the mechanism of dithiolene formation in metal-binding pterins. *J. Biol. Chem.* **2001**, *276*, 36268–36274. [CrossRef]
118. Rudolph, M.J.; Wuebbens, M.M.; Rajagopalan, K.V.; Schindelin, H. Crystal structure of molybdopterin synthase and its evolutionary relationship to ubiquitin activation. *Nat. Struct. Biol.* **2001**, *8*, 42–46. [CrossRef]
119. Wuebbens, M.M.; Rajagopalan, K.V. Mechanistic and mutational studies of *Escherichia coli* molybdopterin synthase clarify the final step of molybdopterin biosynthesis. *J. Biol. Chem.* **2003**, *278*, 14523–14532. [CrossRef]
120. Pitterle, D.M.; Johnson, J.L.; Rajagopalan, K.V. In vitro synthesis of molybdopterin from precursor Z using purified converting factor. Role of protein-bound sulfur in formation of the dithiolene. *J. Biol. Chem.* **1993**, *268*, 13506–13509.
121. Lake, M.W.; Wuebbens, M.M.; Rajagopalan, K.V.; Schindelin, H. Mechanism of ubiquitin activation revealed by the structure of a bacterial MoeB-MoaD complex. *Nature* **2001**, *414*, 325–329. [CrossRef]
122. Leimkühler, S.; Rajagopalan, K.V. A Sulfurtransferase is required in the transfer of cysteine sulfur in the in vitro synthesis of molybdopterin from precursor z in *Escherichia coli*. *J. Biol. Chem.* **2001**, *276*, 22024–22031. [CrossRef] [PubMed]
123. Llamas, A.; Mendel, R.R.; Schwarz, G. Synthesis of adenylated molybdopterin: An essential step for molybdenum insertion. *J. Biol. Chem.* **2004**, *279*, 55241–55246. [CrossRef] [PubMed]
124. Joshi, M.S.; Johnson, J.L.; Rajagopalan, K. V Molybdenum cofactor biosynthesis in *Escherichia coli* mod and mog mutants. *J. Bacteriol.* **1996**, *178*, 4310–4312. [CrossRef] [PubMed]
125. Meyer, O.; Rajagopalan, K.V. Molybdopterin in carbon monoxide oxidase from carboxydotrophic bacteria. *J. Bacteriol.* **1984**, *157*, 643–648. [CrossRef]
126. Reschke, S.; Duffus, B.R.; Schrapers, P.; Mebs, S.; Teutloff, C.; Dau, H.; Haumann, M.; Leimkühler, S. Identification of YdhV as the first molybdoenzyme binding a Bis-Mo-MPT cofactor in *Escherichia coli*. *Biochemistry* **2019**, *58*, 2228–2242.
127. Johnson, J.L.; Bastian, N.R.; Rajagopalan, K. V Molybdopterin guanine dinucleotide: A modified form of molybdopterin identified in the molybdenum cofactor of dimethyl sulfoxide reductase from *Rhodobacter sphaeroides* forma specialis denitrificans. *Proc. Natl. Acad. Sci. USA* **1990**, *87*, 3190–3194. [CrossRef]
128. Reschke, S.; Sigfridsson, K.G.V.; Kaufmann, P.; Leidel, N.; Horn, S.; Gast, K.; Schulzke, C.; Haumann, M.; Leimkühler, S. Identification of a bis-molybdopterin intermediate in molybdenum cofactor biosynthesis in *Escherichia coli*. *J. Biol. Chem.* **2013**, *288*, 29736–29745. [CrossRef]
129. Johannes, J.; Bluschke, A.; Jehmlich, N.; Von Bergen, M.; Boll, M. Purification and characterization of active-site components of the putative p-cresol methylhydroxylase membrane complex from *Geobacter metallireducens*. *J. Bacteriol.* **2008**, *190*, 6493–6500. [CrossRef]
130. Peters, F.; Heintz, D.; Johannes, J.; Van Dorsselaer, A.; Boll, M. Genes, enzymes, and regulation of para-cresol metabolism in *Geobacter metallireducens*. *J. Bacteriol.* **2007**, *189*, 4729–4738. [CrossRef]
131. Neumann, M.; Stöcklein, W.; Leimkühler, S. Transfer of the molybdenum cofactor synthesized by *Rhodobacter capsulatus* MoeA to XdhC and MobA. *J. Biol. Chem.* **2007**, *282*, 28493–28500. [CrossRef]

132. Thomé, R.; Gust, A.; Toci, R.; Mendel, R.; Bittner, F.; Magalon, A.; Walburger, A. A sulfurtransferase is essential for activity of formate dehydrogenases in *Escherichia coli*. *J. Biol. Chem.* **2012**, *287*, 4671–4678. [CrossRef] [PubMed]
133. Arnoux, P.; Ruppelt, C.; Oudouhou, F.; Lavergne, J.; Siponen, M.I.; Toci, R.; Mendel, R.R.; Bittner, F.; Pignol, D.; Magalon, A.; et al. Sulphur shuttling across a chaperone during molybdenum cofactor maturation. *Nat. Commun.* **2015**, *6*, 6148. [CrossRef] [PubMed]
134. Schwanhold, N.; Iobbi-Nivol, C.; Lehmann, A.; Leimkühler, S. Same but different: Comparison of two system-specific molecular chaperones for the maturation of formate dehydrogenases. *PLoS ONE* **2018**, *13*. [CrossRef] [PubMed]

Review

Metal–Dithiolene Bonding Contributions to Pyranopterin Molybdenum Enzyme Reactivity

Jing Yang [1], John H. Enemark [2] and Martin L. Kirk [1,*]

[1] Department of Chemistry and Chemical Biology, The University of New Mexico, MSC03 2060, Albuquerque, NM 87131-0001, USA; yangjing@unm.edu

[2] Department of Chemistry Biochemistry, University of Arizona, Tucson, AZ 85721, USA; jenemark@email.arizona.edu

* Correspondence: mkirk@unm.edu; Tel.: +1-505-277-5992

Received: 2 February 2020; Accepted: 2 March 2020; Published: 5 March 2020

Abstract: Here we highlight past work on metal–dithiolene interactions and how the unique electronic structure of the metal–dithiolene unit contributes to both the oxidative and reductive half reactions in pyranopterin molybdenum and tungsten enzymes. The metallodithiolene electronic structures detailed here were interrogated using multiple ground and excited state spectroscopic probes on the enzymes and their small molecule analogs. The spectroscopic results have been interpreted in the context of bonding and spectroscopic calculations, and the pseudo-Jahn–Teller effect. The dithiolene is a unique ligand with respect to its redox active nature, electronic synergy with the pyranopterin component of the molybdenum cofactor, and the ability to undergo chelate ring distortions that control covalency, reduction potential, and reactivity in pyranopterin molybdenum and tungsten enzymes.

Keywords: metal–dithiolene; pyranopterin molybdenum enzymes; fold-angle; tungsten enzymes; electronic structure; pseudo-Jahn–Teller effect; thione; molybdenum cofactor; Moco

---

1. Introduction

It is now well-established that all known molybdenum-containing enzymes [1–3], with the sole exception of nitrogenase, contain a common pyranopterin dithiolene (PDT) (Figure 1) organic cofactor (originally called molybdopterin (MPT)), which coordinates to the Mo center of the enzymes through the sulfur atoms of the dithiolene fragment. To date, the PDT component [4] of the molybdenum cofactor (Moco) is the only known occurrence of dithiolene ligation in biological systems. This cofactor is also found in anaerobic tungsten enzymes, and it may be one of the most ancient cofactors in biology [5]. The study of metal–dithiolene compounds (metallodithiolenes) has undergone a recent renaissance, with their synthesis, geometric structure, spectroscopy, bonding, and electronic structure having been recently highlighted [4,6–20]. Here, we briefly review the discovery of metallodithiolene compounds [13,21]. This history is followed by a more extensive discussion of key investigations into the myriad roles of the dithiolene ligands in the structure, bonding and reactivity of metal compounds, using multiple spectroscopic techniques, as well as theoretical calculations. Throughout this review, the key implications of these results for Mo and W enzymes are discussed.

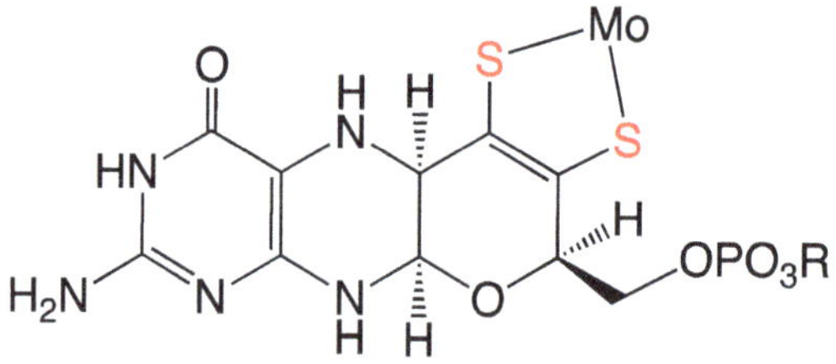

Figure 1. The reduced tetrahydro form of the pyranopterin dithiolene (PDT) coordinated to Mo in the molybdenum cofactor (Moco). In the enzymes, the Mo ion can redox cycle between the Mo^{IV}, Mo^{V}, and Mo^{VI} oxidation states.

In the early 1960s, several research groups reported intensely colored square planar metal complexes with chelating sulfur-donor ligands that could stabilize metal compounds in a range of formal oxidation states related by one-electron oxidation-reduction (i.e., redox) reactions (Figure 2) [22–24]. McCleverty gave these novel ligands the general name "dithiolene" in order to emphasize their delocalized electronic structures [25]. These ligands are also described as being "non-innocent" due to the participation of the dithiolene ligands in the multiple one-electron reactions of their metal complexes and the inability to assign a specific oxidation state to the metal ion or the dithiolene ligands [11].

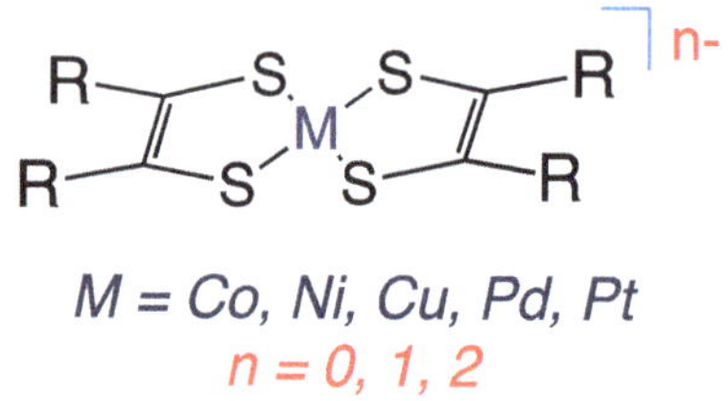

Figure 2. Square planer metallodithiolene complexes. R = CN, CH_3, Ph, CF_3.

Importantly, these non-innocent dithiolene ligands can modulate the nature of the covalent bonding with transition metal ions via the various redox states accessible to the dithiolene (Figure 3) [13]. The ene-1,2-dithiolate is the reduced form of the ligand and possesses six π-electrons. This ligand form is both a σ-donor and π-donor that usually forms strong covalent bonds with an oxidized transition metal ion, as is observed in the active sites of most pyranopterin Mo and W enzymes (e.g., Mo(V)/Mo(VI)-dithiolene bonds). The radical anion form with five π-electrons is usually found in molecules chelated by multiple dithiolene ligands, where extended delocalization of the π-electrons and mixed-valency assists in the stabilization of the metal–ligand bonds. The fully oxidized 1,2-dithione form of the ligand possesses only four π-electrons and can be described by two resonance structures (e.g., the 1,2-dithione and 1,2-dithiete). The low-lying empty π* orbitals of the S=C bonds in the dithione can accept π-electron density from electron-rich low-valent transition metals [16,17], thereby stabilizing such compounds. However, dithione-containing low-valent metal complexes are encountered much less frequently than high-valent transition metal ions coordinated by reduced forms of dithiolene ligands.

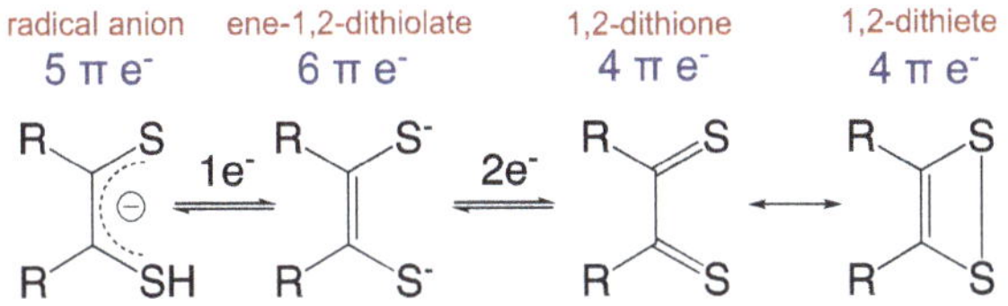

Figure 3. Dithiolene redox states and resonance structures for the oxidized dithione/dithiete forms. (Adapted with permission from *Inorganic Chemistry*, 2016, 55, 785–793. Copyright (2016) American Chemical Society).

In 1982, Johnson and Rajagopalan proposed that Moco consisted of the Mo ion coordinated by the dithiolene fragment of the PDT (Figure 1), from the results of an elegant series of degradative, analytical and spectroscopic studies of sulfite oxidase [26]. This proposed structure was subsequently confirmed by X-ray crystallography [27,28], and numerous examples are now known [29]. Molybdenum and tungsten enzymes are the only known examples of dithiolene coordination in biology, and given the "non-innocent" behavior of dithiolene ligands in simple metal compounds, one may ask what role does dithiolene coordination play in molybdenum enzymes? Through a series of examples involving small molecules and enzymes, we will address this important question and how it relates to control of metal–ligand covalency, reduction potentials, and reactivity in pyranopterin Mo and W enzymes.

2. Mo–Dithiolene Bonding

2.1. Early Descriptions of Mo–Dithiolene Bonding

Some insight into the role of dithiolene coordination in enzymes is provided by the organometallic compounds of the general formula $Cp_2M(bdt)$, where Cp is $C_5H_5^-$, and M is either Mo, V or Ti. The fold angle of the dithiolene ligand depends on the formal d-electron count of the metal, and this angle ranges from nearly planar (9°) for Mo (d^2), to 35° for V (d^1), and 46° for Ti (d^0) (Figure 4). Lauher and Hoffman [30] related this increase in the fold angle with decreased d-electron count to donation from the filled out-of-plane $S_\pi{}^+$ orbital to the in-plane metal d-orbital (Figure 5). For the molybdenum enzymes, these model compound results imply that the Mo–dithiolene fold angle in Moco could be related to the formal oxidation state of the Mo atom, with Mo(VI) (d^0) sites possessing a relatively large fold angle and Mo(V) (d^1) and Mo(IV) (d^0) sites possessing smaller fold angles. Accurate fold angles are difficult to determine for large protein molecules, but values ranging from 6–33° have been calculated for various molybdenum enzymes [31]. The binding of substrate or inhibitors, and/or dynamic conformational changes in the protein, are expected to modulate the active site chelate fold angle and thereby affect enzyme reactivity [4,32].

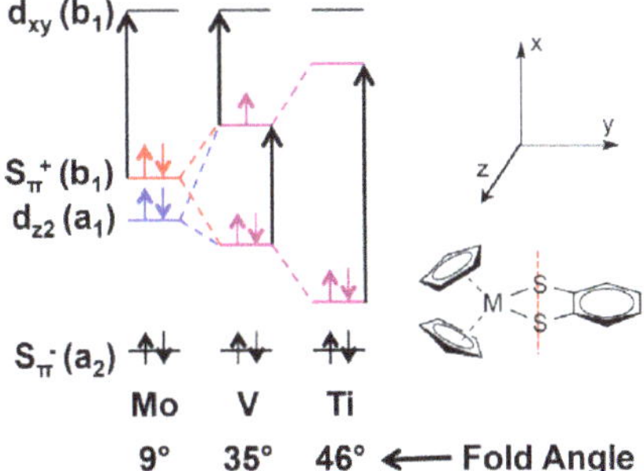

Figure 4. Fold angle distortions as a function of redox orbital electron occupancy in a series of Cp_2M^{IV}(bdt) complexes. (Adapted with permission from *J. Am. Chem. Soc.* 2018, 140, 14777–14788. Copyright (2018) American Chemical Society).

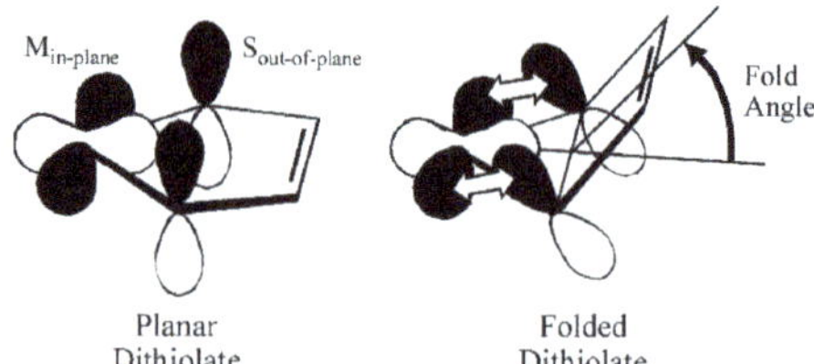

Figure 5. Pictorial description of how the ligand fold angle modulates the degree of mixing between the dithiolene out-of-plane S orbitals ($S_\pi{}^+$) and the in-plane Mo(xy) redox orbital. The chelate ring fold is along the dithiolene S–S vector. (Adapted with permission from *Proc. Natl. Acad. Sci. USA*. 2003, 100, 3719–3724. Copyright (2003) National Academy of Sciences.

2.2. Spectroscopic Investigations of Mo–Dithiolene Bonding

2.2.1. Electron Paramagnetic Resonance (EPR) Spectroscopy

An important spectroscopic signature of molybdenum enzymes, such as xanthine oxidase and sulfite oxidase, is a unique Mo(V) electron paramagnetic resonance (EPR) spectrum. The EPR spectra of the enzymes display a relatively large average g-value (g_{ave} = 1.97) and relatively small ^{95,97}Mo hyperfine interactions (*hfi*) compared to the EPR spin-Hamiltonian parameters from typical inorganic Mo(V) complexes that possess hard N, O, and Cl donor ligands. The unique EPR parameters for molybdenum enzymes have been ascribed to covalent delocalization of electron density between the Mo(V) center and the sulfur atoms of the coordinated pyranopterin dithiolene unit [33]. The oxo-Mo(V) model compound Tp*MoO(bdt) (Figure 6, where Tp* is hydrotris-(3,5-dimethyl-1-pyrazolyl)borate and bdt is 1,2-benzenedithiolate)) displays Mo(V) EPR spin-Hamiltonian parameters that are very similar to those observed in the enzymes. This supports the proposal of dithiolene coordination in Mo enzymes [34], which has been confirmed by X-ray crystal structures [2]. Recent multidimensional variable frequency pulsed EPR studies of sulfite oxidase, where the sulfur atoms of Moco have been isotopically labeled with ^{33}S (I = 3/2), have provided *direct* experimental evidence for delocalization of Mo(V) spin density onto the S atoms of the dithiolene fragment of Moco [35,36]. Density functional theory (DFT) computations show spin polarization effects and strong covalent intermixing between the in-plane metal d_{xy} orbital and out-of-plane p_z orbitals of the PDT dithiolene S atoms, which provide a mechanism for the observation of a significant ^{33}S hyperfine interaction [12,36].

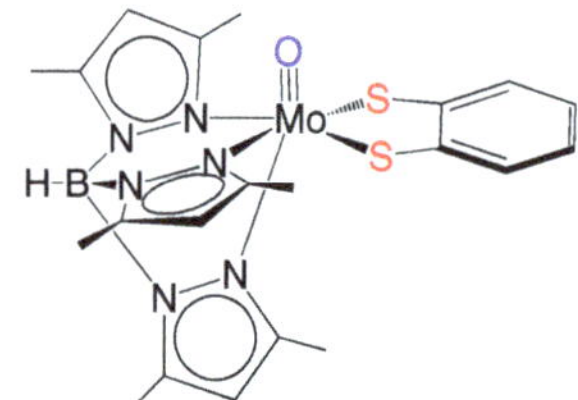

Figure 6. The Tp*Mo^{V}O(bdt) model. Note that the apical oxo ligand can be changed to a terminal sulfido or nitrosyl to probe the electronic structure of the Mo–dithiolene unit. The bdt ligand can also be conveniently interchanged with a large variety of other dithiolenes.

2.2.2. Electronic Absorption and Resonance Raman Spectroscopies

Experimental investigation of the electronic structures of the Mo centers of enzymes is difficult because of the intense absorptions from other chromophores (e.g., the *b*-type heme in sulfite oxidase and iron sulfur centers and FAD in xanthine oxidase) [37–41]. However, the effects of dithiolene coordination on electronic structure have been investigated for model oxo-Mo(V) compounds (Figure 6) by electronic absorption, XAS, magnetic circular dichroism (MCD), and resonance Raman (rR) spectroscopies [12,14–17,32,33,42–51]. For Tp*MoO(bdt), the electronic absorptions at 19,400 cm^{-1} (Band 4) and 22,100 cm^{-1} (Band 5) are assigned to S → Mo charge transfer bands (Figure 7A) [12]. These assignments have been confirmed by rR spectroscopy (Figure 7A,B), which shows three resonantly enhanced vibrations at 362.0, 393.0, and 931.0 cm^{-1}. The lower frequency vibrations (ν_1 and ν_6) can be assigned to symmetric S–Mo–S stretching and bending vibrations, and the 931.0 cm^{-1} frequency (ν_3) is primarily the Mo≡O stretch. Figure 7C shows a molecular orbital diagram that is consistent with the spectroscopic data of Figure 7A,B. Band 5 of Figure 7A is assigned as $\psi_{op}{}^{a''} \rightarrow \psi_{xz}{}^{a''}$, $\psi_{yz}{}^{a'}$ (blue arrow, Figure 7C), a transition which formally results in the promotion of an electron from an out-of-plane dithiolene molecular orbital to the nearly degenerate Mo $d_{xz,yz}$-based orbitals, which are strongly antibonding with respect to the apical Mo≡O bond. This band assignment is supported by the rR enhancement of ν_3 (squares) with excitation into Band 5 (Figure 7A,B). The preferential enhancement of vibrations ν_1 (diamonds) and ν_6 (circles) upon excitation at 514.5 nm (Figure 7A,B) supports assignment of Band 4 as the electronic transition $\psi_{ip}{}^{a''} \rightarrow \psi_{xy}{}^{a'}$ (red arrow, Figure 7C), which promotes an electron from the antisymmetric in-plane dithiolene orbital ($\psi_{ip}{}^{a''}$) to the half-filled in-plane Mo d_{xy} ($\psi_{xy}{}^{a'}$) orbital. The intensity of this electronic transition illustrates the covalency of in-plane metal–dithiolene bonding and suggests that such a pseudo-σ-mediated process could play a role in one-electron transfer steps of enzyme catalysis.

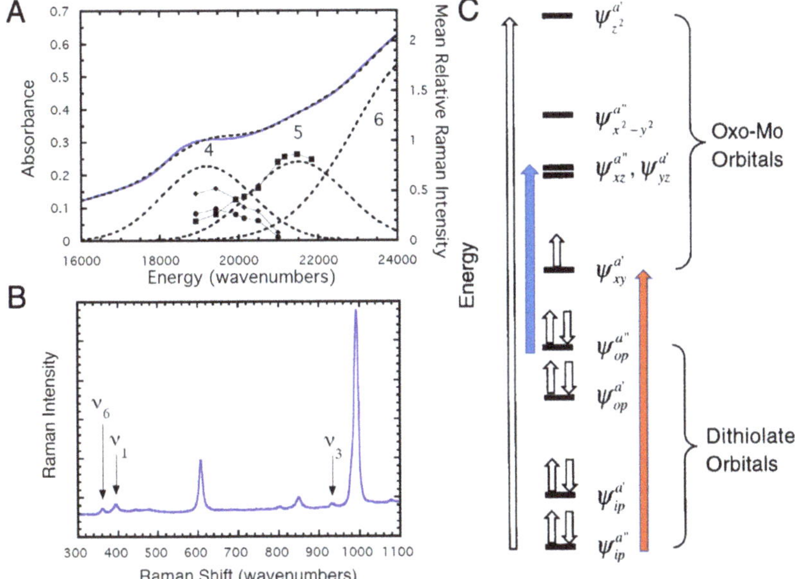

Figure 7. (**A**) Solid state resonance Raman profiles and 5K mull electronic absorption spectrum for $Tp^{*}Mo^{V}O(bdt)$. (**B**) Resonance Raman spectrum for $Tp^{*}Mo^{V}O(bdt)$ (293K) using 514.5 nm excitation (75 mW). (**C**) General molecular orbital diagram for $Tp^{*}Mo^{V}O$(dithiolene) complexes. The z-axis is oriented along the Mo≡O bond and the energies of the molecular orbitals are not drawn to scale. Transitions are described in the text. (Adapted with permission from *Inorganic Chemistry*, 1999, 38, 1401. Copyright (1999) American Chemical Society).

3. Synergistic Interactions between the Dithiolene and Pterin Components of the PDT

Electronic coupling between the dithiolene and the pterin components of the PDT is most prevalent in the dihydropyranopterin form of the PDT [4,15,20,29,52]. This coupling is dramatically reduced in a tetrahydropyranopterin due to the loss of extended π-conjugation in these systems. Two-electron oxidation of the tetrahydro pyranopterin component of the PDT can result in an unusual asymmetric dithiolene known as the "thiol–thione" form that leads to bond and electronic asymmetry in the metal–dithiolene core [4,15,52]. As depicted in Figure 8, the two-electron oxidized 10,10a-dihydropyranopterin can undergo an induced internal redox reaction upon protonation at the N-5 position that involves a subsequent charge transfer between the dithiolene chelate and the piperazine ring of the pterin. This protonation results in a dominant monoanionic thiol–thione chelate form of the ligand when bound to Mo or W. This thiol–thione character can also occur in the absence of protonation by the concept of resonance, which may also be described as configurational mixing between the thiol–thione and dithiol states. This type of thiol–thione chelate has been observed and studied in a small molecule Mo(IV) systems [4,15,20,52]. In these systems, excited state thiol–thione character was shown to be admixed into the ground state configuration using a variety of spectroscopic and computational probes of the electronic structure. The analysis of the data indicates that a two-electron oxidized pterin is inherently electron withdrawing, allowing for a low-lying dithiolene → pterin intraligand charge transfer (ILCT) state to mix with the ground state to provide a variable degree of thiol–thione character in the electronic ground state.

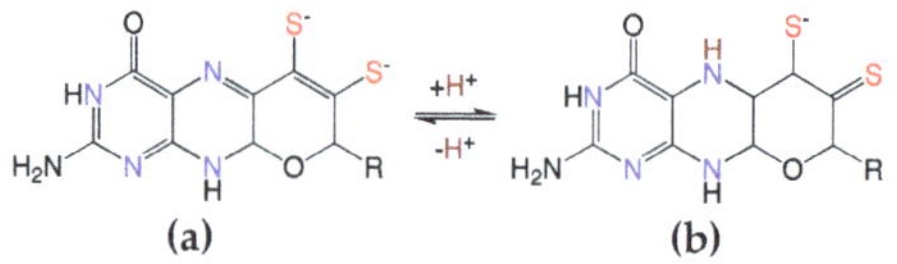

Figure 8. Oxidized PDT ligands: dihydropyranopterin (**a**) and protonated dihydropyranopterin (**b**) yielding the thiol/thione.

Definitive spectroscopic signatures are associated with the presence of a dihydropterin form of the PDT ligand. It is observed that the dithiolene → pterin intraligand charge transfer (ILCT) band is intense (E = 20,000–27,500 cm^{-1}; $\varepsilon \sim$ 10,000–16,000 M^{-1} cm^{-1}) [52], and there is considerable resonance enhancement of numerous Raman vibrations that can be assigned as originating from pterin and dithiolene C = C and C = N vibrations. Key resonance enhanced vibrational modes that can be used to characterize the presence of dihydropterin thiol–thione character in the enzymes include the 1508 cm^{-1} and 1549 cm^{-1} pyranopterin–dithiolene stretching frequencies that were observed in this Mo(IV) cyclized pyranopterin dithiolene model compound. This oxidized pyran ring closed form of the PDT has yet to be definitively observed in any pyranopterin Mo enzyme, but its presence would have profound implications on the electronic structure of the Mo site. Namely, the change in ligand charge from −2 to −1 leads to an asymmetric reduction in the charge donated by the monoanionic ligand compared to the dianionic dithiolene. Charge effects on oxygen atom transfer catalysis have recently been explored in model compounds showing dramatic rate enhancements in the oxidative half reaction that leads to substrate reduction [53]. This reactivity correlates with a large shift in the Mo(VI/V) reduction potential between cationic $[Tpm^*MoO_2Cl]^+$ (−660 mV vs. Fc^+/Fc) and charge neutral Tp^*MoO_2Cl (−1010 mV vs. Fc^+/Fc) [53]. The same effect on redox potential and reactivity would be expected in enzymes that could adopt an oxidized PDT with a thiol–thione configuration. The presence of a thiol–thione form of the PDT in an enzyme would also have a considerable impact on the active site electronic structure, and enable the pyranopterin to play a more significant role in catalysis by fine-tuning the Mo redox potential and providing a π-pathway for electron transfer regeneration of the active site [52]. Additionally, the asymmetry in the dithiolene (thiol/thione) charge donation would be expected to result in a significant *trans* effect or *trans* influence on oxo or sulfido ligands that are coordinated to the Mo or W ion and oriented *trans* to the thione sulfur.

4. The Electronic Buffer Effect and Fold Angle Distortions

4.1. Photoelectron Spectroscopy (PES) Studies

A common structural feature of the large group of pyranopterin Mo enzymes that catalyze a wide range of oxidation/reduction reactions in carbon, sulfur, and nitrogen metabolism is coordination by the sulfur atoms of one (or two) unique dithiolene groups derived from the side chain of a novel substituted pterin (PDT, Figure 1). Given the electronic lability of the dithiolene, a possible role of dithiolene coordination in molybdoenzymes is to buffer the influence of other ligands and changes in the formal oxidation state of the metal. Gas-phase photoelectron spectroscopy (PES) is a powerful tool for probing metal–ligand covalency in isolated molecules. Gas-phase ultraviolet PES of the molybdenum model complexes with the general formula Tp*MoE(tdt) (Figure 6, where E = O, S, or NO, and tdt = 3,4-toluenedithiolate), exhibit nearly identical first ionization energies (6.88–6.95 eV) even though there is a dramatic difference in the electronic structure properties of the axial ligand. Collectively, these results have provided direct experimental evidence for the "electronic buffer" effect of dithiolene ligands [54].

Additional evidence for the electronic buffer effect of dithiolene ligands has been provided by gas-phase core and valence electron ionization energy measurements of the series of molecules $Cp_2M(bdt)$ (Figure 4, Cp = η^5-cyclopentadienyl, M = Ti, V, Mo, and bdt = benzene-1,2-dithiolate). Comparison of the gas-phase core and valence ionization energy shifts provides a unique quantitative energy measure of valence orbital overlap interactions between the metal and the sulfur orbitals that is separated from the effects of charge redistribution. The results explain the large amount of sulfur character in the redox-active orbitals and the electronic buffering of oxidation state changes in metal–dithiolene systems. The experimentally determined orbital interaction energies also reveal a previously unidentified overlap interaction of the predominantly sulfur HOMO of the bdt ligand with the filled π orbitals of the Cp ligands, suggesting that direct dithiolene interactions with other ligands bound to the metal could be significant for other metallodithiolene systems in chemistry and biology [55].

4.2. A Large Fold Angle Distortion in a Mo(IV)–Dithione Complex

Mo(IV)–dithione complexes are much rarer than Mo(V)/Mo(VI)-dithiolene complexes. Recently, a detailed spectroscopic and computational study was performed on a novel Mo(IV)–dithione complex, $MoO(SPh)_2(^iPr_2Dt^0)$ (where $^iPr_2Dt^0$ = *N,N'*-isopropylpiperazine-2,3-dithione) [17]. The structure of this unusual molecule was determined by x-ray crystallography and displays a remarkably large dithiolene fold angle (η = 70°). This large fold angle was compared to that observed in more than 75 other metallodithiolene complexes found in the Cambridge crystallographic database, where fold angles were found to range from 0.3° to 37.3° with an average value for η of 12.5° [17]. The large fold angle distortion in the metallodithiolene ring of $MoO(SPh)_2(^iPr_2Dt^0)$ is reflected in its unusual electron absorption spectrum. The combination of an electron rich Mo(IV) center and electron donating thiolate (SPh) ligands results in the presence of low-energy Mo(IV) $d(x^2\text{-}y^2) \rightarrow$ dithione MLCT and thiolate $\rightarrow$ dithione LL'CT transitions as a result of the strong π-acceptor character of the dithione ligand. These spectral assignments are supported by resonance Raman profiles constructed for the 378 cm^{-1} S–Mo–S symmetric stretch and the 945 cm^{-1} Mo≡O stretch in addition to the results of TDDFT computations. The donor–acceptor nature of the complex was revealed in a molecular orbital fragments analysis using a donor fragment, $[(PhS)_2Mo(IV)]^{2+}$ (F1) and an acceptor fragment, $[^iPr_2Dt^0]$ (F2). The analysis showed that 21% of the F1 HOMO was mixed into the F2 fragment LUMO at a 70° fold angle. In contrast, only 5% of the F1 HOMO was mixed into F2 fragment LUMO in a planer configuration (η = 0°), correlating the effective π-acceptor ability of the dithione with the ligand fold angle. The effects of this HOMO-LUMO mixing also affects the HOMO-LUMO gap, with the HOMO-LUMO gap increasing at larger fold angles (Figure 9). The increased covalency that results from the fold angle distortion represents an example of a strong pseudo-Jahn–Teller effect, vide infra, involving vibronic coupling between the ground state and a low-energy excited state in the non-distorted (η = 0°) geometry of this molecule. A scan of the potential energy surface as a function of this fold angle distortion coordinate results in an asymmetric double well potential (Figure 10), with the global minimum representing a ground state geometry with the dithione ligand fold distorted toward the apical oxo ligand. Thus, an oxidized dithione form of the PDT present in an enzyme active site would be expected to possess a very large ligand fold angle, unless the polypeptide enforces a more planer fold angle geometry.

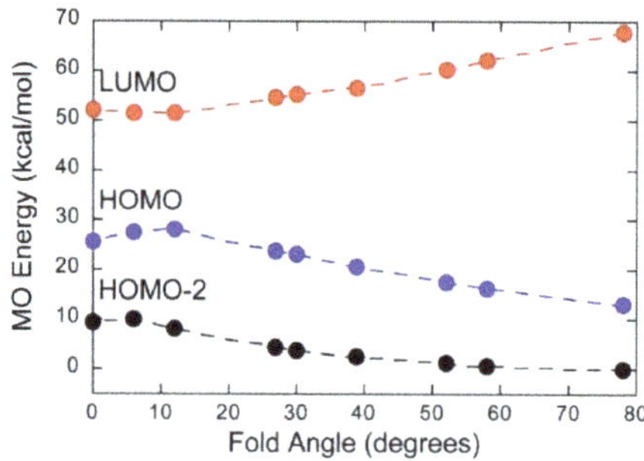

Figure 9. Frontier orbital energies as a function of fold angle in $Mo^{IV}O(SPh)_2(^iPr_2Dt^0)$, which possesses a dithione π-acceptor ligand. (Adapted with permission from *Inorganic Chemistry*, 2016, 55, 785–793. Copyright (2016) American Chemical Society).

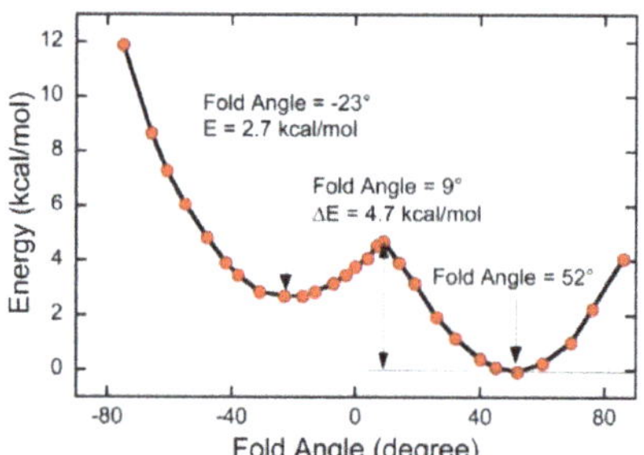

Figure 10. A double well in the ground state potential energy surface of $Mo^{IV}O(SPh)_2(^iPr_2Dt^0)$ as a function of the ligand fold angle. (Adapted with permission from *Inorganic Chemistry*, 2016, 55, 785–793. Copyright (2016) American Chemical Society).

4.3. Low-Frequency Pyranopterin Dithiolene Vibrational Modes in Xanthine Oxidase/Dehydrogenase

Low-frequency dithiolene distortions that are coupled to large electron density changes at the Mo ion represent an example of the electronic buffer effect [54], and have been probed in bovine xanthine oxidase (XO) and *R. capsulatus* xanthine dehydrogenase (XDH) using resonance Raman spectroscopy [40]. Computations have shown that exciting into a low-energy Mo(IV) → product metal-to-ligand charge transfer (MLCT) band results in a large degree of change transfer from the Mo(IV) HOMO to the product LUMO, resulting in an excited state with significant Mo(V) hole character (e.g., Mo(IV)–P^0 → Mo(V)–P·). Thus, the optical charge transfer process mimics the instantaneous one-electron oxidation of the Mo ion, which is encountered in the electron transfer reactions of the enzymes.

The Mo(IV) → 2,4-TV and Mo(IV) → 4-TV (2,4-TV = 2,4-thioviolapterin; 4-TV = 4-thioviolapterin) MLCT bands are red-shifted relative to the Mo(IV) → violapterin MLCT band [39,40,56–59]. The red-shift of the Mo^{IV}–2,4-TV and Mo^{IV}–4-TV MLCT bands eliminates spectral overlap with the absorption envelope of the 2Fe–2S spinach ferredoxin clusters and FAD. The elimination of the

FAD fluorescence background and spurious signals deriving from 2Fe–2S vibrations contributing to the Raman spectrum allow for the acquisition of very high-quality resonance Raman data. Multiple low-frequency (200–400 cm^{-1}) Raman vibrations are observed to be enhanced when using laser excitation on resonance with the Mo(IV) → product MLCT band [40], and these have been assigned as a vibrational mode involving dithiolene folding, Mo≡O rocking, and pyranopterin motions (Band A: Mo^{IV}–4-TV = 234 cm^{-1}; Mo^{IV}–2,4-TV = 236 cm^{-1}), a ring distortion vibration that possesses both Mo–SH and pyranopterin motions (Band B: Mo^{IV}–4-TV = 290 cm^{-1}; Mo^{IV}–2,4-TV = 286 cm^{-1}), the symmetric S–Mo–S dithiolene core stretching vibration (Band C: Mo^{IV}–4-TV = 326 cm^{-1}; Mo^{IV}–2,4-TV = 326 cm^{-1}), and the corresponding asymmetric S–Mo–S dithiolene stretch (Band D: Mo^{IV}–4-TV = 351 cm^{-1}; Mo^{IV}–2,4-TV = 351 cm^{-1}) (Figure 11). Thus, the instantaneous generation of a hole on the Mo center (Mo(IV)–P^0 → Mo(V)–P·) by photoexcitation is felt by the dithiolene chelate and extends all the way to the amino terminus of the PDT. The most resonantly enhanced mode in this spectral region is Band C, the symmetric S–Mo–S dithiolene core stretching, and the frequency of this mode and Band D are similar to those observed in Tp*MoO(bdt) [12,32], which were assigned as the chelate ring symmetric S–Mo–S stretching and bending vibrations, respectively. Band A is significant, since it possesses dithiolene ring folding character indicating that electron density changes at Mo are buffered by a distortion along this low-frequency coordinate, as has been observed in the various model systems described in this review. These observations strongly support an electron transfer role for the PDT in catalysis, with the dithiolene contributing to the Mo–S covalency necessary for increasing the electronic coupling matrix element for electron transfer (H_{DA}) and to affect the Mo reduction potential via the covalency in the Mo–$S_{dithiolene}$ bonds.

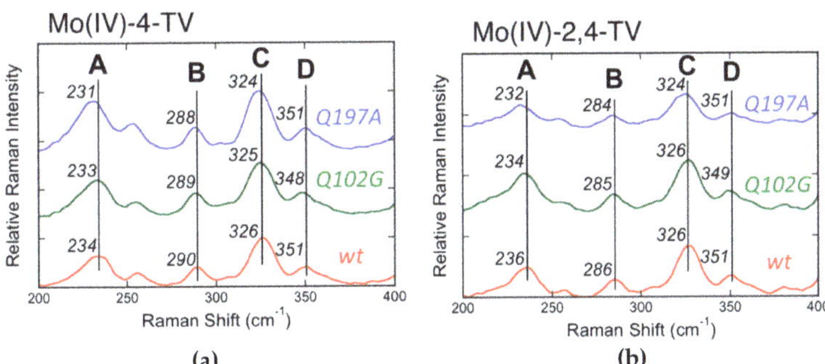

Figure 11. Low-frequency rR spectra for wt, Q102G, and Q197A XDH, Mo^{IV}–4-TV (**a**) and Mo^{IV}–2,4-TV (**b**). Raman spectra were collected on resonance with the Mo(IV) → P MLCT band using 780 nm laser excitation (Adapted with permission from *Inorganic Chemistry*, 2017, 56, 6830–6837. Copyright (2017) American Chemical Society).

5. Vibrational Control of Covalency

A combination of MCD, electronic absorption, electron paramagnetic resonance, resonance Raman, and photoelectron spectroscopies has been used in conjunction with theory to reveal vibrational control of metal–ligand covalency in a series of $Cp_2M(bdt)$ complexes (M = Ti, V, Mo; Cp = η^5-C_5H_5) [60] (Figure 4). The work is important because it has allowed for a detailed understanding of how redox orbital electron occupancy (Ti(IV) = d^0, V(IV) = d^1, Mo(IV) = d^2,) affects the nature of the M–dithiolene bonding scheme at parity of the ligand set and at parity of charge. In this series of complexes, large changes in the metallodithiolene fold angle and electronic structure are observed as electrons are successively removed from the redox orbital (Figure 4). These electron occupancy effects on the fold angle distortion are now understood in terms of the pseudo-Jahn–Teller effect (PJT). PJT-derived molecular distortions originate from the mixing of the electronic ground state (Ψ_0) with specific excited states (Ψ_i) [61,62]. The ground state–excited state energy gap (2Δ), the matrix elements (F_{0i}) of the vibronic contribution to the force constant (F), and the primary non-vibronic force constant (K_0) all govern the degree of the ligand fold distortion according to:

$$F_{0i} = \left\langle \Psi_0 \middle| \frac{\partial H}{\partial Q} \middle| \Psi_i \right\rangle \quad (1)$$

$$F^2 > \Delta \cdot K_0 \quad (2)$$

At the critical threshold defined by Equation (2), the metallodithiolene centers of $Cp_2M(bdt)$ can distort along the dithiolene fold angle coordinate to yield a double well potential energy surface (Figure 12), and the magnitude of the PJT distortion is maximized by a large F, a small Δ, and a small K_0. Thus, the PJT distortion in these $Cp_2M(bdt)$ complexes effectively couples soft fold angle bending modes in the M-dithiolene chelate ring to the inherent electronic structure of the system via the d-electron count. Importantly, the mixing of low-energy charge transfer states into the ground state by the PJT effect controls the covalency of the M–S bonds.

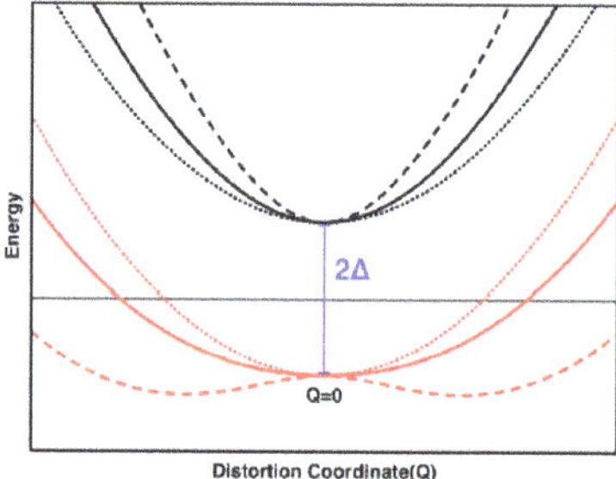

Figure 12. The excited state (black) and ground state (red) potential energy surfaces associated with varying values of F^2 (Dotted: $F^2 = 0$, solid: $F^2 = \Delta \cdot K0$, dashed: $F^2 = 2\Delta \cdot K0$) When the condition $F^2 > \Delta \cdot K0$ is met one observes that the single-well ground state potential energy surface distorts into a double-well potential. The $F^2 > \Delta \cdot K0$ criteria describe a strong PJT effect. (Adapted with permission from *J. Am. Chem. Soc.* 2018, 140, 14777–14788. Copyright (2018) American Chemical Society).

One of the unique aspects of Mo–S and W–S bonding is the small energy gap between filled dithiolene-based orbitals and the lowest energy metal-based orbital, which naturally leads to low-energy charge transfer states that can mix with the ground state. Mode softening along the dithiolene fold coordinate is important in pyranopterin Mo and W enzymes since this leads to a potential energy surface where a large range of dithiolene fold angles may be sampled without paying a prohibitive energy penalty. This effect is maximized when $F^2 \cong \Delta \cdot K_0$. Thus, a low-energy pathway is operative that can minimize energetically unfavorable reorganizational energy contributions along the reaction coordinate, which accompany redox changes at the metal ion. As mentioned previously, these fold angle distortions have been shown to be kinematically coupled to low frequency pyranopterin modes in XO and contribute to low-energy barriers for electron transfer regeneration of the active site. However, in the enzymes there may be either a competing or additive relationship between active site distortions that are driven via the d-electron count of the metal ion and distortions that are imposed by the protein. Vibronic coupling effects that derive from different occupancy numbers for the redox-active orbital will function to modulate the enzyme reduction potential in the oxidative and reductive half reactions of pyranopterin Mo and W enzymes, and this occurs by modulating the degree of metal–ligand covalency via low-frequency distortions at the active site.

6. Conclusions

This review focuses on the electronic structures, molecular structures, and spectroscopic properties of well-characterized metallodithiolene compounds in order to provide deep insight into the role(s) of metal–dithiolene bonding in pyranopterin dithiolene containing enzymes (Figure 1). The discovery, in the early 1960s, that transition metal dithiolene compounds undergo a series of one-electron oxidation-reduction reactions (Figure 2), provided the first evidence for the "non-innocence" of dithiolene ligands and the highly covalent nature of metal–dithiolene bonding. Additional links between metal–dithiolene covalency and electronic and molecular structure were posited from theoretical studies of bent metallocene−dithiolene compounds (Figure 4) by Lauher and Hoffman in 1976 [30], who related metal−dithiolene chelate ring "folding" with the metal ion d-electron configuration. Investigations of Mo dithiolene compounds by electronic absorption, resonance Raman, and EPR spectroscopies showed that S → Mo charge transfer bands dominate the visible spectrum and that there is substantial delocalization of spin density onto the S atoms of the dithiolene. Recent comprehensive studies of bent metallocene−dithiolene compounds have shown that low-energy ligand fold distortions arise from a pseudo-Jahn−Teller (PJT) effect, which involves vibronic coupling of the electronic ground state with electronic excited states to control metal−ligand covalency [60] (Section 5). This vibronic coupling process may play critical roles in the catalytic cycles of pyranopterin Mo and W enzymes by dynamic and/or static modulation of redox potentials and providing a superexchange pathway for electron transfer through the PDT framework. However, greater understanding of how geometric and electronic structure control reactivity, and define function in Mo and W enzymes, will require linking the concepts that have been developed for metallodithiolenes to the emerging results from studies of well-characterized compounds that mimic the pterin component of PDT (Section 3). Exploring the synergistic interactions between the dithiolene and pterin components of the PDT and the metal ion will be challenging, but such research promises to provide important insights into these critically important enzymes.

Author Contributions: J.H.E., J.Y., and M.L.K. collectively conceived and drafted this article. All authors have read and agreed to the published version of the manuscript.

Funding: M.L.K.'s research contributions to this article were funded by the National Institutes of Health (R01-GM-057378).

Conflicts of Interest: The authors declare no conflict of interest.

References

1. Hille, R.; Schulzke, C.; Kirk, M.L. *Molybdenum and Tungsten Enzymes*; The Royal Society of Chemistry: Cambridge, UK, 2017.
2. Hille, R.; Hall, J.; Basu, P. The Mononuclear Molybdenum Enzymes. *Chem. Rev.* **2014**, *114*, 3963–4038. [CrossRef]
3. Kirk, M.L.; Stein, B. The Molybdenum Enzymes. In *Comprehensive Inorganic Chemistry II*, 2nd ed.; Jan, R., Kenneth, P., Eds.; Elsevier: Amsterdam, The Netherlands, 2013; pp. 263–293. [CrossRef]
4. Nieter Burgmayer, S.J.; Kirk, M.L. The Role of the Pyranopterin Dithiolene Component of Moco in Molybdoenzyme Catalysis. In *Metallocofactors that Activate Small Molecules: With Focus on Bioinorganic Chemistry*; Ribbe, M.W., Ed.; Springer International Publishing: Cham, Switzerland, 2019; pp. 101–151. [CrossRef]
5. Weiss, M.C.; Sousa, F.L.; Mrnjavac, N.; Neukirchen, S.; Roettger, M.; Nelson-Sathi, S.; Martin, W.F. The physiology and habitat of the last universal common ancestor. *Nat. Microbiol.* **2016**, *1*, 116. [CrossRef]
6. Fourmigue, M. Mixed cyclopentadienyl/dithiolene complexes. *Coord. Chem. Rev.* **1998**, *180*, 823–864. [CrossRef]
7. Faulmann, C.; Cassoux, P. Solid-State Properties (Electronic, Magnetic, Optical) of Dithiolene Complex-Based Compounds. *Prog. Inorg. Chem.* **2004**, *52*, 399–490.

8. Deplano, P.; Pilia, L.; Espa, D.; Mercuri, M.L.; Serpe, A. Square-planar d(8) metal mixed-ligand dithiolene complexes as second order nonlinear optical chromophores: Structure/property relationship. *Coord. Chem. Rev.* **2010**, *254*, 1434–1447. [CrossRef]
9. Sproules, S.; Wieghardt, K. o-Dithiolene and o-aminothiolate chemistry of iron: Synthesis, structure and reactivity. *Coord. Chem. Rev.* **2010**, *254*, 1358–1382. [CrossRef]
10. Sproules, S.; Wieghardt, K. Dithiolene radicals: Sulfur K-edge X-ray absorption spectroscopy and Harry's intuition. *Coord. Chem. Rev.* **2011**, *255*, 837–860. [CrossRef]
11. Eisenberg, R.; Gray, H.B. Noninnocence in Metal Complexes: A Dithiolene Dawn. *Inorg. Chem.* **2011**, *50*, 9741–9751. [CrossRef]
12. Inscore, F.E.; McNaughton, R.; Westcott, B.L.; Helton, M.E.; Jones, R.; Dhawan, I.K.; Enemark, J.H.; Kirk, M.L. Spectroscopic evidence for a unique bonding interaction in oxo-molybdenum dithiolate complexes: Implications for sigma electron transfer pathways in the pyranopterin dithiolate centers of enzymes. *Inorg. Chem.* **1999**, *38*, 1401–1410. [CrossRef]
13. Kirk, M.L.; McNaughton, R.L.; Helton, M.E. The Electronic Structure and Spectroscopy of Metallo-Dithiolene Complexes. In *Progress in Inorganic Chemistry: Synthesis, Properties, and Applications*; Stiefel, E.I., Karlin, K.D., Eds.; John Wiley and Sons: Hoboken, NJ, USA, 2004; Volume 52, pp. 111–212.
14. Peariso, K.; Helton, M.E.; Duesler, E.N.; Shadle, S.E.; Kirk, M.L. Sulfur K-edge spectroscopic investigation of second coordination sphere effects in oxomolybdenum-thiolates: Relationship to molybdenum-cysteine covalency and electron transfer in sulfite oxidase. *Inorg. Chem.* **2007**, *46*, 1259–1267. [CrossRef]
15. Matz, K.G.; Mtei, R.P.; Leung, B.; Burgmayer, S.J.N.; Kirk, M.L. Noninnocent Dithiolene Ligands: A New Oxomolybdenum Complex Possessing a Donor Acceptor Dithiolene Ligand. *J. Am. Chem. Soc.* **2010**, *132*, 7830–7831. [CrossRef]
16. Mtei, R.P.; Perera, E.; Mogesa, B.; Stein, B.; Basu, P.; Kirk, M.L. A Valence Bond Description of Dizwitterionic Dithiolene Character in an Oxomolybdenum-Bis(dithione) Complex. *Eur. J. Inorg. Chem.* **2011**, *2011*, 5467–5470. [CrossRef]
17. Yang, J.; Mogesa, B.; Basu, P.; Kirk, M.L. Large Ligand Folding Distortion in an Oxomolybdenum Donor Acceptor Complex. *Inorg. Chem.* **2016**, *55*, 785–793. [CrossRef]
18. Yang, J.; Kersi, D.K.; Richers, C.P.; Giles, L.J.; Dangi, R.; Stein, B.W.; Feng, C.; Tichnell, C.R.; Shultz, D.A.; Kirk, M.L. Ground State Nuclear Magnetic Resonance Chemical Shifts Predict Charge-Separated Excited State Lifetimes. *Inorg. Chem.* **2018**, *57*, 13470–13476. [CrossRef]
19. Yang, J.; Kersi, D.K.; Giles, L.J.; Stein, B.W.; Feng, C.J.; Tichnell, C.R.; Shultz, D.A.; Kirk, M.L. Ligand Control of Donor-Acceptor Excited-State Lifetimes. *Inorg. Chem.* **2014**, *53*, 4791–4793. [CrossRef]
20. Matz, K.G.; Mtei, R.P.; Rothstein, R.; Kirk, M.L.; Burgmayer, S.J.N. Study of Molybdenum(4+) Quinoxalyldithiolenes as Models for the Noninnocent Pyranopterin in the Molybdenum Cofactor. *Inorg. Chem.* **2011**, *50*, 9804–9815. [CrossRef]
21. Kirk, M.L. *Spectroscopic and Electronic Structure Studies of Mo Model Compounds and Enzymes*; The Royal Society of Chemistry: Cambridge, UK, 2016; pp. 13–67.
22. Schrauzer, G.N.; Mayweg, V. Reaction of Diphenylacetylene with Nickel Sulfides. *J. Am. Chem. Soc.* **1962**, *84*, 3221. [CrossRef]
23. Gray, H.B.; Billig, E.; Williams, R.; Bernal, I. Spin-Free Square Planar Cobaltous Complex. *J. Am. Chem. Soc.* **1962**, *84*, 3596. [CrossRef]
24. Davison, A.; Holm, R.H.; Edelstein, N.; Maki, A.H. Preparation and Characterization of 4-Coordinate Complexes Related by Electron-Transfer Reactions. *Inorg. Chem.* **1963**, *2*, 1227. [CrossRef]
25. McCleverty, J.A. Metal 1,2-Dithiolene and Related Complexes. *Prog. Inorg. Chem.* **1968**, *10*, 49–221.
26. Johnson, J.; Rajagopalan, K. Structural and Metabolic Relationship Between the Molybdenum Cofactor and Urothione. *Proc. Natl. Acad. Sci. USA* **1982**, *79*, 6856–6860. [CrossRef]
27. Chan, M.K.; Mukund, S.; Kletzin, A.; Adams, M.W.W.; Rees, D.C. Structure of a Hyperthermophilic Tungstopterin Enzyme, Aldehyde Ferredoxin Oxidoreductase. *Science* **1995**, *267*, 1463–1469. [CrossRef]
28. Romao, M.J.; Archer, M.; Moura, I.; Moura, J.J.G.; Legall, J.; Engh, R.; Schneider, M.; Hof, P.; Huber, R. Crystal Structure of the Xanthine Oxidase Related Aldehyde Oxidoreductase from *D. gigas*. *Science* **1995**, *270*, 1170–1176. [CrossRef]

29. Rothery, R.A.; Stein, B.; Solomonson, M.; Kirk, M.L.; Weiner, J.H. Pyranopterin conformation defines the function of molybdenum and tungsten enzymes. *Proc. Natl. Acad. Sci. USA* **2012**, *109*, 14773–14778. [CrossRef]
30. Lauher, J.W.; Hoffmann, R. Structure and Chemistry of Bis(Cyclopentadienyl)-MLn Complexes. *J. Am. Chem. Soc.* **1976**, *98*, 1729–1742. [CrossRef]
31. Joshi, H.K.; Cooney, J.J.A.; Inscore, F.E.; Gruhn, N.E.; Lichtenberger, D.L.; Enemark, J.H. Investigation of metal-dithiolate fold angle effects: Implications for molybdenum and tungsten enzymes. *Proc. Natl. Acad. Sci. USA* **2003**, *100*, 3719–3724. [CrossRef]
32. Inscore, F.E.; Knottenbelt, S.Z.; Rubie, N.D.; Joshi, H.K.; Kirk, M.L.; Enemark, J.H. Understanding the origin of metal-sulfur vibrations in an oxo-molybdenurn dithiolene complex: Relevance to sulfite oxidase. *Inorg. Chem.* **2006**, *45*, 967. [CrossRef]
33. Peariso, K.; Chohan, B.S.; Carrano, C.J.; Kirk, M.L. Synthesis and EPR characterization of new models for the one-electron reduced molybdenum site of sulfite oxidase. *Inorg. Chem.* **2003**, *42*, 6194–6203. [CrossRef]
34. Cleland, W.E.; Barnhart, K.M.; Yamanouchi, K.; Collison, D.; Mabbs, F.E.; Ortega, R.B.; Enemark, J.H. Syntheses, Structures, and Spectroscopic Properties of 6-Coordinate Mononuclear Oxo-Molybdenum(V) Complexes Stabilized by the Hydrotris(3,5-Dimethyl-1-Pyrazolyl)Borate Ligand. *Inorg. Chem.* **1987**, *26*, 1017–1025. [CrossRef]
35. Enemark, J.H. Consensus structures of the Mo(V) sites of sulfite-oxidizing enzymes derived from variable frequency pulsed EPR spectroscopy, isotopic labelling and DFT calculations. *Dalton Trans.* **2017**, *46*, 13202–13210. [CrossRef]
36. Klein, E.L.; Belaidi, A.A.; Raitsimring, A.M.; Davis, A.C.; Kramer, T.; Astashkin, A.V.; Neese, F.; Schwarz, G.; Enemark, J.H. Pulsed Electron Paramagnetic Resonance Spectroscopy of S-33-Labeled Molybdenum Cofactor in Catalytically Active Bioengineered Sulfite Oxidase. *Inorg. Chem.* **2014**, *53*, 961–971. [CrossRef] [PubMed]
37. Jones, R.M.; Inscore, F.E.; Hille, R.; Kirk, M.L. Freeze-Quench Magnetic Circular Dichroism Spectroscopic Study of the "Very Rapid" Intermediate in Xanthine Oxidase. *Inorg. Chem.* **1999**, *38*, 4963–4970. [CrossRef] [PubMed]
38. Yang, J.; Dong, C.; Kirk, M.L. Xanthine oxidase-product complexes probe the importance of substrate/product orientation along the reaction coordinate. *Dalton Trans.* **2017**, *46*, 13242–13250. [CrossRef]
39. Dong, C.; Yang, J.; Reschke, S.; Leimkühler, S.; Kirk, M.L. Vibrational Probes of Molybdenum Cofactor–Protein Interactions in Xanthine Dehydrogenase. *Inorg. Chem.* **2017**, *56*, 6830–6837. [CrossRef]
40. Dong, C.; Yang, J.; Leimkühler, S.; Kirk, M.L. Pyranopterin Dithiolene Distortions Relevant to Electron Transfer in Xanthine Oxidase/Dehydrogenase. *Inorg. Chem.* **2014**, *53*, 7077–7079. [CrossRef]
41. Helton, M.E.; Pacheco, A.; McMaster, J.; Enemark, J.H.; Kirk, M.L. An MCD Spectroscopic Study of the Molybdenum Active Site in Sulfite Oxidase: Insight into the Role of Coordinated Cysteine. *J. Inorg. Biochem.* **2000**, *80*, 227–233. [CrossRef]
42. Sugimoto, H.; Sato, M.; Asano, K.; Suzuki, T.; Mieda, K.; Ogura, T.; Matsumoto, T.; Giles, L.J.; Pokhrel, A.; Kirk, M.L.; et al. A Model for the Active-Site Formation Process in DMSO Reductase Family Molybdenum Enzymes Involving Oxido Alcoholato and Oxido Thiolato Molybdenum(VI) Core Structures. *Inorg. Chem.* **2016**, *55*, 1542–1550. [CrossRef]
43. Sugimoto, H.; Sato, M.; Giles, L.J.; Asano, K.; Suzuki, T.; Kirk, M.L.; Itoh, S. Oxo-carboxylato-molybdenum(VI) complexes possessing dithiolene ligands related to the active site of type II DMSOR family molybdoenzymes. *Dalton Trans.* **2013**, *42*, 5927–15930. [CrossRef]
44. Sugimoto, H.; Tatemoto, S.; Suyama, K.; Miyake, H.; Mtei, R.P.; Itoh, S.; Kirk, M.L. Monooxomolybdenum(VI) Complexes Possessing Olefinic Dithiolene Ligands: Probing Mo–S Covalency Contributions to Electron Transfer in Dimethyl Sulfoxide Reductase Family Molybdoenzymes. *Inorg. Chem.* **2010**, *49*, 5368–5370. [CrossRef]
45. Sugimoto, H.; Tatemoto, S.; Suyama, K.; Miyake, H.; Itoh, S.; Dong, C.; Yang, J.; Kirk, M.L. Dioxomolybdenum(VI) Complexes with Ene-1,2-dithiolate Ligands: Synthesis, Spectroscopy, and Oxygen Atom Transfer Reactivity. *Inorg. Chem.* **2009**, *48*, 10581–10590. [CrossRef]
46. Burgmayer, S.J.N.; Kim, M.; Petit, R.; Rothkopf, A.; Kim, A.; BelHamdounia, S.; Hou, Y.; Somogyi, A.; Habel-Rodriguez, D.; Williams, A.; et al. Synthesis, characterization, and spectroscopy of model molybdopterin complexes. *J. Inorg. Biochem.* **2007**, *101*, 1601–1616. [CrossRef]

47. Kirk, M.L.; Peariso, K. Ground and excited state spectral comparisons of models for sulfite oxidase. *Polyhedron* **2004**, *23*, 499. [CrossRef]
48. Helton, M.E.; Gebhart, N.L.; Davies, E.S.; McMaster, J.; Garner, C.D.; Kirk, M.L. Thermally Driven Intramolecular Charge Transfer in an Oxo-Molybdenum Dithiolate Complex. *J. Am. Chem. Soc.* **2001**, *123*, 10389–10390. [CrossRef]
49. Davie, S.R.; Rubie, N.D.; Hammes, B.S.; Carrano, C.J.; Kirk, M.L.; Basu, P. Geometric control of reduction potential in oxomolybdenum centers: Implications to the serine coordination in DMSO reductase. *Inorg. Chem.* **2001**, *40*, 2632. [CrossRef]
50. McNaughton, R.L.; Helton, M.E.; Rubie, N.D.; Kirk, M.L. The oxo-gate hypothesis and DMSO reductase: Implications for a psuedo-sigma bonding interaction involved in enzymatic electron transfer. *Inorg. Chem.* **2000**, *39*, 4386. [CrossRef]
51. Helton, M.; Gruhn, N.; McNaughton, R.; Kirk, M. Control of oxo-molybdenum reduction and ionization potentials by dithiolate donors. *Inorg. Chem.* **2000**, *39*, 2273–2278. [CrossRef]
52. Gisewhite, D.R.; Yang, J.; Williams, B.R.; Esmail, A.; Stein, B.; Kirk, M.L.; Burgmayer, S.J.N. Implications of Pyran Cyclization and Pterin Conformation on Oxidized Forms of the Molybdenum Cofactor. *J. Am. Chem. Soc.* **2018**, *140*, 12808–12818. [CrossRef]
53. Paudel, J.; Pokhrel, A.; Kirk, M.L.; Li, F. Remote Charge Effects on the Oxygen-Atom-Transfer Reactivity and Their Relationship to Molybdenum Enzymes. *Inorg. Chem.* **2019**, *58*, 2054–2068. [CrossRef]
54. Westcott, B.L.; Gruhn, N.E.; Enemark, J.H. Evaluation of Molybdenum-Sulfur Interactions in Molybdoenzyme Model Complexes by Gas-Phase Photoelectron Spectroscopy. The "Electronic Buffer" Effect. *J. Am. Chem. Soc.* **1998**, *120*, 3382–3386. [CrossRef]
55. Wiebelhaus, N.J.; Cranswick, M.A.; Klein, E.L.; Lockett, L.T.; Lichtenberger, D.L.; Enemark, J.H. Metal-Sulfur Valence Orbital Interaction Energies in Metal-Dithiolene Complexes: Determination of Charge and Overlap Interaction Energies by Comparison of Core and Valence Ionization Energy Shifts. *Inorg. Chem.* **2011**, *50*, 11021–11031. [CrossRef]
56. Davis, M.; Olson, J.; Palmer, G. The Reaction of Xanthine Oxidase with Lumazine: Characterization of the Reductive Half-reaction. *J. Biol. Chem.* **1984**, *259*, 3526–3533. [PubMed]
57. Pauff, J.M.; Cao, H.; Hille, R. Substrate Orientation and Catalysis at the Molybdenum Site in Xanthine Oxidase Crystal structures in complex with xanthine and lumazine. *J. Biol. Chem.* **2009**, *284*, 8751–8758. [CrossRef] [PubMed]
58. Hemann, C.; Ilich, P.; Stockert, A.L.; Choi, E.Y.; Hille, R. Resonance Raman studies of xanthine oxidase: The reduced enzyme–Product complex with violapterin. *J. Phys. Chem. B* **2005**, *109*, 3023–3031. [CrossRef] [PubMed]
59. Hemann, C.; Ilich, P.; Hille, R. Vibrational spectra of lumazine in water at pH 2–13: Ab initio calculation and FTIR/Raman spectra. *J. Phys. Chem. B* **2003**, *107*, 2139–2155. [CrossRef]
60. Stein, B.W.; Yang, J.; Mtei, R.; Wiebelhaus, N.J.; Kersi, D.K.; LePluart, J.; Lichtenberger, D.L.; Enemark, J.H.; Kirk, M.L. Vibrational Control of Covalency Effects Related to the Active Sites of Molybdenum Enzymes. *J. Am. Chem. Soc.* **2018**, *140*, 14777–14788. [CrossRef]
61. Bersuker, I.B. *The Jahn-Teller Effect*; Cambridge University Press: Cambridge, UK, 2006.
62. Bersuker, I.B. Pseudo-Jahn-Teller Effect-A Two-State Paradigm in Formation, Deformation, and Transformation of Molecular Systems and Solids. *Chem. Rev.* **2013**, *113*, 1351–1390. [CrossRef]

Review

Structure: Function Studies of the Cytosolic, Mo- and NAD^+-Dependent Formate Dehydrogenase from *Cupriavidus necator*

Russ Hille [1,*], Tynan Young [2], Dimitri Niks [1], Sheron Hakopian [3], Timothy K. Tam [2], Xuejun Yu [4], Ashok Mulchandani [5] and Gregor M. Blaha [1,*]

1 Department of Biochemistry, University of California, Riverside, CA 92521, USA; dimitri.niks@ucr.edu
2 Department of Biochemistry and the Biochemistry and Molecular Biology Graduate Program, University of California, Riverside, CA 92521, USA; tynan.young@email.ucr.edu (T.Y.); ttam004@ucr.edu (T.K.T.)
3 Department of Biochemistry and the Environmental Toxicology Graduate Program, University of California, Riverside, CA 92521, USA; shako001@ucr.edu
4 Bioengineering Graduate Program, University of California, Riverside, CA 92521, USA; xyu010@ucr.edu
5 Department of Chemical and Environmental Engineering, University of California, Riverside, CA 92521, USA; adani@engr.ucr.edu
* Correspondence: russ.hille@ucr.edu (R.H.); gregor.blaha@ucr.edu (G.M.B.); Tel.: +1-951-827-6354 (R.H.); +1-951-827-4294 (G.M.B.)

Received: 19 May 2020; Accepted: 28 June 2020; Published: 6 July 2020

Abstract: Here, we report recent progress our laboratories have made in understanding the maturation and reaction mechanism of the cytosolic and NAD^+-dependent formate dehydrogenase from *Cupriavidus necator.* Our recent work has established that the enzyme is fully capable of catalyzing the reverse of the physiological reaction, namely, the reduction of CO_2 to formate using NADH as a source of reducing equivalents. The steady-state kinetic parameters in the forward and reverse directions are consistent with the expected Haldane relationship. The addition of an NADH-regenerating system consisting of glucose and glucose dehydrogenase increases the yield of formate approximately 10-fold. This work points to possible ways of optimizing the reverse of the enzyme's physiological reaction with commercial potential as an effective means of CO_2 remediation. New insight into the maturation of the enzyme comes from the recently reported structure of the FdhD sulfurase. In *E. coli*, FdhD transfers a catalytically essential sulfur to the maturing molybdenum cofactor prior to insertion into the apoenzyme of formate dehydrogenase FdhF, which has high sequence similarity to the molybdenum-containing domain of the *C. necator* FdsA. The FdhD structure suggests that the molybdenum cofactor may first be transferred from the sulfurase to the C-terminal cap domain of apo formate dehydrogenase, rather than being transferred directly to the body of the apoenzyme. Closing of the cap domain over the body of the enzymes delivers the Mo-cofactor into the active site, completing the maturation of formate dehydrogenase. The structural and kinetic characterization of the NADH reduction of the FdsBG subcomplex of the enzyme provides further insights in reversing of the formate dehydrogenase reaction. Most notably, we observe the transient formation of a neutral semiquinone FMNH·, a species that has not been observed previously with holoenzyme. After initial reduction of the FMN of FdsB by NADH to the hydroquinone (with a k_{red} of 680 s^{-1} and K_d of 190 μM), one electron is rapidly transferred to the Fe_2S_2 cluster of FdsG, leaving FMNH·. The Fe_4S_4 cluster of FdsB does not become reduced in the process. These results provide insight into the function not only of the *C. necator* formate dehydrogenase but also of other members of the NADH dehydrogenase superfamily of enzymes to which it belongs.

Keywords: nicotinamide adenine dinucleotide (NADH); electron transfer; enzyme kinetics; enzyme structure; formate dehydrogenase; carbon assimilation

1. Introduction

The molybdenum- and tungsten-dependent formate dehydrogenases have drawn increased attention over the past 5–10 years due to the demonstration that under the appropriate conditions, most, if not all, are able to catalyze the reverse reaction, reduction of CO_2 to formate, under the appropriate conditions. Indeed, some enzymes of this family, which include the closely related formylmethanofuran dehydrogenases of the Wood–Ljungdahl pathway, function physiologically to reduce CO_2 to formate in what is probably the most evolutionarily ancient mechanism of carbon fixation.

Cupriavidus necator H16 (previously known as *Ralstonia eutropha*) has four formate dehydrogenases, of which, two are cytosolic enzymes that utilize NAD^+ as oxidizing substrate [1,2]. One of these contains molybdenum and is encoded by the *fdsGBACD* operon, the other possesses tungsten and is encoded by the *fdwAB* operon (presumably enlisting additional subunits from the *fds* operon) [3]. The molybdenum-containing enzyme was originally isolated and characterized by Bowien and coworkers, who showed that the mature holoenzyme belongs to the NADH dehydrogenase superfamily of enzymes [4–6]. The FdsA, FdsB, and FdsG subunits are homologous to corresponding subunits in the cytosolic arm of NADH dehydrogenase [7–10]. FdsA is homologous to the Nqo3 subunit of the NADH dehydrogenase from *Thermus thermophilus* [8,9], although the latter lacks a molybdenum center as found in the former [7]. The homology between FdsA and Nqo3 includes the presence of a histidine ligand to one of the Fe_4S_4 clusters. The C-terminal domain of FdsA, containing the molybdenum center, is also homologous to the crystallographically characterized FdhF formate dehydrogenase of *E. coli*, with the molybdenum-coordinating Cys 378 of FdsA equivalent to Sec 140 in FdhF [11]. The Fe_4S_4- and FMN-containing FdsB subunit is homologous to the *T. thermophilus* Nqo1 subunit and like Nqo1 also possesses a binding site for NADH/NAD^+. The FdsG subunit is homologous to the *T. thermophilus* Nqo2 subunit and has a single Fe_2S_2 cluster. We report here recent work from our laboratories on both mechanistic and structural aspects of the *C. necator* enzyme.

2. Catalysis of CO_2 Reduction by the *C. necator* Formate Dehydrogenase

Under physiological conditions, reducing equivalents enter the *C. necator* formate dehydrogenase holoenzyme at its molybdenum center and leave at the FMN, reducing NAD^+ to NADH; electron transfer between the molybdenum and flavin, which are separated by some 55 Å [10], is mediated by the intervening iron–sulfur clusters. Although originally reported to be unable to catalyze the CO_2 by NADH, we have recently shown that the enzyme is indeed capable of doing so when CO_2 (not bicarbonate) is used as substrate [12]. This is consistent with a mechanism for formate oxidation involving direct hydride transfer of the C_α-H to the $Mo^{VI} = S$ group of the $L_2Mo^{VI}S$(S-Cys) molybdenum center (where L is the bidentate enedithiolate-coordinated pyranopterin cofactor found in molybdenum- and tungsten-containing enzymes, present in this enzyme as the guanine dinucleotide) to L_2Mo^{IV}(SH)(S-Cys), with CO_2 rather than bicarbonate as the immediate product of the reaction [13]. The steady-state kinetic parameters have been determined in both the forward and reverse directions and are shown in Table 1.

Table 1. Steady-state kinetic parameters for the *C. necator* formate dehydrogenase.

	Forward	Reverse
k_{cat}	201 s^{-1}	10 s^{-1}
$K_m^{formate}$	130 μM	–
K_m^{NAD+}	310 μM	–
K_m^{CO2}	–	2700 μM
K_m^{NADH}	–	46 μM

The Haldane relationship for these parameters for an enzyme, such as formate dehydrogenase operating via a ping-pong mechanism with separate sites for the reductive and oxidative half-reactions, is as follows:

$$K_{eq} = \frac{\frac{\left(k_{cat}^{forward}\right)}{\left(K_m^{formate}\right)} \cdot \frac{\left(k_{cat}^{forward}\right)}{\left(K_m^{NAD+}\right)}}{\frac{\left(k_{cat}^{reverse}\right)}{\left(K_m^{CO_2}\right)} \cdot \frac{\left(k_{cat}^{reverse}\right)}{\left(K_m^{NADH}\right)}} = \frac{\frac{\left(201\ s^{-1}\right)}{(130\ \mu M)} \cdot \frac{\left(201\ s^{-1}\right)}{(310\ \mu M)}}{\frac{\left(10\ s^{-1}\right)}{(2700\ \mu M)} \cdot \frac{\left(10\ s^{-1}\right)}{(46\ \mu M)}} = 1250 \quad (1)$$

The value 1250 compares favorably with the K_{eq} calculated from the 100 mV difference between $\Delta E_{0'}$ for the NAD^+/NADH and CO_2/formate couples of 2100, the disparity reflecting a ~10% uncertainty in $k_{cat}{}^{forward}$ and $k_{cat}{}^{reverse}$, which are squared terms in numerator and denominator, respectively, of Equation (1).

The reaction can be pushed significantly in the direction of CO_2 reduction by the addition of an NADH regeneration system [14]. As shown in Figure 1 left, addition of a catalytic amount of formate dehydrogenase to a solution that is 29.5 mM in CO_2 (at 30 °C) and 300 μM in NADH results in the formation of 120 μM formate (as quantified by ion chromatography) and 130 μM NAD^+—in other words, the reaction is tightly coupled. When the experiment is repeated with the addition of 50 mM glucose and a catalytic amount of glucose dehydrogenase for NADH regeneration, the amount of formate generated increases approximately 10-fold to 1.0 mM (Figure 1 right). This illustrates the potential commercial use of the enzyme for generation of formate from CO_2 using a suitable source of reducing equivalents.

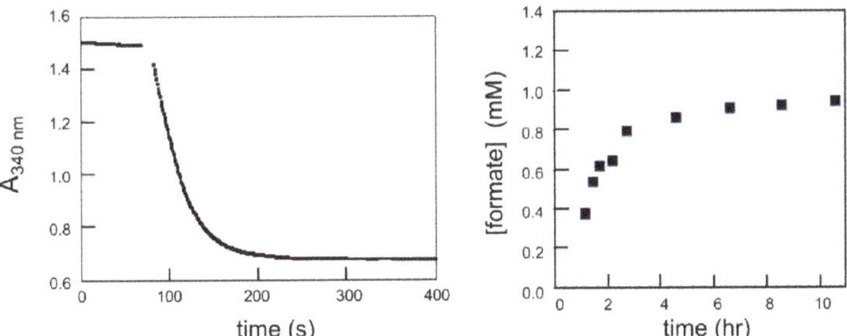

Figure 1. Catalysis of CO_2 reduction to formate by *C. necator* formate dehydrogenase. (**Left**) the reaction of 0.2 μM holoenzyme with 29.5 mM CO_2(aq) and 300 μM NADH, 20 mM Bis-Tris propane, pH 6.3, 30 °C. The absorbance change reflects the consumption of 130 μM NADH. The amount of formate accumulated as determined by ion chromatography was 120 μM [12]. (**Right**) the accumulation of formate under the same conditions upon addition of 50 mM glucose and catalytic glucose dehydrogenase to the reaction mix. Adapted with permission from *Biochemistry* **2019**, *58*, 1861–1868 [14]. Copyright (2019) American Chemical Society.

3. Insertion of the Molybdenum Cofactor into Formate Dehydrogenase and Other Members of the DMSO Reductase Family

As is seen with all members of the DMSO reductase family of molybdenum enzymes, the active site molybdenum center is deeply buried in the C-terminal domain of the FdsA subunit, and the means by which it is introduced into the apoenzyme is not presently understood. Given the extensive structural homology of this domain to the *E. coli* FdhF, one can consider the overall topology of the latter which consists of three interlaced domains that constitute the body of the protein, and a contiguous C-terminal domain that constitutes a "cap" over the cofactor in the holoenzyme (Figure 2A [11]). It is now well-accepted that all the Mo- and W-containing formate dehydrogenases possess a M=S group that is inserted into the metal coordination sphere as the final step of cofactor maturation [15]. Sulfur forms a more covalent bond with molybdenum and tungsten than does the more electronegative oxygen. As a result, there is less formal negative charge on the sulfur, e.g., making it better able to accept a hydride in the course of the reaction. A similar argument has been made in the case of xanthine oxidase and related enzymes, which also require a Mo=S [10]. The sulfur transferase catalyzing this reaction (the product of the *fdhD* gene in *E. coli*, *fdsC* in the *C. necator* operon) is also thought to play a role in the insertion process, and the X-ray crystal structure of FdhD has recently been reported with two equivalents of GDP bound at the presumed position of the maturing molybdenum center [16]. Modeling the cofactor into the structure yields a complex with the apex of the molybdenum coordination sphere pointing into the channel through which the catalytically essential sulfur is delivered (via a cysteine desulfurase), with the *bis*(MGD) portion of the cofactor presenting a concave basket to the surface of the complex; the pyranopterin portion of both cofactors is solvent-exposed with the principal interactions with FdhD principally involving the guanine dinucleotide extensions of the cofactor (Figure 2B). In this orientation, the cofactor cannot be transferred directly to the body of the apo FdhF, where it is also oriented concave outwards with the two guanine dinucleotide arms extended into the protein, not out toward solvent. On the other hand, the molybdenum center in FdhF interacts with the cap domain of the protein principally via its pyranopterin rings, interacting minimally with the guanine dinucleotide arms. This being the case, it is possible to dock the cap (with bound cofactor) to the dimeric $FdhD_2{\cdot}(GDP)_2$ complex in such a way that the GDP arms overlap; both proteins are in a position to interact optimally with the cofactor sandwiched between them. If this interaction is physiologically significant, the implication is that FdhD passes the now sulfurated and mature cofactor not to the body of apo FdhF but to its cap, which then closes over the body swinging the cofactor into position in the active site. In this way the highly unstable cofactor is never released into free solution. There have been a number of efforts in the past to identify regions on the surface of DMSO reductase family enzymes that interact with the cofactor insertion machinery. If the above analysis is correct, then this surface, on the face of the C-terminal cap domain that interfaces with the body of the enzyme, is buried in the structure of the holoenzyme. In support of this model, the C-terminal cap domains of the apo forms of both *E. coli* trimethylamine-*N*-oxide reductase TorA [17] and *E. coli* periplasmic nitrate reductase NapA [18] in complex with their respective chaperones TorD and NapD (that recognize the proteins' N-terminal twin-Arg signal sequence that targets them to the periplasm) have been reported to assume an open position in readiness to accept the mature molybdenum cofactor. This indicates that the C-terminal cap domain is indeed able to adopt the type of open configuration that is proposed here.

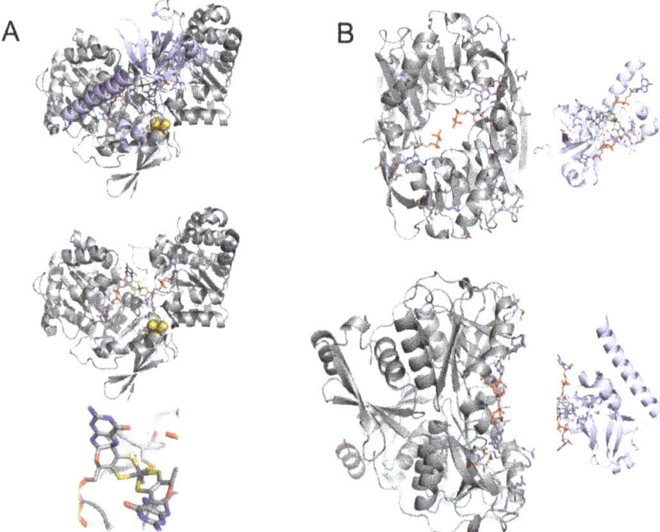

Figure 2. A comparison of the structures of FdhD and FdhF. Panel (**A**) top, the structure of FdhF (PDB 1FDO), with the body of the enzyme in *gray* and the C-terminal cap domain in *blue*. Middle, the body

of FdhF with the C-terminal cap domain removed, revealing the buried molybdenum center. Bottom, an enlargement of the molybdenum center, illustrating its disposition in the body of FdhF. Panel (**B**) top, views of the putative interfaces of the dimeric FdhD in complex with two equivalents of GDP (*gray*; PDB 4PDE) and the C-terminal cap domain of FdhF with the molybdenum center shown (*blue*). Bottom, the structures shown rotated toward one another illustrating the overlap between the GDP's of the FdhD structure and the molybdenum center of FdhF.

4. The Crystal Structure of FdsBG

The X-ray crystal structure of the FdsBG subcomplex of the *C. necator* formate dehydrogenase has also been determined at a resolution of 2.3 Å [19]. This fragment of the holoenzyme contains FMN and a Fe_4S_4 cluster in the FdsB subunit and a Fe_2S_2 cluster in FdsG; it lacks the FdsA subunit that contains the molybdenum center and additional iron–sulfur clusters. As expected, the structure of each subunit is closely related to its cognate subunits in the NADH dehydrogenases from *T. thermophilus* and *Aquifex aeolicus*, FdsB being homologous to the *T. thermophilus* Nqo2 and *A. aeolicus* NuoF subunits and contains FMN and a Fe_4S_4 cluster, and FdsG to the *T. thermophilus* Nqo1 and *A. aeolicus* NuoE subunits and contains a Fe_2S_2 cluster [20]. Compared to the *A. aeolicus* NuoF, FdsB has an RMSD of 1.48 Å for 394 C_α atoms and contains a Rossmann-like fold [21] encompassing the FMN binding site, a ubiquitin-like and a four-helix domain containing the Fe_4S_4 cluster [22] (Figure 3); all these structural elements are shared with NuoF. The Fe_4S_4 cluster is in a mostly hydrophobic environment close to the protein surface and is bound by $C443^B$, $C446^B$, $C449^B$, and $C489^B$ (superscript B indicating the residue is in FdsB). The principal structural differences between FdsB and NuoF are in surface loops of the protein. However, the N-terminal portion of FdsB consists of a thioredoxin-like fold [20] not seen in NuoF or Nqo1. This domain lacks the Fe_2S_2 cluster typical of thioredoxins owing to mutation of the iron-coordinating cysteines ($P10^B$, $A15^B$, $S45^B$, and $F49^B$). This resembles the C-terminal portion of FdsG (RMSD 1.97 Å for 71 $C\alpha$ atoms of the core of the domain) that contains the Fe_2S_2 cluster (vide infra).

The FdsG subunit is homologous to the NuoE subunit of *A. aeolicus* NADH dehydrogenase, consisting of an N-terminal four-helical bundle (residues 29^G–74^G) and a C-terminal thioredoxin-like domain (residues 79^G–159^G) that possesses the subunit's Fe_2S_2 cluster. These two domains are connected by a four-amino acid linker and are rotated by ~26° relative to the orientation seen in NuoE.

It is to be noted that while the N-terminal domains of FdsG and NuoE both have four helices, the first helix of FdsG runs parallel, but in NuoE, it runs across the second and third helices. The C-terminal thioredoxin-like domain of FdsG resembles the N-terminal thioredoxin-like domain of FdsB, but unlike the cofactorless FdsB domain, the FdsG domain contains a spinach ferredoxin-like Fe_2S_2 cluster [19]. This iron–sulfur cluster is again in a hydrophobic environment near the surface, coordinated by $C86^G$, $C91^G$, $C127^G$, and $C131^G$. Like *A. aeolicus* NuoE (but unlike *T. thermophilus* Nqo1), the C-terminus of FdsG is truncated at its C-terminus and lacks a disulfide bond.

At the FdsB–FdsG interface, two helices of the N-terminal four-helical bundle domain of FdsG contact the Rossmann-like domain of FdsB, as seen in the subunit interface between NuoE and NuoF. FdsG's C-terminal thioredoxin-like domain interacts with the same surfaces of the ubiquitin-like and Rossmann-like domains of FdsB as do the corresponding structural elements of NuoE and NuoF. The different placement of the ubiquitin and of the 183^B–190^B loop in FdsB as compared to NuoE results in an increase in the distance between the Fe_2S_2 cluster of FdsG and the Fe_4S_4 cluster of FdsB, being of ~1.0 Å as compared to the separation seen in NuoEF.

At a distance of nearly 21 Å (edge-to-edge), direct electron transfer between the two Fe/S clusters is expected to be extremely slow and unlikely to be physiologically relevant. On the other hand, each cluster is within 12 Å of the FMN isoalloxazine ring (Figure 4). Given the model that has been constructed for the holoenzyme based on the structures of FdhF and Nqo123, the Fe_4S_4 cluster of FdsB clearly lies on-path for electron transfer between the molybdenum center and FMN of the formate dehydrogenase, while the Fe_2S_2 cluster of FdsG is off-path. In the NADH dehydrogenases, it has been suggested that this cluster may serve to temporarily store electrons during electron transfer out of the FMN once it has been reduced by NADH (vide infra). The FMN in FdsB is surrounded by a network of interactions that is also conserved in other NADH dehydrogenase-like enzymes [20,23] and is part of the solvent-accessible cavity that constitutes the NAD^+/NADH binding site. The C8-methyl of the FMN points towards the Fe_4S_4 cluster, with the C_4=O facing the Fe_2S_2 cluster (Figure 4).

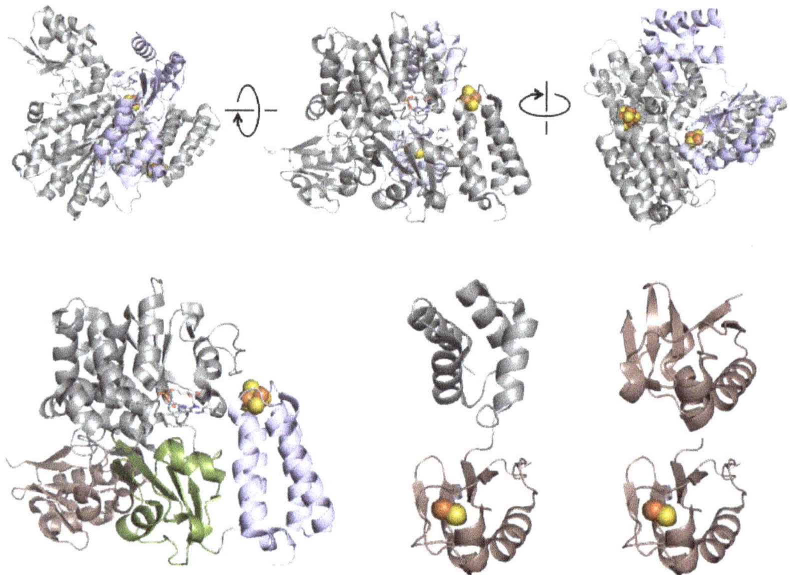

Figure 3. The structure of FdsBG. (**Top**) the overall structure of FdsBG, with FdsB in *gray* and FdsG in *blue* (PDB 6VW8). The center structure is rotated about the horizontal or vertical axis as indicated to give the structures to the left and right, respectively. The two iron–sulfur clusters are rendered as CPK spheres and the FMN as CPK sticks. (**Bottom Left**) the domain structure of FdsB, with the thioredoxin-like domain in *red*, the Rossmann fold in *gray*, the ubiquitin-like domain in *green*, and the four-helix domain in *blue*; (**Center**) the domain structure of FdsG, with the N-terminal domain in gray and the thioredoxin-like domain in *red*; (**Right**) a comparison of the thioredoxin-like domains of FdsB (*upper*) and FdsG (*lower*).

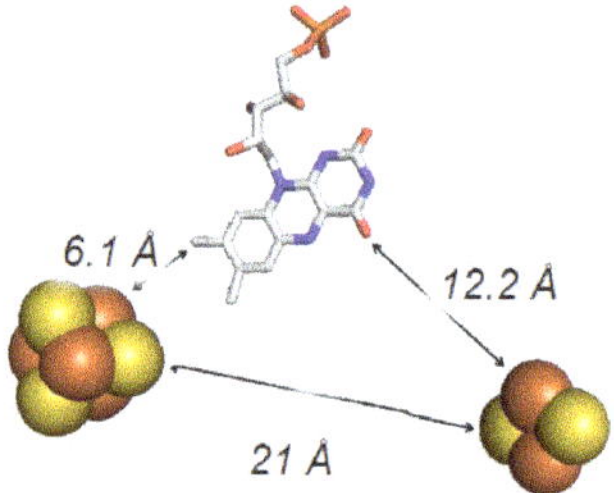

Figure 4. Disposition of the redox-active centers of FdsBG. From left to right the Fe_4S_4 and FMN of FdsB and the Fe_2S_2 of FdsG (PDB 6VW8).

5. EPR Characterization of the Iron–Sulfur Clusters of FdsBG

Extended incubation of FdsBG complex with sodium dithionite (pH 7.0) yields two signals (as seen in Figure 5 top). The first is seen at liquid nitrogen temperatures as high as 200 K, with $g_{1,2,3}$ = 2.000, 1.948, and 1.920 and linewidths of 1.4, 1.7, and 1.6 mT, respectively. These values are in good agreement with previously reported parameters for the cluster designated as Fe/S_1 for holoenzyme (Figure 5 middle, dotted spectrum) [13]. The second signal is seen only below 20 K and has $g_{1,2,3}$ = 2.039, 1.955, and 1.891, and linewidths of 4.5, 1.4, and 5.3 mT (Figure 5 middle, dashed spectrum). This signal has not been previously seen with the holoenzyme [13], owing to its low intensity, broad linewidths, and overlap with the much stronger Fe/S_3 signal, and is designated Fe/S_5. In the *A. aeolicus* NuoEF and *T. thermophilus* Nqo12 NADH dehydrogenases, only the Fe_2S_2 clusters exhibit EPR signals at 77 K, while the Fe_4S_4 signals appear only below 50 K [24]. Further, the g-values $g_{1,2,3}$ = 2.004, 1.945, and 1.917 for the N1a Fe_2S_2 cluster of the 24 kDa subunit of bovine complex I [25] (the homolog to FdsG) are in very good agreement with the Fe/S_1 signal seen here in FdsBG, and accordingly, the Fe/S_1 signal has been assigned to the Fe_2S_2 cluster of FdsG. The signal assigned to the N3 Fe_4S_4 cluster in the NADH dehydrogenase systems, with g-values of $g_{1,2,3}$ = 2.037, 1.945, and 1.852, is also in good agreement with the Fe/S_5 signal of the FdsBG complex, and we accordingly assign the Fe/S_5 signal to the Fe_4S_4 cluster of FdsB. The Fe/S_1 signal seen with holoenzyme was initially assigned to the His-coordinated Fe_4S_4 cluster of FdsA (Figure 5 bottom, dotted spectrum) [13], but that is clearly incorrect as FdsBG exhibits the signal. The Fe/S_5 signal has similar g-values to those reported for the previously observed Fe/S_3, but the signal-giving cluster couples to the molybdenum center and must be the proximal Fe_4S_4 cluster to the molybdenum center in FdsA; the Fe/S_5 signal must, therefore, be a new signal not seen previously with holoenzyme. The assignment of EPR signal Fe/S_5 to the Fe_4S_4 cluster of FdsB is consistent with our previously published structural model of holoformate dehydrogenase placing the Fe_4S_4 cluster near FdsA in the intact holoenzyme and on-path between the molybdenum and FMN of the enzyme [10].

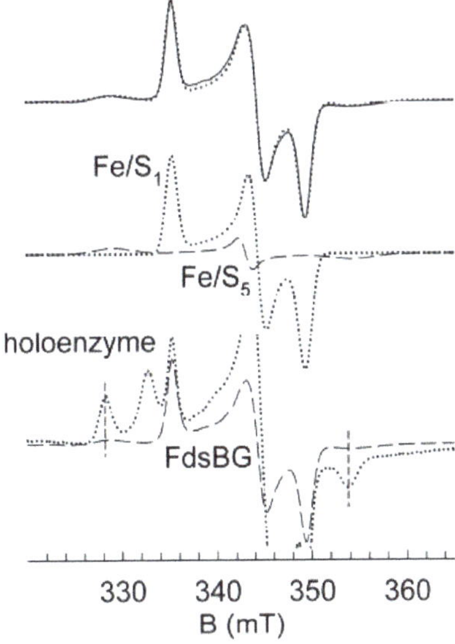

Figure 5. EPR spectra of the iron–sulfur clusters of the FdsBG complex. (**Top**) the observed iron–sulfur EPR spectrum (solid) and simulated composite spectrum (dashed) of dithionite-reduced FdsBG, collected at 9 K with modulation amplitude of 8 Gauss and microwave power of 2 µW. The sample was prepared by incubation of 125 µM of FdsBG complex in 100 mM potassium phosphate, pH 7.0 with 2 mM buffered sodium dithionite under anaerobic conditions for 1 h at room temperature prior to freezing. (**Middle**) the individual signals resulting from the simulation of the composite spectrum above. The spectrum corresponding to the previously assigned Fe/S_1 (dotted) and an additional Fe/S_5 signal (dashed) are resolved only in the FdsBG complex. (**Bottom**) a comparison of the spectra seen with FdsBG (dashed) and holoenzyme, at 20 K (dotted). The dashed lines mark the location of g_1 and g_3 features of the Fe/S_3 component seen in the spectrum of the reduced holoenzyme (see text) [19].

6. Rapid-Reaction Kinetics of FdsBG Reduction by NADH

The rapid-reaction kinetics of FdsBG reduction by NADH have also been examined. This is the reverse of the physiological reaction for the formate dehydrogenase, but is the physiological direction seen in the NADH dehydrogenases. A typical kinetic transient is shown in the inset of Figure 6 left. The reaction is biphasic, with the fast phase of the reaction completing within 120 ms. A plot of the observed k_{fast} as a function of NADH is hyperbolic (Figure 6 left), yielding a limiting k_{red} of 680 s^{-1} and K_d^{NADH} of 190 μM at 5 °C. We note that this rate is more than sixfold faster than the limiting rate of reduction of holoenzyme at high formate, some 20-fold faster than k_{cat} for formate oxidation (13), and 350-fold faster than k_{cat} for CO_2 reduction [12] under comparable conditions.

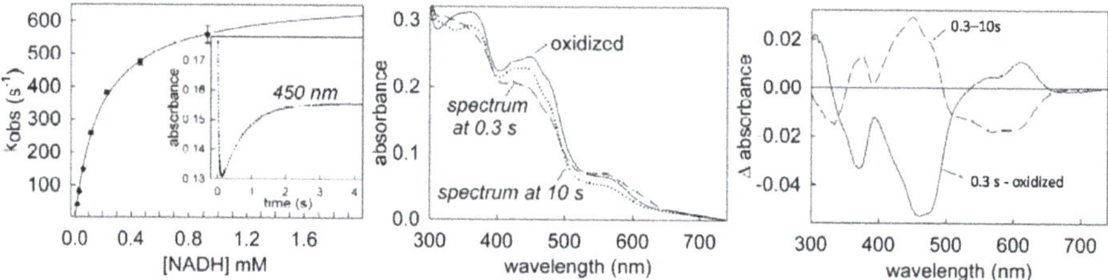

Figure 6. The kinetics of FdsBG reduction by NADH. (**Left**) a plot of k_{fast} vs. NADH, yielding a limiting k_{red} of 680 s^{-1} and a K_d^{NADH} of 190 μM. The *inset* shows a typical biphasic transient, as described in the text. The reaction conditions were 100 mM potassium phosphate, pH 7.0, 5 °C. (**Center**) spectra seen at the times indicated in the reaction of 10 μM FdsBG with 5 μM NADH, as obtained with a photodiode array detector on the stopped-flow. (**Right**) difference spectra obtained using the spectra obtained at the times indicated. The formation of FMNH· in the fast phase of the reaction is reflected in the positive feature in the 500–650 nm region in the *dotted* difference spectrum and its loss during the slow disproportionation phase of the reaction by the negative feature in the same region in the *dashed* difference spectrum [19].

When the reaction was repeated using substoichiometric NADH (so that the $FdsBG_{1e}$- formed on disproportionation will not be further reduced by reaction with a second equivalent of NADH), the absorbance increase seen at the end of the fast phase of the reaction implies formation of the neutral semiquinone FMNH· (Figure 6 center, dashed). This species has not seen in previous work with holoenzyme and must be the result of transfer of a single electron from the fully reduced flavin hydroquinone (formed on its initial two-electron reduction by NADH) to the Fe_2S_2 cluster of FdsG iron, leaving the flavin as FMNH· (and giving rise to the Fe/S_1 EPR signal; see further below and Scheme 1). On a second, slower timescale, two equivalents of the $FdsBG_{2e}$- formed upon reaction with NADH disproportionate to give one equivalent each of $FdsBG_{1e}$- and $FdsBG_{3e}$-. To the extent this occurs, $FdsBG_{1e}$- will have its Fe_2S_2 cluster (Fe/S_1) reduced and its FMN oxidized, whereas $FdsBG_{3e}$- will have both Fe/S_1 and the FMN reduced.

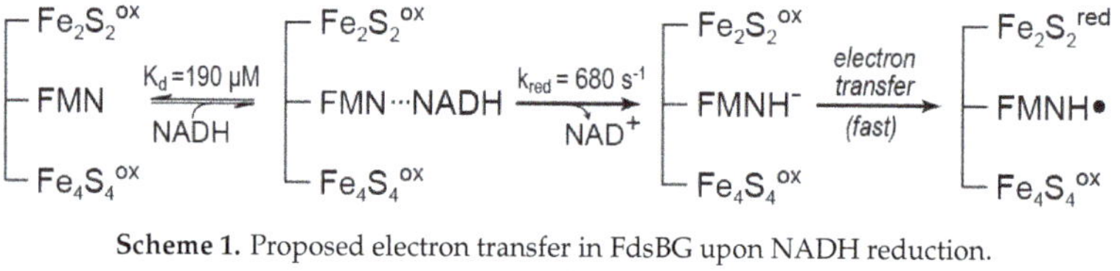

$$[Fe_2S_2^{ox},\ FMN,\ Fe_4S_4^{ox}] \xrightarrow[NADH]{K_d = 190\ \mu M} [Fe_2S_2^{ox},\ FMN\cdots NADH,\ Fe_4S_4^{ox}] \xrightarrow[NAD^+]{k_{red} = 680\ s^{-1}} [Fe_2S_2^{ox},\ FMNH^-,\ Fe_4S_4^{ox}] \xrightarrow[(fast)]{electron\ transfer} [Fe_2S_2^{red},\ FMNH\bullet,\ Fe_4S_4^{ox}]$$

Scheme 1. Proposed electron transfer in FdsBG upon NADH reduction.

The UV/Visible signature of FMNH· [26] is seen as the positive feature in the 500–650 nm region in the difference spectrum between fully oxidized enzyme and that collected at 0.3 s after the initial reaction of FdsBG with NADH (Figure 6 right, black spectrum). The extended negative feature in the 300–500 nm region is due to reduction of the Fe/S clusters. The subsequent disappearance of the

FMNH· formed in the fast phase of the reaction is reflected in the pronounced negative feature in the 500–650 nm region seen in the 0.3–10 s difference spectrum (Figure 6 right, dashed difference spectrum). Here, the two broad, negative peaks centered at 570 and 605 nm reflect the presence of FMNH· at 0.3 s but its loss over the next 10 s. Formation of FMNH· has independently been confirmed in a freeze-quench EPR experiment. Figure 7 shows the spectrum seen at 150 K when FdsBG is reacted with NADH and frozen after ~40 ms (Top), which exhibits the EPR signatures of both FMNH· and Fe/S_1. In addition to demonstrating the formation of FMNH·, this experiment establishes that it is the Fe_2S_2 cluster of FdsG that becomes reduced in forming FMNH·. Subtracting out the Fe/S_1 signal from the measured EPR spectrum yields an isotropic EPR signal with a $g_{iso} = 2.003$ and a linewidth of ~1.9 mT, reflecting formation of the neutral rather than anionic semiquinone (Figure 7 middle) [27,28]. No additional iron–sulfur signal attributable to the Fe/S_4 signal is seen at 9 K, even when the experiment is repeated in the presence of excess NADH, indicating that it does not get reduced in the course of reduction of FdsBG by NADH (Figure 7 bottom).

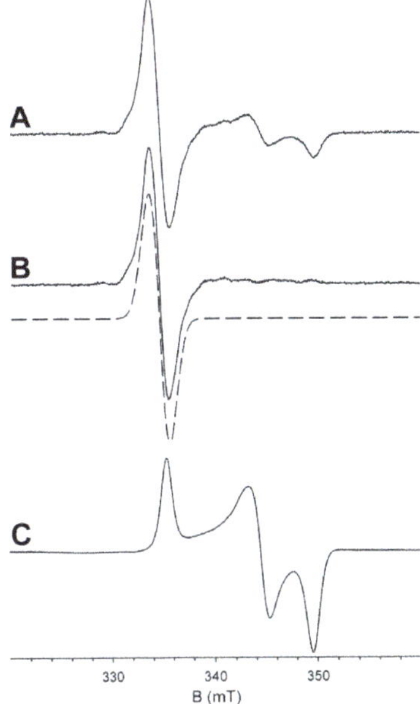

Figure 7. EPR of the neutral flavin semiquinone of FdsBG. (**Top**) The EPR spectrum seen by rapid-freeze quench (quenching time ~40 ms) on reaction of 40 μM FdsBG with 0.8 mM NADH at 0 °C. The spectrum was obtained at 150 K with modulation amplitude of 8 Gauss and microwave power of 0.4 mW. (**Middle**) the spectrum of FADH·, obtained by subtracting out the Fe/S1 contribution (solid line) along with a simulation (dashed line). (**Bottom**) The EPR spectrum seen on mixing 160 μM of FdsBG with 60 μM NADH at room temperature for 10 s prior to freezing. The EPR spectrum was collected at 9 K with modulation amplitude of 8 Gauss and microwave power of 2 μW, and it exhibits only the Fe/S_1 signal with no contribution from Fe/S_4. All samples were prepared in 100 mM potassium phosphate, pH 7.0, under anaerobic conditions [19].

7. The Thioredoxin-Like Domain of FdsB

The N-terminal thioredoxin-like domain of FdsB is extremely similar to the Fe_2S_2 ferredoxin domain from *A. aeolicus* [21], which like the *C. necator* formate dehydrogenase is known to dimerize. The largest contact interface between the two FdsBG protomers within the asymmetric unit of the crystal is between the two FdsB thioredoxin-like domains, suggesting that these contribute significantly to the dimerization of the holoenzyme.

The buried area is small (~560 Å^2), however, with only a few specific interactions across the interface, which are not highly conserved among the NADH dehydrogenase family. In addition, FdsBG remains monomeric throughout purification. It thus appears that the dimer seen in the asymmetric unit is simply the result of crystal packing [29,30].

8. Electron Transfer in FdsBG

During normal turnover with holoformate dehydrogenase, reducing equivalents from the oxidation of formate pass from the active site molybdenum center through a chain of iron–sulfur clusters to the FMN, which ultimately reduces NAD$^+$. FdsBG is able to reduce NAD$^+$, as are the corresponding subcomplexes from NADH dehydrogenases [31,32] and the NAD$^+$-reducing hydrogenase [33]. Consistent with the Fe_4S_4 cluster of FdsB being on-path between the molybdenum center and FMN, it lies close to the C8-methyl of FMN and to highly conserved surface residues that are implicated in the interaction of FdsB with FdsA (i.e., E441^B–S487^B, I451^B–G452^B, and K292^B–L298^B). By contrast, the Fe_2S_2 cluster of FdsG lies further from the region implicated in FdsA binding. Nevertheless, as summarized in Scheme 1, the present evidence indicates that the initial electron transfer event out of the FMN in FdsB upon reduction by NADH is to the off-path Fe_2S_2 cluster, which has the higher reduction potential relative to the on-path Fe_4S_4 cluster. In the case of the NADH dehydrogenases, a similar electron transfer to the Fe_2S_2 cluster is thought to minimize formation of neutral flavin semiquinone, FMNH·, thereby reducing the accumulation of reactive oxygen species [20,22].

9. Concluding Remarks

The cytosolic and NAD$^+$-dependent formate dehydrogenase is fully capable of catalyzing the reduction of CO_2 to formate using NADH as a source of reducing equivalents. The values for the forward and reverse steady-state kinetic parameters are consistent with the expected Haldane relationship. The addition of an NADH-regenerating system consisting of glucose and glucose dehydrogenase increases the yield of formate approximately 10-fold, suggesting the commercial potential of the enzyme as an effective means of CO_2 remediation.

A consideration of the recently reported structure of the FdhD sulfurase that transfers a catalytically essential sulfur to the maturing molybdenum cofactor prior to insertion into apoenzyme suggests that the cofactor may first be transferred to the C-terminal cap domain of apo formate dehydrogenase, which then closes over the body of the enzyme to assemble the active site, rather than transferring the cofactor directly to the body of the protein.

The X-ray crystal structure of the FdsBG fragment of the *C. necator* formate dehydrogenase is reviewed, as are the kinetics of its reduction by NADH. Notably, the neutral semiquinone FMNH· is observed transiently in the course of the reaction of FdsBG with NADH. In a reaction analogous to the physiological reduction of NADH dehydrogenases by NADH, after initial reduction of the FMN of FdsB by NADH to the hydroquinone (with a k_{red} of 680 s^{-1} and K_d of 190 μM), one electron is rapidly transferred to the Fe_2S_2 cluster of FdsG, leaving FMNH·, as characterized by both UV/visible spectroscopy and EPR. The Fe_4S_4 cluster of FdsB does not become reduced in the process.

Finally, we note that the cryo-EM structure of the *R. capsulatus* formate dehydrogenase, which is very similar to the *C. necator* enzyme that has been the focus here, has very recently been reported at a resolution of 3.3 Å [34]. The overall structure of the FdsBG fragment of the *R. capsulatus* protein is very much in agreement with that described here for the *C. necator* protein, and the overall structure is also in good agreement with that predicted previously on the basis of the structures of the *E. coli* FdhF formate dehydrogenase and the *T. thermophilus* NADH dehydrogenase [10]. Not previously anticipated is the proximity (9.5 Å edge-to-edge) of the Fe_4S_4 clusters designated A4 in the two FdsA subunits of the dimeric Fds(ABGD)$_2$ dimer, which suggests reducing equivalents are able to pass easily between protomers (a situation also seen in the even more complex formylmethanofuran dehydrogenase from *Methanothermobacter wolfeii* [35]). Interestingly, the A4 Fe_4S_4 cluster is the one possessing histidine as a ligand and presumably having a higher rather than lower reduction potential.

Importantly, the cryo-EM work clearly demonstrates that the small FdsD subunit is an integral component of the mature holoenzyme, interacting intimately with the C-terminal domain of FdsA and apparently stabilizing the inserted molybdenum cofactor. The position of FdsD is not inconsistent with the model for cofactor incorporation considered here, although its role in cofactor insertion remains to be elucidated. Very interestingly, evidence is presented that cryo-EM is able to distinguish between oxidized and reduced iron–sulfur clusters in complex systems. If found to be generally applicable, this would constitute a very significant advantage of the cryo-EM method over other structural methods and provide information presently obtainable by spatially resolved anomalous dispersion refinement [36].

Author Contributions: R.H. supervised the EPR and kinetic work described here, and prepared the manuscript. T.Y., S.H., and T.K.T. analyzed the X-ray crystallographic work with FdsBG under the supervision of G.M.B., who also contributed to preparation of the manuscript. D.N. performed the EPR and kinetic work described. X.Y. performed the work involving CO_2 reduction to formate under the supervision of A.M., who also contributed to preparation of the manuscript. All authors have read and agreed to the published version of the manuscript.

Funding: This research was funded by a grant from the U.S. Department of Energy (DE-SC0010666 to R.H.).

Acknowledgments: We thank the staff of the following beamlines for their assistance in data collection: 5.0.1 and 5.0.2 of the Berkeley Center of Structural Biology at Advanced Light Source (ALS), 24-ID-C of the Northeastern Collaborative Access Team at Advanced Photon Source (APS), and BL7-1 of the Stanford Synchrotron Radiation Lightsource (SSRL). The Berkeley Center for Structural Biology is supported in part by the Howard Hughes Medical Institute. The Advanced Light Source is a Department of Energy Office of Science User Facility under Contract No. DE-AC02-05CH11231. The Pilatus detector on 5.0.1 was funded under NIH grant S10OD021832. The ALS-ENABLE beamlines are supported in part by the National Institutes of Health, National Institute of General Medical Sciences, grant P30 GM124169. The Northeastern Collaborative Access Team beamlines are funded by the National Institute of General Medical Sciences from the National Institutes of Health (P30 GM124165). The Advanced Photon Source is a U.S. Department of Energy (DOE) Office of Science User Facility operated for the DOE Office of Science by Argonne National Laboratory under Contract No. DE-AC02-06CH11357. The Stanford Synchrotron Radiation Lightsource, SLAC National Accelerator Laboratory, is supported by the U.S. Department of Energy, Office of Science, Office of Basic Energy Sciences under Contract No. DE-AC02-76SF00515. The SSRL Structural Molecular Biology Program is supported by the DOE Office of Biological and Environmental Research and by the National Institutes of Health, National Institute of General Medical Sciences (including P41GM103393). The contents of this publication are solely the responsibility of the authors and do not necessarily represent the official views of NIGMS or NIH.

Conflicts of Interest: The authors declare no conflict of interest.

References

1. Pohlmann, A.; Fricke, W.F.; Reinecke, F.; Kusian, B.; Liesegang, H.; Cramm, R.; Eitinger, T.; Ewering, C.; Potter, M.; Schwartz, E.; et al. Genome sequence of the bioplastic-producing "Knallgas" bacterium *Ralstonia eutropha* H16. *Nat. Biotechnol.* **2006**, *24*, 1257–1262. [CrossRef] [PubMed]
2. Little, G.T.; Ehsaan, M.; Arenas-Lopez, C.; Jawed, K.; Winzer, K.; Kovacs, K.; Minton, N.P. Complete genome sequence of *Cupriavidus necator* H16 (DSM 428). *Microbiol. Resour. Announc.* **2019**, *8*. [CrossRef] [PubMed]
3. Cramm, R. Genomic view of energy metabolism in *Ralstonia eutropha* H16. *J. Mol. Microbiol. Biotechnol.* **2009**, *16*, 38–52. [CrossRef]
4. Friedebold, J.; Bowien, B. Physiological and biochemical characterization of the soluble formate dehydrogenase, a molybdoenzyme from *Alcaligenes eutrophus*. *J. Bacteriol.* **1993**, *175*, 4719–4728. [CrossRef] [PubMed]
5. Friedebold, J.; Mayer, F.; Bill, E.; Trautwein, A.X.; Bowien, B. Structural and immunological studies on the soluble formate dehydrogenase from *Alcaligenes eutrophus*. *Biol. Chem. Hoppe Seyler* **1995**, *376*, 561–568. [CrossRef] [PubMed]
6. Oh, J.I.; Bowien, B. Structural analysis of the Fds operon encoding the NAD^+-linked formate dehydrogenase of *Ralstonia eutropha*. *J. Biol. Chem.* **1998**, *273*, 26349–26360. [CrossRef] [PubMed]
7. Sazanov, L.A.; Hinchliffe, P. Structure of the hydrophilic domain of respiratory complex I from *Thermus thermophilus*. *Science* **2006**, *311*, 1430–1436. [CrossRef]
8. Berrisford, J.M.; Sazanov, L.A. Structural basis for the mecahnism of respiratory complex I. *J. Biol. Chem.* **2009**, *284*, 29773–29783. [CrossRef]
9. Efremov, R.G.; Baradaran, R.; Sazanov, L.A. The architecture of respiratory complex I. *Nature* **2010**, *465*, 441–445. [CrossRef]

10. Hille, R.; Hall, J.; Basu, P. The mononuclear molybdenum enzymes. *Chem. Rev.* **2014**, *114*, 3963–4038. [CrossRef]
11. Boyington, J.C.; Gladyshev, V.N.; Khangulov, S.V.; Stadtman, T.C.; Sun, P.D. Crystal structure of formate dehydrogenase H: Catalysis involving Mo, molybdopterin, selenocysteine, and an Fe_4S_4 cluster. *Science* **1997**, *275*, 1305–1308. [CrossRef] [PubMed]
12. Yu, X.; Niks, D.; Mulchandani, A.; Hille, R. Efficient reduction of CO_2 by the molybdenum-containing formate dehydrogenase from *Cupriavidus necator* (*Ralstonia eutropha*). *J. Biol. Chem.* **2017**, *292*, 16872–16879. [CrossRef] [PubMed]
13. Niks, D.; Duvvuru, J.; Escalona, M.; Hille, R. Spectroscopic and kinetic properties of the molybdenum-containing, NAD^+-dependent Formate Dehydrogenase from *Ralstonia eutropha*. *J. Biol. Chem.* **2016**, *291*, 1162–1174. [CrossRef] [PubMed]
14. Yu, X.; Niks, D.; Ge, X.; Liu, H.; Hille, R.; Mulchandani, A. Synthesis of formate from CO_2 gas catalyzed by an O_2-tolerant NAD^+-dependent formate dehydrogenase and glucose dehydrogenase. *Biochemistry* **2019**, *58*, 1861–1868. [CrossRef] [PubMed]
15. Thomé, R.; Gust, A.; Toci, R.; Mendel, R.; Bittner, F.; Magalon, A.; Walburger, A. A sulfurtransferase is essential for activity of formate dehydrogenase in *Escherichia coli*. *J. Biol. Chem.* **2012**, *287*, 4671–4678. [CrossRef]
16. Arnoux, P.; Ruppelt, C.; Oudouhou, F.; Lavergne, J.; Siponen, M.I.; Toci, R.; Mendel, R.R.; Bittner, F.; Pignol, D.; Magalon, A.; et al. Sulphur shuttling across a chaperone during molybdenum cofactor maturation. *Nat. Commun.* **2015**, *6*, 6148. [CrossRef]
17. Dow, J.M.; Grabel, F.; Sargent, F.; Palmer, T. Characterization of a pre-export enzyme-chaperone complex on the twin-arginine transport pathway. *Biochem. J.* **2013**, *452*, 57–66. [CrossRef]
18. Dow, J.M.; Grahl, S.; Ward, R.; Evans, R.; Byron, O.; Norman, D.G.; Palmer, T.; Sargent, F. Characterization of a periplasmic nitrate reductase in complex with its biosynthetic chaperone. *FEBS J.* **2014**, *281*, 246–260. [CrossRef]
19. Young, T.; Niks, D.; Hakopian, S.; Tam, T.K.; Yu, X.; Hille, R.; Blaha, G.M. Crystallographic and kinetic analyses of the FdsBG subcomplex of the cytosolic formate dehydrogenase from *Cupriavidus necator*. *J. Biol. Chem.* **2020**, *295*, 6570–6585. [CrossRef]
20. Schulte, M.; Frick, K.; Gnandt, E.; Jurkovic, S.; Burschel, S.; Labatzke, R.; Aierstock, K.; Fiegen, D.; Wohlwend, D.; Gerhardt, S.; et al. A mechanism to prevent production of reactive oxygen species by *Escherichia coli* respiratory complex I. *Nat. Commun.* **2019**, *10*, 2551. [CrossRef]
21. Marchler-Bauer, A.; Bo, Y.; Han, L.; He, J.; Lanczycki, C.J.; Lu, S.; Chitsaz, F.; Derbyshire, M.K.; Geer, R.C.; Gonzales, N.R.; et al. CDD/SPARCLE: Functional classification of proteins via subfamily domain architectures. *Nucleic Acids Res.* **2017**, *45*, D200–D203. [CrossRef]
22. Hinchliffe, P.; Sazanov, L.A. Organization of iron-sulfur clusters in respiratory complex I. *Science* **2005**, *309*, 771–774. [CrossRef] [PubMed]
23. Shomura, Y.; Taketa, M.; Nakashima, H.; Tai, H.; Nakagawa, H.; Ikeda, Y.; Ishii, M.; Igarashi, Y.; Nishihara, H.; Yoon, K.S.; et al. Structural basis of the redox switches in the NAD(+)-reducing soluble [NiFe]-hydrogenase. *Science* **2017**, *357*, 928–932. [CrossRef]
24. Ohnishi, T.; Ohnishi, S.T.; Salerno, J.C. Five decades of research on mitochondrial NADH-quinone oxidoreductase (complex I). *Biol. Chem.* **2018**, *399*, 1249–1264. [CrossRef] [PubMed]
25. Reda, T.; Barker, C.D.; Hirst, J. Reduction of the iron-sulfur clusters in mitochondrial NADH:ubiquinone oxidoreductase (complex I) by Eu^{II}-DTPA, a very low potential reductant. *Biochemistry* **2008**, *47*, 8885–8893. [CrossRef]
26. Edmondson, D.E.; Tollin, G. Semiquinone formation in flavo- and metalloflavoproteins. *Top. Curr. Chem.* **1983**, *108*, 109–138. [PubMed]
27. Edmondson, D.E.; Ackrell, B.A.; Kearney, E.B. Identification of neutral and anionic 8 alpha-substituted flavin semiquinones in flavoproteins by electron spin resonance spectroscopy. *Arch. Biochem. Biophys.* **1981**, *208*, 69–74. [CrossRef]
28. Palmer, G.; Mueller, F.; Massey, V. Electron Paramagnetic Resonance Studies on Flavoprotein Radicals. In *Third International Symposium of Flavins and Flavoproteins*; Kamin, H., Ed.; University Park Press: Baltimore, MD, USA, 1971; pp. 123–140.
29. Bliven, S.; Lafita, A.; Parker, A.; Capitani, G.; Duarte, J.M. Automated evaluation of quaternary structures from protein crystals. *PLoS Comput. Biol.* **2018**, *14*, e1006104. [CrossRef]
30. Krissinel, E.; Henrick, K. Inference of macromolecular assemblies from crystalline state. *J. Mol. Biol.* **2007**, *372*, 774–797. [CrossRef]

31. Yano, T.; Sled, V.D.; Ohnishi, T.; Yagi, T. Expression and characterization of the flavoprotein subcomplex composed of 50-kDa (NQO1) and 25-kDa (NQO2) subunits of the proton-translocating NADH-quinone oxidoreductase of *Paracoccus denitrificans*. *J. Biol. Chem.* **1996**, *271*, 5907–5913. [CrossRef]
32. Galante, Y.M.; Hatefi, Y. Purification and molecular and enzymic properties of mitochondrial NADH dehydrogenase. *Arch. Biochem. Biophys.* **1979**, *192*, 559–568. [CrossRef]
33. Lauterbach, L.; Idris, Z.; Vincent, K.A.; Lenz, O. Catalytic properties of the isolated diaphorase fragment of the NAD-reducing [NiFe]-hydrogenase from *Ralstonia eutropha*. *PLoS ONE* **2011**, *6*, e25939. [CrossRef] [PubMed]
34. Radon, C.; MIttelStädt, G.; Duffus, B.R.; Bürger, J. Hartmann, T.; Mielke, T.; Teutloff, C.; Leimkühler, S.; Wendler, P. Cryo-EM structures reveal intricate Fe-S cluster arrangement and charging in *Rhodobacter capsulatus* formate dehydrogenase. *Nat. Commun.* **2020**, *11*, 1912. [CrossRef] [PubMed]
35. Wagner, T.; Ermler, U.; Shima, S. The methanogenic CO2 redcuing-and-fixing enzyme is bifunctional and contains 46 [4Fe-4S] clusters. *Science* **2016**, *354*, 114–117. [CrossRef]
36. Spatzao, T.; Schlesier, J.; Burger, E.-M.; Sippel, D.; Zhang, L.; Andrade, S.L.A.; Rees, D.C.; Einsle, O. Nitrogenase FeMoCo investigated by spatially resolved anomalous dispersion refinement. *Nat. Commun.* **2016**, *7*, 10902. [CrossRef]

Article

Susceptibility of the Formate Hydrogenlyase Reaction to the Protonophore CCCP Depends on the Total Hydrogenase Composition

Janik Telleria Marloth and Constanze Pinske *

Institute for Biology/Microbiology, Martin-Luther-University Halle-Wittenberg, 06108 Halle, Germany; janik.telleria-marloth@student.uni-halle.de

* Correspondence: constanze.pinske@mikrobiologie.uni-halle.de; Tel.: +49-(0)-345-5526-353

Received: 23 April 2020; Accepted: 27 May 2020; Published: 28 May 2020

Abstract: Fermentative hydrogen production by enterobacteria derives from the activity of the formate hydrogenlyase (FHL) complex, which couples formate oxidation to H_2 production. The molybdenum-containing formate dehydrogenase and type-4 [NiFe]-hydrogenase together with three iron-sulfur proteins form the soluble domain, which is attached to the membrane by two integral membrane subunits. The FHL complex is phylogenetically related to respiratory complex I, and it is suspected that it has a role in energy conservation similar to the proton-pumping activity of complex I. We monitored the H_2-producing activity of FHL in the presence of different concentrations of the protonophore CCCP. We found an inhibition with an apparent EC_{50} of 31 μM CCCP in the presence of glucose, a higher tolerance towards CCCP when only the oxidizing hydrogenase Hyd-1 was present, but a higher sensitivity when only Hyd-2 was present. The presence of 200 mM monovalent cations reduced the FHL activity by more than 20%. The Na^+/H^+ antiporter inhibitor 5-(*N*-ethyl-*N*-isopropyl)-amiloride (EIPA) combined with CCCP completely inhibited H_2 production. These results indicate a coupling not only between Na^+ transport activity and H_2 production activity, but also between the FHL reaction, proton import and cation export.

Keywords: formate hydrogenlyase; hydrogen metabolism; energy conservation; MRP (multiple resistance and pH)-type Na^+/H^+ antiporter; CCCP—carbonyl cyanide *m*-chlorophenyl-hydrazone; EIPA—5-(*N*-ethyl-*N*-isopropyl)-amiloride

1. Introduction

Anaerobic or fermentative growth in the absence of oxygen requires that conservation of energy is at its utmost efficiency. The reaction of the formate hydrogenlyase (FHL) complex contributes indirectly to the generation of a proton gradient at the cytoplasmic membrane by consumption of protons (H^+) from the cytoplasm, diffusion of H_2-gas across the membrane and subsequent oxidation by the two periplasmic H_2-oxidizing hydrogenases (Hyd-1 and Hyd-2) [1] (Figure 1). Hyd-1 forms a redox loop, and Hyd-2 contributes directly to proton transfer from the cytoplasm to the quinone during oxidation of H_2 [1,2]. The *hyaB* and *hybC* genes encode the catalytic subunits of Hyd-1 and Hyd-2, respectively [3]. This kind of intracellular syntrophy was initially described for *Desulfovibrio* species [4] and more recently for *Acetobacterium woodii* [5]. The FHL complex of *Escherichia coli* is active during mixed acid fermentation to disproportionate formate, which, when accumulated intracellularly, has cytotoxic effects [6]. The two active sites of the FHL complex comprise the [NiFe]-hydrogenase 3 (HycE protein) and the molybdenum- and selenium-containing formate dehydrogenase H (FdhH protein), which function together with three electron-transferring iron-sulfur carrying subunits located in the cytoplasm. Membrane attachment to the membrane-integral HycC and HycD subunits is required for coupling of these half-reactions [7,8]. Our previous studies have shown that the FHL complex can work in reverse in resting cells, a reaction of great biotechnological interest for H_2 and CO_2 storage as formate [8,9]. A second FHL complex with a predicted five instead of two membrane subunits is encoded in the *hyf*-operon on the *E. coli* chromosome, but, presumably due to its low level of gene expression, the activity of this enzyme complex is not detectable under our growth conditions [10,11].

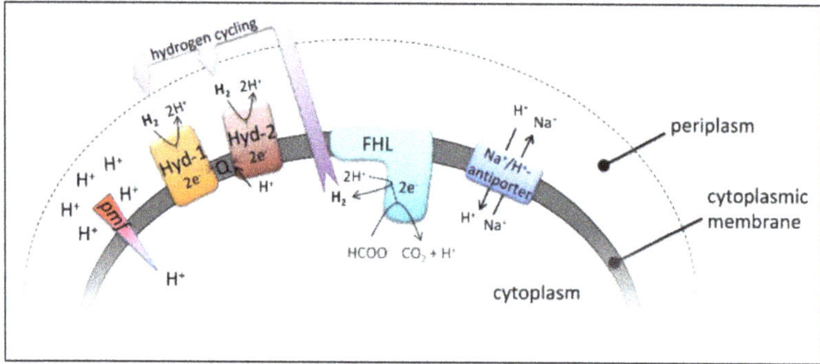

Figure 1. Relevant enzyme complexes of this study, their reactions and their membrane localization. Hyd-1 and Hyd-2 indicate the periplasmic H_2-oxidizing hydrogenases, which channel electrons into the quinone pool (Q), FHL—the cytoplasmic formate hydrogenlyase—and the membrane bound Na^+/H^+ antiporters. Further details are highlighted in the main text.

When the phylogenetic relationship between the FHL complex and complex I was discovered, energy conservation by proton-pumping of the complex was also assumed [12,13]. The hydrogenase 3 (HycE protein) belongs to the class 4 hydrogenases and is similar to the Ech (Energy-converting hydrogenase) hydrogenases, which were shown to generate proton gradients during catalysis [14]. However, while studying FHL bidirectionality as the basis for analyzing the influence of ionophores, we unexpectedly found that the protonophore carbonyl cyanide *m*-chlorophenyl-hydrazone (CCCP) showed only a 40% inhibition of the H_2 production activity and a 30% increase of the reverse formate production reaction from H_2 and CO_2 [8].

It was speculated that due to the small difference in standard redox potentials of formate ($E_0' = -432$ mV, pH 7.0) and H_2 ($E_0' = -414$ mV, pH 7.0), redox energy for proton translocation becomes available only when the physiological conditions shift to a more acidic pH [15]. However, the reversibility of an artificially coupled hydrogenase and formate dehydrogenase reaction from *Desulfovibrio* via an electrode was able to mimic the FHL reaction in both directions in vitro without dependence of the reaction on energy provision [16]. Instead, a controversially discussed model of a multienzyme transport supercomplex comprising F_1F_0-ATPase, the K^+-transporting TrkA symporter and the FHL(-2) complex was suggested for energetic coupling [17,18]; however, its existence still requires verification (reviewed in [19]). Therefore, both the necessity for and the role of the FHL complex's membrane attachment in energy conservation remains to be resolved.

In the present work we have systematically re-examined the CCCP effect on FHL activity, whereby we combined different CCCP concentrations with use of different carbon sources to initiate FHL activity in strains either bearing or lacking the H_2-oxidising Hyds. We found that ultimately all strains could be inhibited by CCCP, but strains lacking Hyd-1 required lower, and those lacking Hyd-2 required higher, concentrations of CCCP for inhibition. Based on the relationship of the membrane subunit HycC to Na^+/H^+ (MRP)-type antiporters, we further addressed the question of whether the FHL reaction is driven by or is contributing to the *pmf* and if it might be linked to the sodium motive force. We present strong evidence for a dependence of the FHL reaction on proton influx and sodium export.

2. Results and Discussion

2.1. CCCP Inhibits H_2 Production to Different Degrees Depending on the Hydrogenase Composition

Cell suspensions initiate H_2 production after the anaerobic addition of either formate or glucose, the latter requires metabolizing by the cells to formate prior to H_2 production. Consequently, the mixed-acid fermentation yields 2 mol formate per mol glucose (reviewed in [3]). Formate is the direct substrate for the FHL complex, but its import via the formate channel FocA could mask the effects seen. FocA behaves as a pH-dependent channel in vitro, while its activity is mostly determined by its interaction with the pyruvate formate lyase protein PflB in vivo [20]. It is feasible that formate export across the membrane contributes to *pmf* generation by proton symport, but the import mechanism is unresolved, except that it only occurs when formate-consuming reactions in the cytoplasm can take place [21]. Therefore, the apparent 1.3 mol H^+ that are translocated across the membrane per mole of formate oxidized during the FHL reaction could be indirect [22]. On the other hand, the transport of glucose across the membrane is PTS (phosphotransferase system)-dependent, where phosphoenolpyruvate provides the phosphoryl group for the incoming glucose in a *pmf*-independent manner [23]. Our parallel experimental design circumvents the fact that indirect effects through the transport of the carbon source into the cells are observed.

The parental strain MC4100, the isogenic *hyaB* deletion strain CP630, which lacks Hyd-1 activity, the isogenic *hybC* deletion strain CP631, which lacks Hyd-2 activity, or the *hyaB hybC* double deletion strain CP734 were grown under the optimal conditions for FHL activity. Subsequently, cells were harvested and suspensions used on a modified Clark-type electrode to monitor H_2 production. The strains showed initial activities very similar to each other in the presence of the same carbon source, but generally higher activity was observed in the presence of formate than glucose (Table 1). A strain lacking the genes for the FHL complex (HDK103) did not produce H_2 under these conditions, supporting the observation that the second FHL is inactive under the conditions tested here [10].

Table 1. H_2 production rates in nmol $H_2 \cdot min^{-1} \cdot mg^{-1}$ ± standard deviation from at least three independent biological samples.

Strain [1]	Glucose	Formate
MC4100	69 ± 12	160 ± 47
CP630 (Δ*hyaB*)	74 ± 23	173 ± 13
CP631 (Δ*hybC*)	101 ± 15	145 ± 20
CP734 (Δ*hyaB* Δ*hybC*)	114 ± 24	121 ± 13

[1] Strains were grown in TGYEP, pH 6.5 medium for 16 h at 30 °C, cells harvested and suspended in 50 mM Tris/HCl, pH 7.0, prepared and assayed on the Clark-type electrode.

When CCCP concentrations of 100 μM were employed, the effect this had on the H_2 production was strain dependent. A previous study from our lab had shown that 100 μM CCCP caused a slight inhibition of the H_2 production rate in a strain with a similar Hyd composition to CP734 [8]. If indeed proton translocation is directly performed by the FHL complex, then CCCP would be expected to cause an increase of the activity, not a decrease. Nevertheless, 100 μM CCCP caused complete inhibition of H_2 production by the parental strain MC4100 independent of the employed carbon source. The slope of the H_2 production even inverted (i.e., H_2 oxidation) after CCCP addition (Figure 2), which indicates consumption of the produced H_2 within the electrode system and indicates the activity of H_2-oxidation reactions, which are usually not performed by the FHL complex, but by the H_2-oxidizing Hyd-1 and Hyd-2 enzymes, indicating they are responsible for the difference.

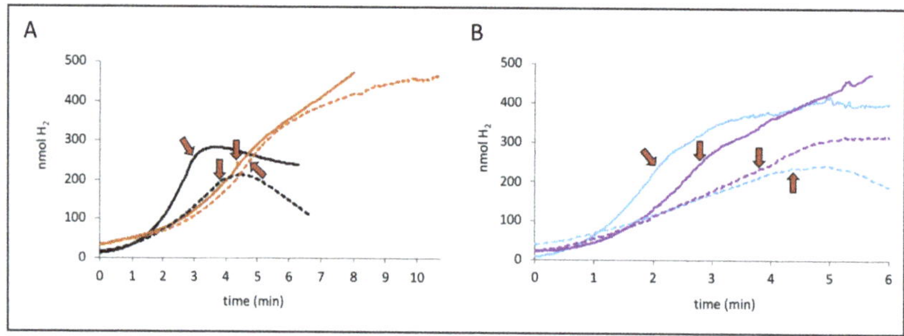

Figure 2. Different degrees of inhibition by protonophore carbonyl cyanide *m*-chlorophenyl-hydrazone (CCCP). Panel A shows H_2 production from the strains MC4100 (parental, black curves) and CP734 (Δ*hyaB* Δ*hybC*, orange curves) while panel B shows the strains CP630 (Δ*hyaB*, blue curves) and CP631 (Δ*hybC*, purple curves) after addition of either 14 mM glucose (dashed lines) or 50 mM formate (solid lines). When H_2 production was linear, a final concentration of 100 μM CCCP was added to the reaction vessel, as indicated by the red arrows.

In the next step, strains lacking the genes encoding the large subunit of either Hyd-1 (Δ*hyaB*, strain CP630) or Hyd-2 (Δ*hybC*, strain CP631) were used in the same experimental setup. While the Hyd-2-deficient strain CP631 (purple curves) mimics the H_2 production curves of the double deletion strain CP734 (Δ*hyaB* Δ*hybC*) after CCCP addition, the Hyd-1 deficient strain CP630 (blue curves) looks very similar to the parental curves and negative slopes after CCCP addition is observed. It can be further observed that the effect of CCCP addition is not immediate, but H_2 production rates shift rather gradually. For the direct dependence of an activity on *pmf*, we have previously observed immediate and complete effects of CCCP addition on the activity [2], suggesting an indirect effect of CCCP occurs here. Notably, the addition of other protonophores like 2,4-dinitrophenol (DNP) up to 1 mM had no effect on the H_2 production activity of the cells, similar to what was observed before at lower concentrations [8]. Taken together, this indicates that the effect of CCCP on FHL activity is indirect.

2.2. Hyd-1 Confers Resistance and Hyd-2 Sensitivity of H_2 Production to CCCP

In order to quantify the effect of CCCP on the FHL activity, we calculated the apparent EC_{50} (half maximal effective concentration) values (Table 2). For that purpose, we analyzed the H_2 production on the electrode, and at about 100 nmol H_2, different concentrations of CCCP were added to the reaction. Figure 3A shows an example of the effect of the CCCP addition on strain CP630 (Δ*hyaB*). A concentration of 1 and 10 μM did not significantly alter the slope of H_2 production, while 50 and 100 μM CCCP clearly slowed down the reaction, and 200 μM led to complete stagnation of H_2 production. A further increase to 400 μM CCCP revealed negative slopes, indicating re-oxidation of H_2 by Hyd-2. Due to slight variations in the initial FHL activities, the reduction of activity after CCCP addition was calculated as the ratio of the activity after CCCP addition to the activity before its addition (Figure 3B). In the case of a negative slope, the ratio after CCCP addition was calculated as zero.

Table 2. Apparent EC_{50} values of CCCP inhibition.

Strain [1]	Glucose	Formate
MC4100	31 μM	81 μM
CP630 (Δ*hyaB*)	0.13 μM	42 μM
CP631(Δ*hybC*)	125 μM	223 μM
CP734 (Δ*hyaB* Δ*hybC*)	98 μM	348 μM

[1] The same cell suspensions were used for the glucose and formate experiments. The same amount of total cell protein (1 mg) was used for the different strains. All values were calculated from at least five different CCCP concentrations.

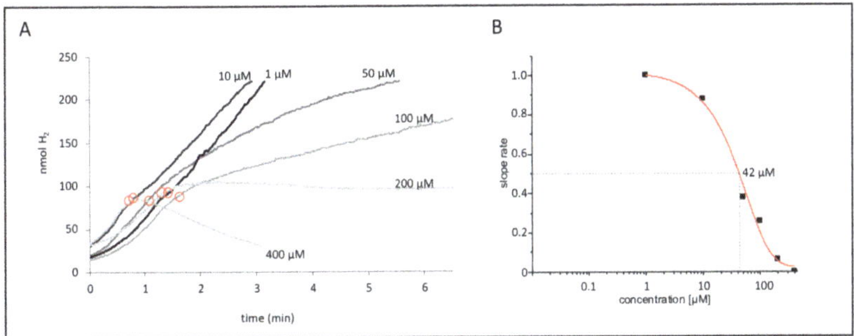

Figure 3. Dose-dependent effect of CCCP. (**A**) Strain CP630 (Δ*hyaB*) was used as a representative strain for CCCP inhibition after initiation of the reaction with 50 mM formate and linear slope; the designated concentrations of CCCP were added at the time points encircled in red. (**B**) The apparent EC_{50} values of this set of activities were calculated from the ratio of slopes after and before CCCP addition and subsequent fitting with the EC_{50} function in SigmaPlot.

An apparent half-maximal inhibition (EC_{50}) of H_2 production can be observed at 31 μM CCCP with glucose or 81 μM CCCP with formate as substrate in strain MC4100 (Table 2). These higher concentrations of CCCP, which are required in the presence of formate compared to glucose, reflect the higher initial activities in the presence of formate (Table 1). The double deletion strain CP734 (Δ*hyaB* Δ*hybC*) only showed reduction of H_2 production at elevated CCCP concentrations, which is reflected in the increased apparent EC_{50} values. For example, we observed complete inhibition of MC4100 (glucose initiated) at 50 μM CCCP but in strain CP734 only at 500 μM CCCP. Looking at the effect of single deletions of the two H_2-oxidizing Hyd in strains CP630 and CP631, respectively, then the Hyd-1 deletion (Δ*hyaB*, strain CP630) caused increased sensitivity to CCCP compared to the parental MC4100, while the Hyd-2 deletion (Δ*hybC*, strain CP631) increased resistance to CCCP similar to the double deletion strain CP734 (Table 2).

Hyd-1 has a lower catalytic activity than Hyd-2 [24,25], while at the same time it has a similar abundance in the membrane under the conditions used to grow the cells here [26]. It is possible that the inactivation of Hyd-1 in strain CP630 results in elevated amounts of Hyd-2 because more maturation enzymes for cofactor construction and more membrane space are available. Hyd-2 is the enzyme that contributes to *pmf* generation [27], and the data show that Hyd-2 but not Hyd-1 catalyzes the oxidation of H_2 at high CCCP concentrations. CCCP enhances Hyd-2 activity by removing the back-pressure of the *pmf* on the enzyme. When assayed directly, the effect is not very prominent because the enzyme works at its catalytic optimum, but in variants that exhibit difficulties in transferring electrons across the membrane, the effect is more pronounced [2]. Electrons from Hyd-2 are concomitantly channeled into the quinone pool for reduction of electron acceptors like fumarate, which can be internally produced during mixed-acid fermentation when glucose is provided to the cells [28]. Therefore, negative slopes occur mostly when Hyd-2 is present and glucose is provided. This does not account for H_2 oxidation in the presence of formate; however, we cannot currently explain this effect (Figure 3A). The higher amounts of Hyd-2 enzyme in combination with its CCCP-elevated activity account for the apparent higher CCCP sensitivity in the absence of Hyd-1. Therefore, the absence of Hyd-2 confers resistance, and the absence of Hyd-1, resulting in a concomitant increase in Hyd-2 levels, causes sensitivity to CCCP.

CCCP is, however, also effective at stalling the H_2 production when Hyd-1 and Hyd-2 are both absent, albeit only at higher concentrations, but independent of the given carbon source. Hence, a secondary effect, exclusively based on FHL activity, must occur. When, in an unrelated study, a proteorhodopsin was expressed in *E. coli* cells, the H_2 production from FHL was increased upon light exposure, which initiates proton-pumping activity of the proteorhodopsin [29] and supports the observed *pmf*-consumption during H_2-evolution observed here. We can exclude a direct inhibition of either the hydrogenase or formate dehydrogenase half-reaction of the FHL complex. When assayed independently of one another with benzyl viologen redox dye, crude extracts from strain CP734 showed a hydrogenase activity of 2.43 ± 0.48 U mg^{-1} in the absence and 2.36 ± 0.42 U mg^{-1} in the presence of CCCP. The formate dehydrogenase activity was similarly unaltered with an activity of 0.26 ± 0.09 U mg^{-1} without and 0.27 ± 0.07 U mg^{-1} with CCCP present.

2.3. CCCP Promotes the Reverse FHL Reaction

Due to the inhibitory effect of CCCP addition on the H_2 production of FHL, we investigated its effect on the reverse FHL reaction, which can be initiated if resting cells are incubated under elevated pressure of H_2 and CO_2 gas mix, as exploited before [8,9]. The reverse reaction was tested in the presence of different concentrations of CCCP, and the resulting formate concentration secreted by the cell suspensions at the equilibrium state was subsequently analyzed using HPLC (Figure 4). While cells produced 23 mM formate in the absence of CCCP in a strain that lacked Hyd-1 and Hyd-2, no formate formation was observed in the parental strain MC4100 under these conditions. The amount of formate initially decreased in the presence of 1, 10 and 50 μM CCCP, was then similar to no addition in the presence of 75 μM CCCP and finally increased to 180 mM formate when 500 μM CCCP was added. The parental strain MC4100 only showed formate production when 200 μM CCCP or more was present but eventually produced 157 mM formate in the presence of 500 μM CCCP (Figure 4). This difference between the two strains clearly showed that when Hyd-1 and Hyd-2 are present (strain MC4100), H_2 oxidation by these enzymes is the preferred route for H_2 oxidation to the FHL reverse reaction. Generally, an increase of the CCCP concentration resulted in an increase of formate production. If the CCCP effect was *pmf*-mediated, then this could be interpreted as CCCP-mediated relief of the *pmf* counter-pressure during the reverse FHL reaction. An alternative explanation is provided by the proton symport of the formate channel FocA during export, which would be enhanced in the presence of CCCP, thus removing cytoplasmic formate from the reaction and, therefore, shifting the equilibrium of the FHL reaction towards formate production.

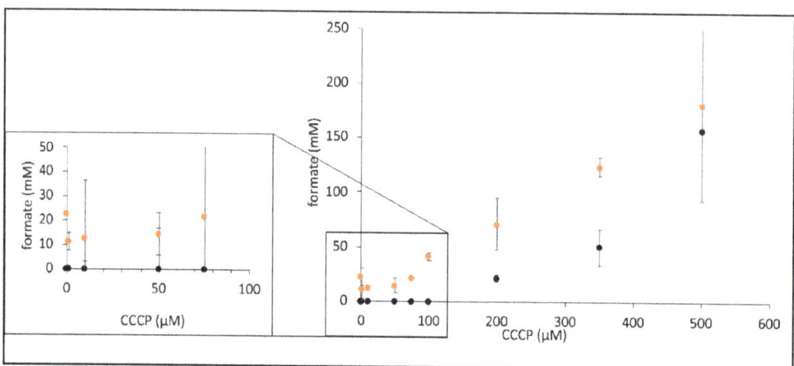

Figure 4. CCCP-dependent increase in formate synthesis by FHL. Strains MC4100 (parental, black dots) and CP734 (Δ*hyaB* Δ*hybC*, orange dots) were triggered for formate production in the presence of different concentration of CCCP as indicated (described in the Materials and Methods section). Each dot represents the average of three biological replicates with standard deviation.

2.4. The DTT Reversal of the CCCP Effect Is Not Mediated by Hyd-2 Activity

It has been observed that the effects of CCCP can be generally relieved by addition of dithiothreitol (DTT) [30] and also specifically restore the activity of FHL with DTT after CCCP inhibition [17]. Initially, it was verified that the addition of DTT on its own had no effect on the H_2 production rate of MC4100 (Figure 5A). Once the H_2 production was CCCP-inhibited, the addition of DTT was able to gradually restore H_2 production up to the rate it had been before CCCP addition (Figure 5A). We also observed that the DTT relief of CCCP inhibition is independent of the strain used, indicating that the presence of Hyd-2 or Hyd-1 is not essential for the reversibility. Furthermore, CCCP was unable to inhibit H_2 production when it was already pre-incubated with DTT (red curve in Figure 5A).

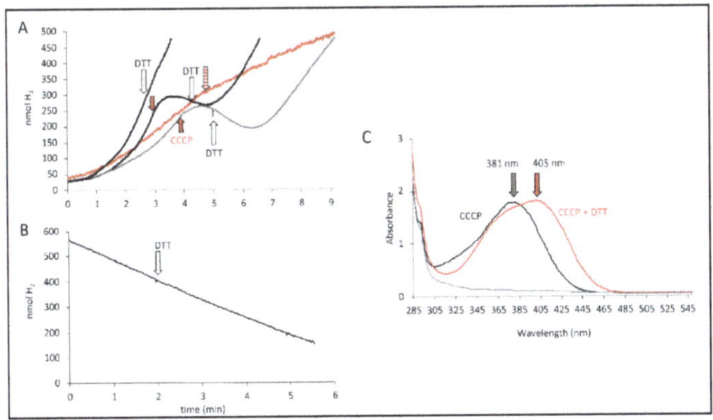

Figure 5. DTT effect. (**A**) Electrode traces of MC4100 cells producing H_2 from either glucose (grey curves) or formate (black curves). CCCP addition (red arrow, 100 μM) and DTT addition (white arrow, 5 mM) are indicated. The red trace shows the combined addition of DTT and CCCP (striped red arrow). (**B**) Electrode trace of strain HDK103 (Δ*hyaB* Δ*hycA-I*) showing the activity of Hyd-2 as H_2 oxidation of H_2-saturated buffer after the addition of 15 mM fumarate. The white arrow indicates the addition of 5 mM DTT. (**C**) Absorption spectra of CCCP solution (black), CCCP solution after DTT addition (red) and DTT alone (grey). The arrows indicate the respective maxima.

We were recently able to show that Hyd-2 contributes to *pmf* generation in an electrogenic fashion that is susceptible to CCCP addition [2,27]. If CCCP addition substantially increased Hyd-2 activity to the amount where it outcompeted the H_2 production of the FHL complex and thus reduced the overall H_2 production, then the DTT addition would inhibit Hyd-2 H_2-oxidizing activity and consequently restore overall H_2 production. Strain HDK103 (Δ*hyaB* Δ*hycA-I*), which only synthesizes active Hyd-2 enzymes, was used to investigate whether DTT addition influenced the H_2-oxidizing activity of Hyd-2. H_2-saturated buffer was applied to the electrode, and cell suspension was added before the reaction was initiated with 15 mM fumarate as electron acceptor and the H_2 trace recorded (Figure 5B). The addition of 5 mM DTT is indicated but did not inhibit H_2 oxidation. Hence, DTT did not inhibit the H_2-oxidizing activity of Hyd-2, and restoration of overall H_2 production is caused otherwise.

It is noteworthy that a redshift of the CCCP spectrum in the presence of DTT was observed, where the absorbance maximum shifts from 381 nm to 405 nm (Figure 5C). The reaction of CCCP derivatives with thiols has been characterized [31] and showed that they are inactivated in the reaction in the presence of DTT. Taken together, this supports the observation made above that the observed effects of CCCP are not merely due to its protonophore activity but rather could be attributed to its interaction with thiols from cysteine residues, possibly those within the FHL complex. Cysteine residues are essential for electron transfer within the FHL complex by coordinating the metal cofactors. Recently, the number of free thiols was determined and was reduced in strains unable to synthesize a functional FHL complex [32]. However, an inactivation of CCCP by DTT and re-formation of the *pmf* by the cell can also not be excluded.

2.5. *Effect of Monovalent Cations on FHL Reactions*

CCCP has been described to collapse both the ΔpH and the transmembrane electrical gradient (ΔΨ) component of the *pmf* [33]. It has been observed that FHL is more active and the *hyc*-genes more strongly expressed under acidic conditions [34,35]. While a ΔpH collapse would increase the availability of protons to the hydrogenase active site in the cytoplasm, and hence increase substrate availability, the $\Delta\Psi$ collapse represents reduction of the outer positive charge and inner negative charge of the membrane. In order to investigate the possibility that the apparent *pmf* dependence could be due to the $\Delta\Psi$ collapse mediated by CCCP, we analyzed the effects of cation addition on H_2 production.

Kinetic experiments on the electrode showed reduction of the H_2 production rate when increasing concentrations of Na^+ ions were present. The rates were reduced by 27% in strain CP734 in the presence of 200 mM sodium compared to in its absence (Figure 6A). When 100 µM CCCP was added to any of the reactions, the reduction was between 40%–50%, regardless of the sodium ion concentration present (Figure 6A, grey dots). Therefore, the presence of NaCl did not enhance the effect of CCCP in this strain.

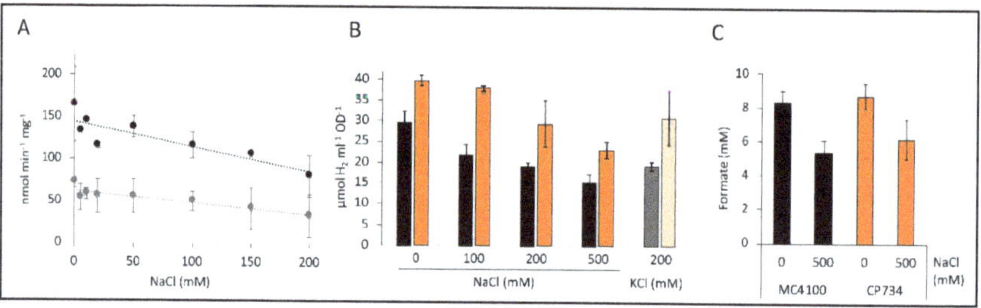

Figure 6. Effect of cations on H_2 production. Panel (**A**) shows the kinetic H_2 production rates of CP734 cells in the presence of different concentrations of NaCl (black dots) and in combination with 100 µM CCCP (gray dots). Reactions were initiated with 50 mM formate. Each point represents the average of two independent biological experiments with the respective standard deviation. The trend lines were calculated using linear regression. Panel (**B**) shows the H_2 content in the headspace of a 7 mL culture tube of MC4100 (black graphs) or CP734 (Δ*hyaB* Δ*hybC*, orange graphs) after anaerobic overnight growth in M9-minimal medium with 0.8% (*w*/*v*) glucose in the presence of NaCl or KCl as indicated. (**C**) The formate concentration of cultures from panel B was quantified using HPLC, as described in the Materials and Methods section. All values represent the average of three independent biological replicates and their standard deviations.

The net charge of the ions remains unaltered through the addition of NaCl or KCl, because the chloride counterion is present. However, the cation/H^+ antiporters counteract the increased salt concentrations that challenge the cell and concomitantly excrete excess cations, which is accompanied by an increased H^+ influx. It was observed that $\Delta\Psi$ decreased over the first 15–20 min after addition of KCl or NaCl [36], hence within the time-frame where H_2 production was monitored. Apparently, it is this disturbance of the electrical gradient that is detrimental to H_2 production by FHL.

Similarly, when cells were already grown in the presence of different concentrations of Na^+ or K^+ ions and subsequently the H_2 content in the headspace sampled by gas chromatography (GC), which indirectly reflects the activity of the FHL complex, a reduction in the presence of increasing concentrations of Na^+ ions could be observed (Figure 6B). Strain CP734 (Δ*hyaB* Δ*hybC*) showed an increased H_2 accumulation by 30%–70% compared to its parent MC4100 under all conditions. This effect is due to the absence of Hyd-1 and Hyd-2 enzymes, which oxidize part of the produced H_2. The H_2 content of strain MC4100 was reduced by 36% and 48% in the presence of 200 mM and 500 mM NaCl, respectively, while strain CP734 showed a similar reduction of 26% and 42% in the presence of 200 and 500 mM NaCl, respectively. This further supports the notion that H_2 oxidation, and hence the activity of the H_2-oxidizing enzymes, was not increased in response to NaCl. To verify whether the effect of Na^+ ions was specific, 200 mM K^+ ions were used instead and showed an identical reduction of the H_2 content by 35% and 22% for MC4100 and CP734, respectively. Therefore, the reduction was not specific for a particular cation. However, Na^+-transport was shown to be K^+ dependent in *E. coli*, albeit not by direct exchange but rather is H^+-antiport coupled [37]. This interconnection of the two cation transport systems could explain the similar effect of both cations.

Headspace H_2 derives from supplemented glucose given to the cells during growth, which is converted to formate as one of the products of mixed-acid fermentation. Elevated concentrations of 500 mM Na^+-ions reduced the amount of produced H_2 (Figure 6B). In order to identify the fate of the carbon, the formate concentration was determined. The cultures had reduced formate amounts by 35% and 29% for MC4100 and CP734, respectively (Figure 6C). Other key metabolites of the mixed acid fermentation like succinate and acetate were similarly reduced under high salt conditions (data not shown). At the same time, the growth of the cultures was reduced by only 15% and was therefore not the main reason for the reduced formate production. This result indicates that under high salt conditions formate is not accumulated because of FHL inhibition, but rather, more generally, the entire mixed-acid fermentation was reduced.

2.6. *The Na^+ Ionophore EIPA Enhances CCCP Inhibition*

The sodium concentration of the cell is maintained by Na^+/H^+ antiporters that reside in the cell membrane and also function in regulation of the intracellular pH. *E. coli* synthesizes three MRP-type antiporter proteins (NhaA, NhaB and ChaA), which are responsible for Na^+/H^+ exchange [38]. Intriguingly, this class of Na^+/H^+ antiporters is homologous to the HycC protein of the FHL membrane domain and to the NuoLMN (*E. coli* nomenclature) proteins, the proton-pumping subunits of the NADH-oxidizing complex I [39]. Some of these antiporters, especially those homologous subunits within the membrane domain of respiratory complex I, can be inhibited by EIPA (5-(*N*-ethyl-*N*-isopropyl)-amiloride) [40]. *Klebsiella pneumoniae* complex I was shown to translocate Na^+ ions instead of protons and some evidence for Na^+ translocation of *E. coli* complex I exists, but this might merely be a secondary effect [41,42]. In addition, in thermophilic organisms like *Thermococcus onnurineus*, energy conservation functions in an intricate mechanism that couples formate oxidation with concomitant H_2 production for the generation of a proton gradient. This proton gradient is subsequently converted into a Na^+ gradient by a Na^+/H^+-antiporter and then employed by a Na^+ ATPase for ATP synthesis [43].

Due to the observed reduction of H_2 production by NaCl, we tested the effect EIPA had on the H_2 production of FHL and found that, on its own or combined with 200 mM NaCl, the effects were negligible as described before [8]. Surprisingly, however, when EIPA and CCCP were given shortly after each other, they were able to inhibit H_2 production completely (Figure 7A). A similar effect was also determined for strain CP734 (Δ*hyaB* Δ*hybC*, Figure 7B); however, inhibition of H_2 production was not as complete as for MC4100. EIPA on its own also had no effect (Figure 7B). Notably, inhibition was also not complete for MC4100 when the reaction was started with formate instead of glucose (Figure 7C), and no effect was seen in strain CP734 (Δ*hyaB* Δ*hybC*) when formate was used to initiate the H_2 production (Figure 7D). This indicates that the activity of the Na^+/H^+ antiporter plays a crucial role during H_2 production from glucose. However, the MRP-type membrane subunit HycC of the FHL complex, which is essential for the reaction, is not inhibited by EIPA. Taken together, this could imply that a H^+ gradient is converted into a Na^+ gradient at the cell membrane during catalysis and that the FHL reaction couples both. Further experiments will be necessary to find the target of the ionophores.

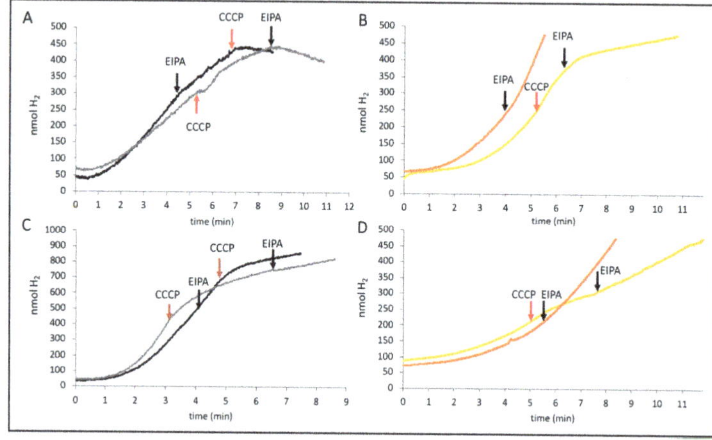

Figure 7. Effect of EIPA combination with CCCP. Cells of MC4100 (black, panels (**A**) and (**C**)) or CP734 (orange, panels (**B**) and (**D**)) were grown anaerobically in TGYEP, pH 6.5, and applied to the electrode as described in the Materials and Methods section. The reaction was initiated with glucose (top panels (**A**) and (**B**)) or formate (bottom panels (**C**) and (**D**)). Where indicated, 50 μM CCCP (red arrows) or 50 μM EIPA (black arrows) were added to the reaction.

3. Materials and Methods

3.1. Strains and Growth Conditions

Strains are listed in Table 3. Strains CP630 and CP631 were constructed using P1*kc*-mediated phage transduction of the Δ*hyaB* and Δ*hybC* alleles of the Keio-collection strains JW0955 and JW2962, respectively, as described [44]. Strains were routinely grown aerobically in LB medium or on LB agar plates at 37 °C. For analysis of FHL activity, the strains were grown anaerobically as standing liquid cultures for 16 h in 50 mL TGYEP medium, pH 6.5 (1% (*w*/*v*) tryptone, 0.5% (*w*/*v*) yeast extract, 0.1 M potassium phosphate buffer, pH 6.5, 0.8% (*w*/*v*) glucose) at 30 °C [45]. Cultures for H_2 headspace and HPLC analyses were grown in phosphate-buffered M9-minimal medium supplemented with 0.8% (*w*/*v*) glucose as described [46]. When indicated, sterile solutions of NaCl or KCl were added. CCCP was used at 100 μM unless otherwise indicated. Cultures were harvested by centrifugation at 4000× g for 15 min at 4 °C, resuspended in 800 μL 50 mM Tris/HCl, pH 7.0, and kept on ice until further use.

Table 3. Strain list.

Strain	Genotype	Reference
MC4100	F⁻ *araD139* Δ(*argF-lac*)*U169* λ *rpsL150 relA1 deoC1 flhD5301* Δ(*fruK-yeiR*)*725*(*fruA25*), *rbsR22*, Δ(*fimB-fimE*)*632*(::IS*1*)	[47]
CP630	Like MC4100, but Δ*hyaB*	This study
CP631	Like MC4100, but Δ*hybC*	This study
CP734	Like MC4100, but Δ*hyaB* Δ*hybC*	[24]
HDK103	Like MC4100, but Δ*hyaB* Δ*hycA-I*	[48]

3.2. Enzymatic Assays

The kinetic activity of the FHL complex was determined on a modified Clark-type electrode equipped with an OXY/ECU module (Oxytherm, Hansatech Instruments, Norfolk, UK) to reverse the polarizing voltage to −0.7 V. A volume of 2 mL degassed 50 mM Tris/HCl buffer, pH 7.0, at 30 °C was added to the chamber prior to adding 50 μL of cell suspension. The reaction was started either with 50 mM formate or with 14 mM glucose. When the effect of sodium ions was studied, the ammonium salt of formate was used to initiate the reaction, which increased the pH by 0.02 units. The amount of H_2 was calibrated as described [49]. EC_{50} values were calculated with Origin Pro 2017G software (OriginLab, Northampton, MA, USA). The protein content of the respective cell suspensions was determined using the method of Lowry in a micro-scale assay [50]. Optical densities and spectra were recorded with a Tecan plate reader.

The H_2 content of the gas headspace of a 15-mL Hungate tube filled with 7 mL of culture was determined using gas chromatography with a GC2010 Plus Gas Chromatograph (Shimadzu, Kyōto, Japan) as described [39].

For the reverse FHL reaction, the cells were initially grown anaerobically in TGYEP, pH 6.5, for 16 h at 30 °C, harvested and resuspended in 50 mM Tris/HCl, pH 7.5. Cell suspensions were further mixed with different concentrations of CCCP, as indicated, and incubated under 2 atm pressure of H_2 and CO_2 1:1 mixture for 16 h at 37 °C. Cells were then removed by filtration through a 0.2 μM filter and supernatants applied to an HPLC system equipped with an Aminex HPX-87H column (Bio-Rad, Hercules, CA, USA), and formate concentrations were determined as previously described [8]. The culture supernatant was centrifuged and subsequently passed through a 0.2 μM filter prior to loading onto the HPLC system.

Calorimetric assays of FdhH activity and hydrogenase activity were carried out with crude extracts from anaerobically grown cells. The cells were harvested, sonicated for 2 min at 20 W with 0.5 s pulses and briefly centrifuged. Anaerobic cuvettes were prepared containing 0.8 mL 50 mM Tris/HCl buffer, pH 7.0, and 4 mM benzyl viologen. The FdhH reaction was started with 15 mM formate, and the hydrogenase reaction was started by adding crude extract to the cuvettes after exchange of the N_2 headspace with H_2. The signal at 600 nm was recorded, and an E_M value of 7400 M^{-1} cm^{-1} was assumed for reduced benzyl viologen. Data were derived from three independent biological replicates. Protein concentration was determined using the Lowry method [50].

4. Conclusions

The data presented here confirm CCCP-dependent inhibition of FHL activity. Here, we quantified this effect for the first time and saw striking differences depending on the presence of active H_2-oxidizing Hyd enzymes. The CCCP-dependent inhibition showed clearly that FHL did not contribute to proton translocation across the membrane, and in contrast, the data suggest it is driven by proton influx. Nevertheless, the data further highlight that the effect of CCCP might not be due to its protonophore activity but might rather be indirect, either by interacting with thiol groups within the complex or by disturbing the charge distribution at the membrane. The absence of an effect of another protonophore on the H_2 production further supports this finding. The sodium/potassium inhibition of the H_2 production showed that in order for it to function most effectively, the FHL complex requires low external cation concentrations. Our data clearly suggests that H_2 production couples H^+ influx with Na^+ efflux. However, evidence shows that it is not the MRP antiporter subunit HycC of the FHL complex that is directly involved, but rather it is the cation/H^+ antiport of the membrane that influences H_2 production.

Author Contributions: C.P. conceived the study; J.T.M. and C.P. collected and analyzed the data, C.P. wrote the original draft and supervised J.T.M. All authors have read and agreed to the published version of the manuscript.

Funding: We acknowledge the financial support within the funding program Open Access Publishing by the German Research Foundation (DFG).

Acknowledgments: We are grateful to Julia Fritz-Steuber (University of Hohenheim, Germany) and Etana Padan (Hebrew University of Jerusalem, Israel), who kindly provided strain EP432. We thank Gary Sawers (Martin-Luther University Halle-Wittenberg, Germany) for fruitful discussions and support in writing this manuscript.

Conflicts of Interest: The authors declare no conflict of interest.

References

1. Pinske, C. Bioenergetic aspects of archaeal and bacterial hydrogen metabolism. *Adv. Microb. Physiol.* **2019**, *74*, 487–514.

2. Lubek, D.; Simon, A.H.; Pinske, C. Amino acid variants of the HybB membrane subunit of *Escherichia coli* [NiFe]-hydrogenase-2 support a role in proton transfer. *FEBS Lett.* **2019**, *156*, 2194–2203. [CrossRef]
3. Pinske, C.; Sawers, R.G. Anaerobic formate and hydrogen metabolism. *EcoSal Plus* **2016**, *7*, ESP-0011-2016. [CrossRef]
4. Odom, J.M.; Peck, H.D. Hydrogen cycling as a general mechanism for energy coupling in the sulfate-reducing bacteria, *Desulfovibrio* sp. *FEMS Microbiol. Lett.* **1981**, *12*, 47–50. [CrossRef]
5. Wiechmann, A.; Ciurus, S.; Oswald, F.; Seiler, V.N.; Müller, V. It does not always take two to tango: "Syntrophy" via hydrogen cycling in one bacterial cell. *ISME J.* **2020**, *14*, 1561–1570. [CrossRef]
6. Sawers, R.G. Formate and its role in hydrogen production in *Escherichia coli*. *Biochem. Soc. Trans.* **2005**, *33*, 42–46. [CrossRef]
7. Sauter, M.; Böhm, R.; Böck, A. Mutational analysis of the operon (*hyc*) determining hydrogenase 3 formation in *Escherichia coli*. *Mol. Microbiol.* **1992**, *6*, 1523–1532. [CrossRef] [PubMed]
8. Pinske, C.; Sargent, F. Exploring the directionality of *Escherichia coli* formate hydrogenlyase: A membrane-bound enzyme capable of fixing carbon dioxide to organic acid. *Microbiologyopen* **2016**, *5*, 721–737. [CrossRef] [PubMed]
9. Roger, M.; Brown, F.; Gabrielli, W.; Sargent, F. Efficient hydrogen-dependent carbon dioxide reduction by *Escherichia coli*. *Curr. Biol.* **2018**, *28*, 140–145. [CrossRef]
10. Redwood, M.D.; Mikheenko, I.P.; Sargent, F.; Macaskie, L.E. Dissecting the roles of *Escherichia coli* hydrogenases in biohydrogen production. *FEMS Microbiol. Lett.* **2008**, *278*, 48–55. [CrossRef]
11. Andrews, S.C.; Berks, B.C.; McClay, J.; Ambler, A.; Quail, M.A.; Golby, P.; Guest, J.R. A 12-cistron *Escherichia coli* operon (*hyf*) encoding a putative proton-translocating formate hydrogenlyase system. *Microbiology* **1997**, *143*, 3633–3647. [CrossRef] [PubMed]
12. Böhm, R.; Sauter, M.; Böck, A. Nucleotide sequence and expression of an operon in *Escherichia coli* coding for formate hydrogenlyase components. *Mol. Microbiol.* **1990**, *4*, 231–243. [CrossRef] [PubMed]
13. Marreiros, B.C.; Batista, A.P.; Duarte, A.M.S.; Pereira, M.M. A missing link between complex I and group 4 membrane-bound [NiFe] hydrogenases. *Biochim. Biophys. Acta* **2012**, *1827*, 198–209. [CrossRef]
14. Welte, C.; Krätzer, C.; Deppenmeier, U. Involvement of Ech hydrogenase in energy conservation of *Methanosarcina mazei*. *FEBS J.* **2010**, *277*, 3396–3403. [CrossRef] [PubMed]
15. McDowall, J.S.; Murphy, B.J.; Haumann, M.; Palmer, T.; Armstrong, F.A.; Sargent, F. Bacterial formate hydrogenlyase complex. *Proc. Natl. Acad. Sci. USA* **2014**, *111*, E3948–E3956. [CrossRef] [PubMed]
16. Sokol, K.P.; Robinson, W.E.; Oliveira, A.R.; Zacarias, S.; Lee, C.-Y.; Madden, C.; Bassegoda, A.; Hirst, J.; Pereira, I.A.C.; Reisner, E. Reversible and selective interconversion of hydrogen and carbon dioxide into formate by a semiartificial formate hydrogenlyase mimic. *J. Am. Chem. Soc.* **2019**, *141*, 17498–17502. [CrossRef]
17. Bagramyan, K.A.; Martirosov, S.M. Formation of an ion transport supercomplex in *Escherichia coli*. An experimental model of direct transduction of energy. *FEBS Lett.* **1989**, *246*, 149–152. [CrossRef]
18. Babu, M.; Bundalovic-Torma, C.; Calmettes, C.; Phanse, S.; Zhang, Q.; Jiang, Y.; Minic, Z.; Kim, S.; Mehla, J.; Gagarinova, A.; et al. Global landscape of cell envelope protein complexes in *Escherichia coli*. *Nat. Biotechnol.* **2018**, *36*, 103–112. [CrossRef]
19. Trchounian, A.; Sawers, R.G. Novel insights into the bioenergetics of mixed-acid fermentation: Can hydrogen and proton cycles combine to help maintain a proton motive force? *IUBMB Life* **2014**, *66*, 1–7. [CrossRef]
20. Doberenz, C.; Zorn, M.; Falke, D.; Nannemann, D.; Hunger, D.; Beyer, L.; Ihling, C.H.; Meiler, J.; Sinz, A.; Sawers, R.G. Pyruvate formate-lyase interacts directly with the formate channel FocA to regulate formate translocation. *J. Mol. Biol.* **2014**, *426*, 2827–2839. [CrossRef]
21. Beyer, L.; Doberenz, C.; Falke, D.; Hunger, D.; Suppmann, B.; Sawers, R.G. Coordination of FocA and pyruvate formate-lyase synthesis in *Escherichia coli* demonstrates preferential translocation of formate over other mixed-acid fermentation products. *J. Bacteriol.* **2013**, *195*, 1428–1435. [CrossRef]
22. Hakobyan, M.; Sargsyan, H.; Bagramyan, K. Proton translocation coupled to formate oxidation in anaerobically grown fermenting *Escherichia coli*. *Biophys. Chem.* **2005**, *115*, 55–61. [CrossRef]
23. Mayer, C.; Boos, W. Hexose/Pentose and Hexitol/Pentitol Metabolism. *EcoSal Plus* **2005**, *1*, 1. [CrossRef] [PubMed]
24. Pinske, C.; Krüger, S.; Soboh, B.; Ihling, C.; Kuhns, M.; Braussemann, M.; Jaroschinsky, M.; Sauer, C.; Sargent, F.; Sinz, A.; et al. Efficient electron transfer from hydrogen to benzyl viologen by the [NiFe]-hydrogenases of *Escherichia coli* is dependent on the coexpression of the iron-sulfur cluster-containing small subunit. *Arch. Microbiol.* **2011**, *193*, 893–903. [CrossRef]

25. Laurinavichene, T.V.; Tsygankov, A.A. H_2 consumption by *Escherichia coli* coupled via hydrogenase 1 or hydrogenase 2 to different terminal electron acceptors. *FEMS Microbiol. Lett.* **2001**, *202*, 121–124. [CrossRef] [PubMed]
26. Sawers, R.G.; Ballantine, S.; Boxer, D. Differential expression of hydrogenase isoenzymes in *Escherichia coli* K-12: Evidence for a third isoenzyme. *J. Bacteriol.* **1985**, *164*, 1324–1331. [CrossRef] [PubMed]
27. Pinske, C.; Jaroschinsky, M.; Linek, S.; Kelly, C.L.; Sargent, F.; Sawers, R.G. Physiology and bioenergetics of [NiFe]-hydrogenase 2-catalyzed H_2-consuming and H_2-producing reactions in *Escherichia coli*. *J. Bacteriol.* **2015**, *197*, 296–306. [CrossRef]
28. Laurinavichene, T.V.; Zorin, N.A.; Tsygankov, A.A. Effect of redox potential on activity of hydrogenase 1 and hydrogenase 2 in *Escherichia coli*. *Arch. Microbiol.* **2002**, *178*, 437–442. [CrossRef]
29. Kuniyoshi, T.M.; Balan, A.; Schenberg, A.C.G.; Severino, D.; Hallenbeck, P.C. Heterologous expression of proteorhodopsin enhances H_2 production in *Escherichia coli* when endogenous Hyd-4 is overexpressed. *J. Biotechnol.* **2015**, *206*, 52–57. [CrossRef]
30. Ridgway, H.F. Source of energy for gliding motility in *Flexibacter polymorphus*: Effects of metabolic and respiratory inhibitors on gliding movement. *J. Bacteriol.* **1977**, *131*, 544–556. [CrossRef]
31. Šturdík, E.; Antalík, M.; Sulo, P.; Baláž, Š.; Ďurčová, E.; Drobnica, Ľ. Relationships among structure, reactivity towards thiols and basicity of phenylhydrazonopropanedinitriles. *Collect. Czechoslov. Chem. Commun.* **1985**, *50*, 2065–2076. [CrossRef]
32. Gevorgyan, H.; Trchounian, A.; Trchounian, K. Formate and potassium ions affect *Escherichia coli* proton ATPase activity at low pH during mixed carbon fermentation. *IUBMB Life* **2020**, *72*, 915–921. [CrossRef] [PubMed]
33. Akiyama, Y. Proton-motive force stimulates the proteolytic activity of FtsH, a membrane-bound ATP-dependent protease in Escherichia coli. *Proc. Natl. Acad. Sci. USA* **2002**, *99*, 8066–8071. [CrossRef]
34. Rossmann, R.; Sawers, R.G.; Böck, A. Mechanism of regulation of the formate-hydrogenlyase pathway by oxygen, nitrate, and pH: Definition of the formate regulon. *Mol. Microbiol.* **1991**, *5*, 2807–2814. [CrossRef]
35. Bagramyan, K.; Mnatsakanyan, N.; Poladian, A.; Vassilian, A.; Trchounian, A. The roles of hydrogenases 3 and 4, and the F_0F_1-ATPase, in H_2 production by *Escherichia coli* at alkaline and acidic pH. *FEBS Lett.* **2002**, *516*, 172–178. [CrossRef]
36. Kashket, E.R. Effects of K^+ and Na^+ on the proton motive force of respiring *Escherichia coli* at alkaline pH. *J. Bacteriol.* **1985**, *163*, 423–429. [CrossRef] [PubMed]
37. Verkhovskaya, M.L.; Verkhovsky, M.I.; Wikström, M. K^+-dependent Na^+ transport driven by respiration in *Escherichia coli* cells and membrane vesicles. *Biochim. Biophys. Acta* **1996**, *1273*, 207–216. [CrossRef]
38. Padan, E.; Landau, M. Sodium-proton (Na^+/H^+) antiporters: Properties and roles in health and disease. *Met. Ions Life Sci.* **2016**, *16*, 391–458.
39. Friedrich, T.; Scheide, D. The respiratory complex I of bacteria, archaea and eukarya and its module common with membrane-bound multisubunit hydrogenases. *FEBS Lett.* **2000**, *479*, 1–5. [CrossRef]
40. Steuber, J. The C-terminally truncated NuoL subunit (ND5 homologue) of the Na^+-dependent complex I from *Escherichia coli* transports Na^+. *J. Biol. Chem.* **2003**, *278*, 26817–26822. [CrossRef] [PubMed]
41. Steuber, J.; Schmid, C.; Rufibach, M.; Dimroth, P. Na^+ translocation by complex I (NADH:quinone oxidoreductase) of *Escherichia coli*. *Mol. Microbiol.* **2000**, *35*, 428–434. [CrossRef]
42. Stolpe, S.; Friedrich, T. The *Escherichia coli* NADH:ubiquinone oxidoreductase (complex I) is a primary proton pump but may be capable of secondary sodium antiport. *J. Biol. Chem.* **2004**, *279*, 18377–18383. [CrossRef] [PubMed]
43. Kim, Y.J.; Lee, H.S.; Kim, E.S.; Bae, S.S.; Lim, J.K.; Matsumi, R.; Lebedinsky, A.V.; Sokolova, T.G.; Kozhevnikova, D.A.; Cha, S.-S.; et al. Formate-driven growth coupled with H_2 production. *Nature* **2010**, *467*, 352–355. [CrossRef]
44. Baba, T.; Ara, T.; Hasegawa, M.; Takai, Y.; Okumura, Y.; Baba, M.; Datsenko, K.; Tomita, M.; Wanner, B.; Mori, H. Construction of *Escherichia coli* K-12 in-frame, single-gene knockout mutants: The Keio collection. *Mol. Syst. Biol.* **2006**, *2*, 2006.0008. [CrossRef]
45. Begg, Y.A.; Whyte, J.N.; Haddock, B.A. The identification of mutants of *Escherichia coli* deficient in formate dehydrogenase and nitrate reductase activities using dye indicator plates. *FEMS Microbiol. Lett.* **1977**, *2*, 47–50. [CrossRef]

46. Sambrook, J.; Russell, D. *Molecular Cloning: A Laboratory Manual*, 3rd ed.; Cold Spring Harbor Laboratory Press: Cold Spring Harbor, NY, USA, 2001.
47. Casadaban, M.J. Transposition and fusion of the *lac* genes to selected promoters in *Escherichia coli* using bacteriophage lambda and Mu. *J. Mol. Biol.* **1976**, *104*, 541–555. [CrossRef]
48. Jacobi, A.; Rossmann, R.; Böck, A. The *hyp* operon gene products are required for the maturation of catalytically active hydrogenase isoenzymes in *Escherichia coli. Arch. Microbiol.* **1992**, *158*, 444–451. [CrossRef]
49. Sargent, F.; Stanley, N.R.; Berks, B.C.; Palmer, T. Sec-independent protein translocation in *Escherichia coli.* A distinct and pivotal role for the TatB protein. *J. Biol. Chem.* **1999**, *274*, 36073–36082. [CrossRef]
50. Lowry, O.; Rosebrough, N.; Farr, A.; Randall, R. Protein measurement with the Folin phenol reagent. *J. Biol. Chem.* **1951**, *193*, 265–275.

Article

Crystal Structures of [Fe]-Hydrogenase from *Methanolacinia paynteri* Suggest a Path of the FeGP-Cofactor Incorporation Process

Gangfeng Huang [1], Francisco Javier Arriaza-Gallardo [1], Tristan Wagner [1,2] and Seigo Shima [1,*]

1 Microbial Protein Structure Group, Max Planck Institute for Terrestrial Microbiology, Karl-von-Frisch-Straße 10, 35043 Marburg, Germany; gangfeng.huang@mpi-marburg.mpg.de (G.H.); francisco.arriaza@mpi-marburg.mpg.de (F.J.A.-G.); twagner@mpi-bremen.de (T.W.)

2 Max Planck Research Group Microbial Metabolism, Max Planck Institute for Marine Microbiology, 28359 Bremen, Germany

* Correspondence: shima@mpi-marburg.mpg.de; Tel.: +49-6421-178100

Received: 18 August 2020; Accepted: 15 September 2020; Published: 17 September 2020

Abstract: [Fe]-hydrogenase (Hmd) catalyzes the reversible heterolytic cleavage of H_2, and hydride transfer to methenyl-tetrahydromethanopterin (methenyl-H_4MPT^+). The iron-guanylylpyridinol (FeGP) cofactor, the prosthetic group of Hmd, can be extracted from the holoenzyme and inserted back into the protein. Here, we report the crystal structure of an asymmetric homodimer of Hmd from *Methanolacinia paynteri* (pHmd), which was composed of one monomer in the open conformation with the FeGP cofactor (holo-form) and a second monomer in the closed conformation without the cofactor (apo-form). In addition, we report the symmetric pHmd-homodimer structure in complex with guanosine monophosphate (GMP) or guanylylpyridinol (GP), in which each ligand was bound to the protein, where the GMP moiety of the FeGP-cofactor is bound in the holo-form. Binding of GMP and GP modified the local protein structure but did not induce the open conformation. The amino-group of the Lys150 appears to interact with the 2-hydroxy group of pyridinol ring in the pHmd–GP complex, which is not the case in the structure of the pHmd–FeGP complex. Lys150Ala mutation decreased the reconstitution rate of the active enzyme with the FeGP cofactor at the physiological pH. These results suggest that Lys150 might be involved in the FeGP-cofactor incorporation into the Hmd protein in vivo.

Keywords: [Fe]-hydrogenase; FeGP cofactor; guanylylpyridinol; conformational changes; X-ray crystallography

1. Introduction

[Fe]-hydrogenase (Hmd) catalyzes the reversible hydride transfer to methenyl-tetrahydromethanopterin (methenyl-H_4MPT^+) from H_2 (Figure 1a) [1,2]. This reaction is involved in the hydrogenotrophic methanogenic pathway [2,3]. [Fe]-hydrogenase forms homodimer and contains an active site cleft at the two dimeric interfaces (Figure 1b) [4,5]. The active-site cleft binds the iron-guanylylpyridinol (FeGP) cofactor as the prosthetic group (Figure 1c) [6]. The FeGP cofactor contains a low spin Fe(II), which is ligated with two CO, one acyl-C and pyridinol nitrogen [5,7–12]. This cofactor is covalently bound to the protein via cysteine-S ligand at the iron site [5,10,13] (Figure 1c). The pyridinol ring is substituted with two methyl- and one guanosine monophosphate (GMP) groups [8]. The GMP part is bound to the mononucleotide-binding site of the Rossmann-fold-like structure of the Hmd protein [4].

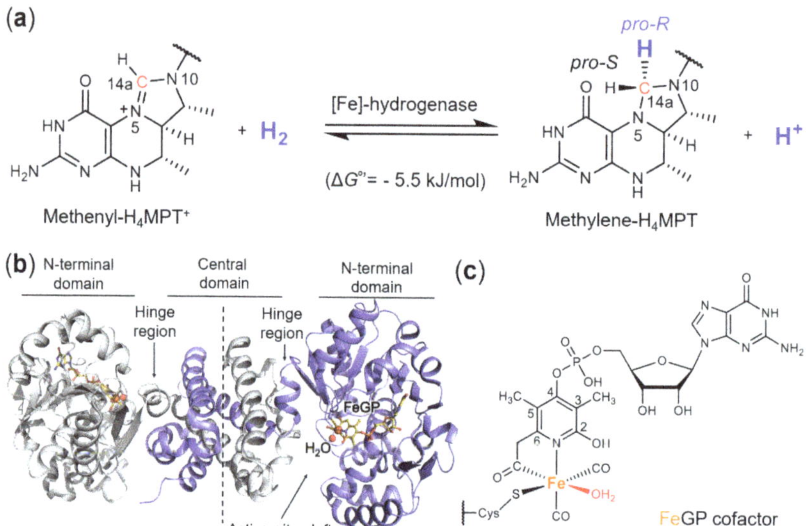

Figure 1. Structure and function of [Fe]-hydrogenase. (**a**) Reaction catalyzed by [Fe]-hydrogenase. The side chain of H_4MPT was omitted. (**b**) Crystal structure of [Fe]-hydrogenase (holoenzyme) from *Methanococcus aeolicus* (PDB: 6HAC). Two monomers are shown as a cartoon model (grey and blue). The FeGP cofactor is depicted as a ball and stick model. A water molecule is bound to the iron in the open resting state. (**c**) Chemical structure of the FeGP cofactor. Cys176 thiolate is covalently bound to the iron of the cofactor.

The FeGP cofactor is extractable from [Fe]-hydrogenase in the presence of 60% methanol, 1-mM 2-mercaptoethanol and 1% NH_3 [14]. The isolated cofactor is stabilized by 2-mercaptoethanol, which makes a complex at the iron site [10,12,15]. The FeGP cofactor can also be isolated in the presence of 50% acetic acid. An acetate ligand binds to the iron site to stabilize the iron complex of the cofactor [12]. The FeGP cofactor is decomposed by UV-A/blue light [16] and hydrogen peroxide [17,18]. A decomposition product of the organic part of the FeGP cofactor is guanylylpyridinol (GP), in which the acyl-methyl substituent was hydrolyzed to a carboxymethyl group [12].

Hmd proteins can be heterologously produced in *Escherichia coli* as an apo-form that does not contain the FeGP cofactor [19]. Crystal structure of the apo-Hmd from *Methanocaldococcus jannaschii* indicated that the active-site cleft of the apoenzyme is in a closed conformation [4]. When the extracted FeGP cofactor is mixed with the apoenzyme, the cofactor binds to the protein in the active-site cleft and generates the active holoenzyme [19]. Crystal structures of holoenzymes without substrate bound have always been observed in an open conformation, in which the active-site cleft is exposed to bulk solvent [5]. The structure of the Hmd from *Methanococcus aeolicus* holoenzyme in complex with methenyl-H_4MPT$^+$ showed that upon binding of methenyl-H_4MPT$^+$, the active-site cleft closes [20]. In the closed tertiary complex, the iron site of the FeGP cofactor is activated by expulsion of the water molecule bound on the iron site. The empty iron coordination site is proposed to be the H_2-binding site [20].

Many mimic complexes of the FeGP cofactor have been synthesized [2,21–26]. Some of the mimics are composed of similar complex structures to the FeGP cofactor containing Fe(II) [27] or Mn(I) [25] as the metal center with two CO and pyridinol but lack the GMP moiety. Several mimic complexes exhibit catalytic activities of H_2 activation and hydride transfer to chemical compounds [22–25]. Recently, reconstituted active semi-synthetic [Fe]-hydrogenases were produced by incorporation of the mimic complexes to the apoenzyme [25,28]. The reconstituted enzyme exhibited only a few percent of the activity of the native enzyme [28]. The reconstitution requires a longer period than that with the FeGP cofactor to achieve full activity [28]. Addition of GMP during the reconstitution of the mimic-complexes increased the enzymatic activity to a certain extent [28].

In this work, we heterologously produced the Hmd apo-form from a mesophilic methanogenic archaeon, *Methanolacinia paynteri* (pHmd, National Center for Biotechnology Information Reference Sequence: WP_048153035.1), which belongs to the *Methanomicrobiales* order. The structures of Hmd from the *Methanomicrobiales* order have not been reported so far. We crystallized pHmd after in vitro reconstitution with the isolated FeGP cofactor from native Hmd purified from *Methanothermobacter marburgensis* [14]. These crystallization trials yielded two unexpected structures. The first one was an asymmetric pHmd homodimer containing one monomer bound with the FeGP cofactor and the other monomer in the apo-form. The second crystal structure was a symmetric homodimer in complex with the broken product of the FeGP cofactor (GP). In addition, co-crystallization experiments of pHmd with GMP and an iron mimic complex (see Section 3) yielded a symmetric homodimer in complex with only GMP. The crystal structures of asymmetric Hmd, and Hmd bound with GMP or GP have not been reported before. Based on the structures, we propose a possible trajectory of the binding process of the FeGP cofactor to the protein to produce the active holoenzyme.

2. Results and Discussion

2.1. Crystal Structure of the Asymmetric Homodimer of pHmd

Reconstitution of the pHmd holoenzyme was performed in the presence of a slight excess of FeGP cofactor (0.18 mM) relative to the pHmd protein (0.13 mM). The specific activity of the reconstituted enzyme was variable in the holo-form obtained in each reconstitution experiment (60–250 U/mg at 40 °C) under standard assay conditions (oxidation of methylene-H_4MPT), which was substantially lower than that of the native Hmd from *M. marburgensis* (~400 U/mg at 40 °C) [14], the reconstituted Hmd from *Methanocaldococcus jannaschii* (jHmd) (~400 U/mg at 40 °C) [14], and the reconstituted Hmd from *Methanococcus aeolicus* (3000 U/mg at 40 °C) [20]. The 2.1 Å crystal structure obtained from this reconstituted preparation contained one homodimer in the asymmetric unit, in which one monomer is bound to the FeGP cofactor and the other is in an apo-form (Figure 2, Table 1). The active-site cleft bound with the FeGP cofactor was in the open conformation and the second monomer without the cofactor was in the closed conformation. The formation of the asymmetric structure indicated that the open/closed conformational change of a monomer of the Hmd homodimer occurs independently from another monomer.

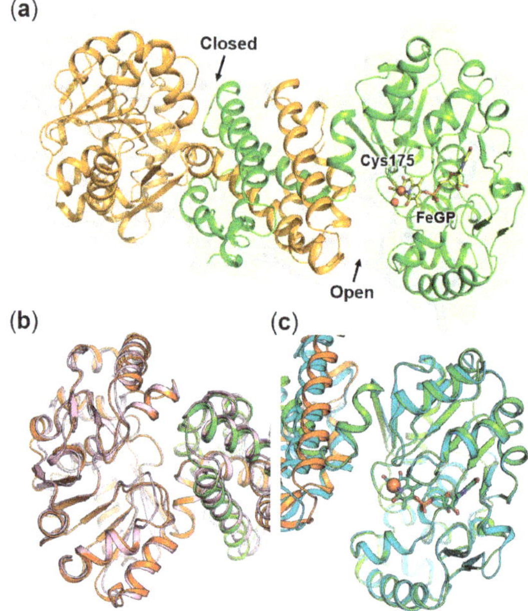

Figure 2. Crystal structure of the asymmetric homodimer of pHmd. (**a**) Overview of the asymmetric homodimer of pHmd. One monomer (left) is the apo-form and the second monomer (right) is bound to the FeGP cofactor. (**b**) Comparisons between the apo-form of the asymmetric pHmd homodimer (orange) and apo-form of jHmd (pink, PDB: 2B0J). The chain from the next monomer of pHmd in the central domain is distinguished by the green color. (**c**) Comparison between the holo-form of the asymmetric pHmd homodimer (green) and holo-form of jHmd (blue, PDB: 3F47). The chain from the next monomer of pHmd in the central domain is distinguished by the orange color. The FeGP cofactor is shown by ball and stick models. For both pictures in (**b**,**c**), the N-terminal domain is superposed.

Table 1. X-ray analysis statistics.

	Asymmetric Homodimer-pHmd	GMP Bound	GP Bound
Data collection			
Wavelength (Å)	0.97980	0.97972	0.97970
Space group	$P3_121$	$P2_1$	C2
Resolution (Å)	45.13–2.10 (2.21–2.10)	43.65–1.55 (1.63–1.55)	46.63–1.70 (1.79–1.70)
Cell dimensions			
a, b, c (Å)	79.48, 79.48, 179.34	67.81, 57.06, 95.13	160.83, 93.27, 83.64
α, β, γ (°)	90, 90, 120	90, 91.43, 90	90, 97.53, 90
Nr. Monomers/asym. Unit	2	2	4
open/closed conformation	1 + 1	0/2	0/4
R_{merge} (%) [a]	5.9 (71.1)	6.8 (35.3)	10.7 (51.7)
R_{pim} (%) [a]	1.9 (22.2)	4.1 (21.3)	6.5 (30.8)
$CC_{1/2}$ [a]	1.000 (0.755)	0.997 (0.653)	0.993 (0.605)
I/σ_I [a]	28.8 (3.6)	12.6 (3.7)	8.0 (2.5)
Completeness (%) [a]	100.0 (100.0)	99.3 (99.7)	99.1 (99.1)
Redundancy [a]	10.8 (11.1)	3.7 (3.7)	3.7 (3.8)
Nr. unique reflections [a]	39,181 (5643)	104,812 (15285)	133,035 (19398)
Refinement			

Table 1. *Cont.*

	Asymmetric Homodimer-pHmd	GMP Bound	GP Bound
Resolution (Å)	39.74–2.10	22.69–1.55	24.89–1.70
Number of reflections	39,126	104,777	132,993
R_{work}/R_{free} [b] (%)	19.5/23.5	18.4/20.9	18.9/21.6
Number of atoms			
Protein	5215	5256	10,490
Ligands/ions	103	78	114
Solvent	224	640	1068
Mean B-value (Å^2)	61.9	22.2	23.1
Molprobity clash score, all atoms	1.49	1.39	0.79
Ramachandran plot			
Favored regions (%)	97.05	96.31	96.31
Outlier regions (%)	0.15	0	0
Rmsd [c] bond lengths (Å)	0.008	0.010	0.010
Rmsd [c] bond angles (°)	1.01	1.10	1.08
PDB code	6YKA	6YKB	6YK9

[a] Values relative to the highest resolution shell are within parentheses. [b] R_{free} was calculated as the R_{work} for 5% of the reflections that were not included in the refinement. [c] rmsd, root mean square deviation.

The N- (residues 1–241) and C- (residues 253–342) terminal domains of the two monomers of the asymmetric homodimer are very similar to each other: 217 Cα superposed with a root mean square deviation (rmsd) of 0.192 Å for the N-terminal domain and 74 Cα superposed with a rmsd of 0.131 Å for the C-terminal domain. The N-terminal domains of the apo- and holo-form monomers of the asymmetric pHmd overlapped with those of apo- and holo-forms of Hmd from *M. jannaschii* (jHmd), respectively (Figure 2b,c). However, when the central domain composed of the dimeric C-terminal domains are superposed, a deviation of the N-terminal domain is observed between pHmd apo- and holo-forms compared to those of apo- and holo-forms of jHmd, respectively (Figure S1). Such variation is caused by the asymmetric structure; the holo conformation impacts the apo conformation at the central domain and vice versa.

According to the similar N-terminal domain structures, the holo-form of pHmd shows an identical binding mode of the FeGP cofactor as observed in the jHmd holoenzyme (Figure S2). In contrast, the apo-forms of pHmd and jHmd showed a slight difference at the loop involved in the FeGP cofactor coordination. One of the differences between the pHmd and jHmd apo-forms is the location of Lys150, which moved outside from the active site in jHmd (Figure S3). This movement might also be attributed to the crystal packing. Lys150 is conserved in the Hmd-encoding genes in the genomes reported with exceptions found in fifteen genomes of *Methanobrevibacter* species (e.g., *M. smithii*), where the lysine position varies to glutamate (Figure S4). Notably, the Hmd activity of the cell extract from the *Methanobrevibacter* species was very low [29,30].

2.2. Crystal Structure of pHmd in Complex with GMP

We obtained crystals from the solutions containing the reconstituted pHmd with an iron complex in the presence of GMP. This mimic complex has been used for reconstitution of semisynthetic jHmd (see Section 3) [28]. The 1.55 Å crystal structure revealed a symmetric pHmd homodimer in the closed conformation, which was fully occupied with GMP but without mimic complex (Figure 3, Table 1). The N-terminal domain of the GMP-binding structure and the apo-form monomer of pHmd in the asymmetric homodimer superposed well (295 Cα superposed with rmsd of 0.416 Å, Figure 3a). The GMP binding site is identical to that of the FeGP cofactor observed in the pHmd–FeGP complex of the asymmetric pHmd. The residues in direct contact with GMP and the GMP moiety of the FeGP cofactor have the same orientations (Figure 3b). The Lys150 side chain adopts the same orientation as in the apo form; however, binding of the GMP slightly rearranged the loop 110–116 and rigidified its surroundings as observed by a lower B-factor profile.

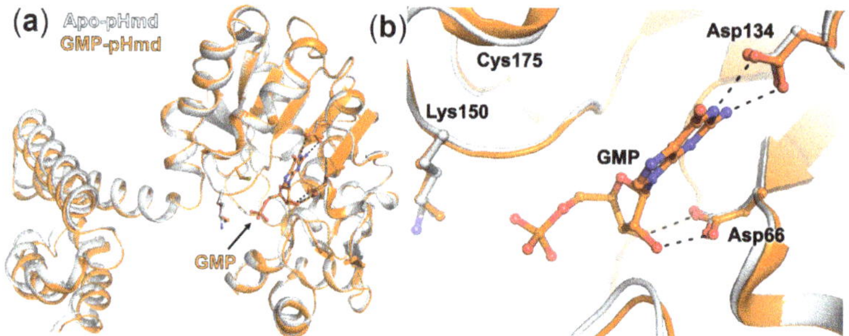

Figure 3. Structural comparison of the apo-form of the pHmd asymmetric-homodimer (grey) and the GMP-bound form (orange) of pHmd. (**a**) Monomers of each structure is shown by cartoon model. The N-terminal domains were superposed. (**b**) Zoom-up view of the GMP binding site. GMP is shown as ball and stick model. Hydrogen bonds are indicated by dashed lines.

2.3. Crystal Structure of pHmd in Complex with GP

From the reconstituted pHmd holoenzyme with the FeGP cofactor, in addition to the asymmetric homodimer crystal, another crystalline form was obtained. This form diffracted to 1.7 Å resolution (Table 1). The asymmetric unit of the crystal contained two homodimers in the closed conformation. Contrary to the other reconstituted holo-Hmd structures, the pyridinol part was only partially visible in the electron density and could be modelled for only one monomer in the four monomers in the asymmetric unit. The FeGP cofactor appeared to be decomposed to GP during crystallization process. The structures of the N-terminal domain and the residues binding GMP and GP superposed well (Figure 4a). Binding of GP induces the local conformational rearrangement of the two loops containing Lys150 and Cys175, respectively, compared to the apo form. In the pHmd–GP structure, the van der Waals interaction between the hydroxyl group of the pyridinol ring and amino group of Lys150 (2.7 Å) stabilizes the pyridinol ring of GP in this structure (Figure 4). The Cys175 side chain was modelled in two conformations of 80% and 20% occupancy in the broad electron density. The interaction between the amino group of Lys150 and the pyridinol does not appear to be optimal for hydrogen-bonding. However, in the incorporation process of the FeGP cofactor, a slight tilt of the pyridinol ring would improve the interaction.

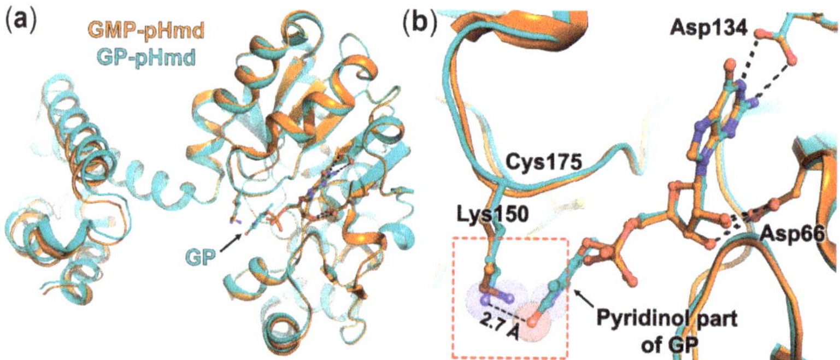

Figure 4. Structural comparison of the GMP-bound form (orange) and the GP-bound form (cyan) of pHmd. (**a**) Monomers of each structure are shown by a cartoon model. The N-terminal domains are superposed. (**b**) Zoomed-in view of the active site. GMP and GP are shown by ball and stick models. Hydrogen bonds are indicated by dashed lines.

The carboxymethyl part of the pyridinol group was not visible, which indicates that the elongated carboxymethyl substituent of the GP does not interact with the protein. This observation contrasts with the structure of hexameric Hmd from *M. marburgensis* in the open conformation obtained in an oxidized broken state. The carboxymethyl group was visible and bound with an Fe atom coordinated by His203 and Cys172, and Asp189 from loop of another dimer (Figure S5a) [18]. The loop containing Asp189 is not conserved in the pHmd sequence. The absence of the loop structure explains why the GP binding mode is different and the iron was removed from the cleft after decomposition (Figure S5b).

The closed conformation of the active-site cleft of the pHmd in complex with GP and also GMP indicated that binding of the GMP and pyridinol parts of the FeGP cofactor does not induce the opening of the active-site cleft. The formation of the Cys175-S–Fe bonding and the resulting conformational change of the loop at Cys175 might trigger the open state. The crystal structure of the reconstituted Cys176Ala-mutated jHmd holoenzyme (Cys176 of jHmd is equivalent to Cys 175 of pHmd) has been reported [10]. This enzymatically inactive jHmd holoenzyme, lacking the Cys176-S–Fe bonding, was crystallized in the open conformation even in the presence of methylene–H_4MPT bound, which indicated that the Cys176Ala mutation hindered the open/closed conformational change essential for the catalytic activity.

In the pHmd holo-form in the asymmetric homodimer, Lys150 is dissociated from the pyridinol ring and the loop containing Lys150 moved slightly away (Figure 5). Lys150 should move further away from the active site in the closed conformation induced by methenyl-H_4MPT^+ binding, because the Lys150 side chain clashes with the phenyl ring part of methenyl-H_4MPT^+ observed in the holoenzyme structure Hmd from *Methanococcus aeolicus* in complex with methenyl-H_4MPT^+ (Figure S6) [20]. In the ternary complex, the Lys150 comes into contact with the side chain of methenyl-H_4MPT^+, mainly via a water network, which might therefore have an additional role in the substrate binding.

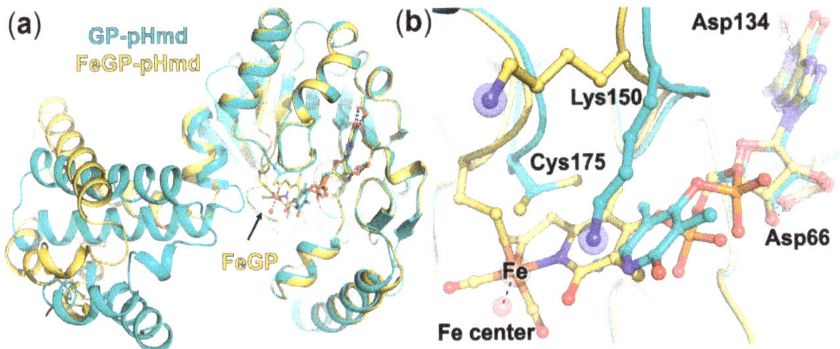

Figure 5. Structural comparison of the GP-bound form (cyan) of pHmd and the FeGP-bound form (yellow) of the pHmd asymmetric homodimer. (**a**) Monomers of each structure are shown by a cartoon model. The N-terminal domains were superposed. (**b**) Zoomed-in view of the active site. GP and FeGP are shown by ball and stick models. In the GP-bound form, the carboxymethyl group of GP is not visible. Hydrogen bonds are indicated by dashed lines.

2.4. Effect of Lys150Ala Substitution on the Reconstitution Rate of pHmd

The structural analysis of pHmd in complex with GP indicated that Lys150 could interact with the FeGP cofactor in the incorporation process. This hypothesis was tested using the pHmd Lys150Ala variant. Prior to the reconstitution assay, we measured the kinetic parameters of the wild-type enzyme and Lys150Ala variant reconstituted under the standard reconstitution condition. Unexpectedly, the Lys150Ala variant exhibited much higher V_{max} and K_m values (640 U/mg and 160 μM) than those of the wild-type enzyme (66 U/mg and 6 μM) in the oxidation reaction of methylene-H_4MPT at pH 6.0 (Figures S7 and S8). The V_{max} and K_m of the reduction reaction of methenyl-H_4MPT^+ with H_2 at pH 7.5 are 1300 U/mg and 62 μM for the wild-type enzyme and 820 U/mg and 110 μM for the K150A variant, respectively. Increase of the K_m values of the Lys150Ala variant is consistent with the observation of the contact of Lys150 with the side chain of methenyl-H_4MPT described in the last section.

The formation of the holoenzyme by binding of the FeGP cofactor was followed by monitoring the increase of enzymatic activity after addition of the FeGP cofactor (4–700 nM) to the assay solution, which contained the 4-nM apoenzyme, and 20-μM methenyl-H_4MPT^+ under H_2 at pH 7.5 or 20-μM methylene–H_4MPT under N_2 at pH 6.0. The time course of the change of the absorbance at 336 nm was recorded (Figure 6a,b,e,f). We calculated the specific activity (U/mg) at each time point (per 2 s) from the slope of the absorbance change (Figure 6c,d,g,h). In these experiments, we assume that the increase of the Hmd activity indicates the increase of the active holoenzyme in the assay by incorporation of the FeGP cofactor into the apoenzymes. Hence, the specific activity is a function of the concentration of the reconstituted enzyme and the residual substrate concentration in the assay. In the assay condition at pH 7.5, which is the physiological pH [31], the wild-type enzyme was quickly reconstituted and reached the maximum activity within 25 s in the presence of 700-nM FeGP cofactor (Figure 6c). The Lys150Ala variant also exhibits activity, but the increase of the specific activity at the same condition was much slower than the case of the wild-type enzyme; to reach the maximal activity, 130 s was required (Figure 6d). These results support the hypothesis that Lys150 contributes to the binding kinetics of the FeGP cofactor to the protein. In the case of the reverse reaction, oxidation of methylene–H_4MPT at pH 6.0, the Lys150Ala mutation did not affect the reconstitution of the holoenzyme (Figure 6e–h). To obtain the reconstitution rate from the data, we simulated the reconstitution curves (Figures S9 and S10). The simulated curves of reconstitution fitted to the observed curve when the reconstitution rate (k_2 in Figure S9) of the FeGP cofactor to the wild and Lys150Ala apoenzymes are assumed as 0.01 ± 0.005 $\mu M^{-1} \cdot s^{-1}$ and 0.0024 ± 0.0003 $\mu M^{-1} \cdot s^{-1}$ at pH 7.5 and 0.10 ± 0.01 $\mu M^{-1} \cdot s^{-1}$ and 0.14 ± 0.03 $\mu M^{-1} \cdot s^{-1}$ at pH 6.0, respectively. These data support the function of Lys150 in the binding kinetics of incorporation of the FeGP cofactor in the physiological pH. Protonation of the 2-hydroxy group of the FeGP cofactor might affect the incorporation rate of the FeGP cofactor into the protein.

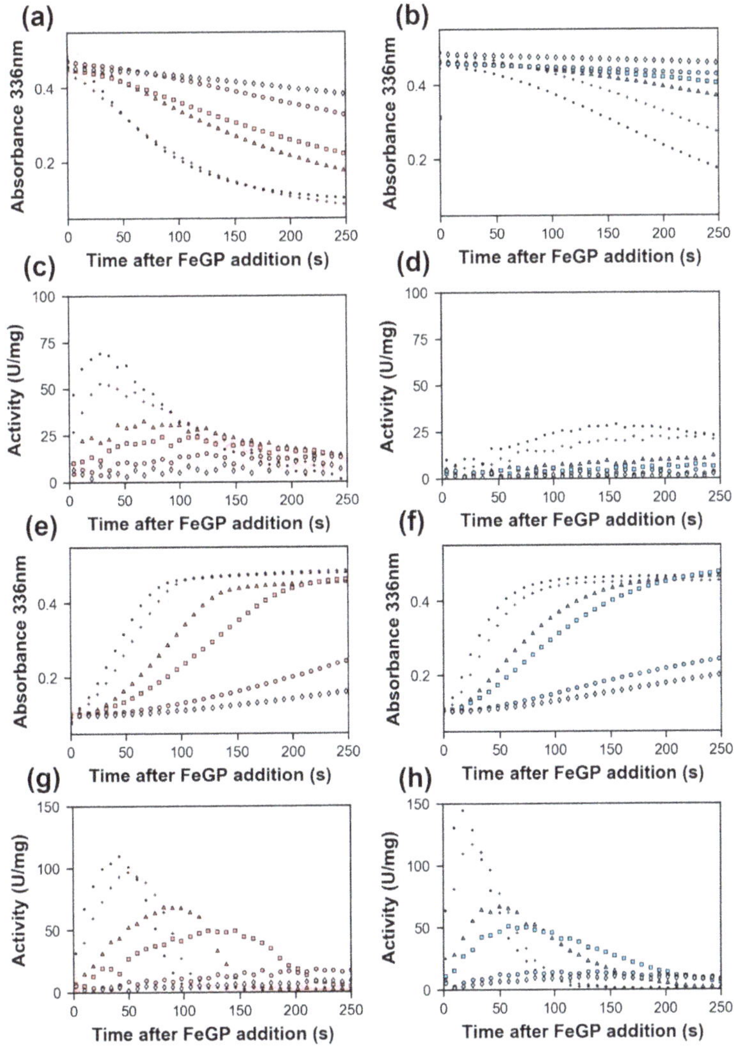

Figure 6. Reconstitution of the [Fe]-hydrogenase holoenzyme from the apoenzyme of the wild-type (**a**,**c**,**e**,**g**) and Lys150Ala variant (**b**,**d**,**f**,**h**) with the FeGP cofactor. Changes of the absorbance at 336 nm according to the reduction of methenyl-H_4MPT^+ (**a**,**b**) and oxidation of methylene-H_4MPT (**e**,**f**) were recorded. The color area indicates the standard deviation of three tests. The enzymatic activity (U/mg) (**c**,**d**,**g**,**h**) of each 2 s was calculated from the data of (**a**,**b**,**e**,**f**), respectively. Concentrations of the FeGP cofactor were 700 (•), 350 (+), 100 (Δ), 50 (□), 10 (○) and 4 nM (◊). The plots of the data of the wild type and the Lys150Ala variant are shown in red and blue, respectively.

3. Materials and Methods

3.1. Chemicals and Reagents

Tetrahydromethanopterin (H_4MPT) and methenyl-H_4MPT^+ were isolated from *M. marburgensis* cells [32]. Methylene-H_4MPT was produced by the reaction of H_4MPT with formaldehyde [14]. The FeGP cofactor was isolated from [Fe]-hydrogenase from *M. marburgensis* as described previously [14]. All other chemical compounds used in this work were purchased from Sigma-Aldrich (Darmstadt, Germany).

3.2. Gene Synthesis of [Fe]-Hydrogenase from Methanolacinia Paynteri

The [Fe]-hydrogenase gene from *M. paynteri* (NCBI Reference Sequence: WP_048153035.1) was modified for the codon usage optimization as shown below and synthesized by GenScript. The DNA synthesized was inserted into the expression vector pET-24b(+) at the *Nde*I and *Sal*I restriction-enzyme digestion sites. Genes of the Lys150Ala variant were synthesized using the template of the wild-type gene.

5′-CATATGACAATAAAGAAGGTAGCTATACTAGGAGCAGGGTGTTATAGGACTCACTCA
GCGACCGGCATTACCAACTTTGCGCGTGCGTGCGAGGTGGCGGAAATGGTTGGTAAACCGG
AGATCGCGATGACCCACAGCACCATTGCGATGGCGGCGGAACTGAAGTACCTGGCGGGCAT
CGACAACATCGTGATTAGCGATCCGAGCTTCGCGGGCGAGTTTACCGTGGTTAAGGACTTCG
ATTACAACGAAGTTATCAAGGCGCACAAAGAGAACCCGGAAACCATCATGCCGAAGATTCGT
GAGAAAGTGAACGAACTGGCGAAAACCGTTCCGAAGCCGCCGAAAGGCGCGATCCACTTTG
TGCACCCGGAGGACCTGGGTCTGAAGGTGACCACCGACGATCGTGAAGCGGTTCGTGACG
CGGATCTGATCATTACCTGGCTGCCGAAGGGTGACATGCAGAAAGGCATCATTGAGAAGTTC
GCGGGTGATATCAAGCAAGGCGCGATCATTACCCACGCGTGCACCATTCCGACCACCCTGTT
CTACAAAATCTTTGAGGAACTGGGCATTGCGGATAAGGTGGAAGTTACCAGCTATCACCCGG
GTGCGGTGCCGGAGATGAAAGGCCAGGTTTACATCGCGGAAGGTTATGCGAGCGAGGAAGC
GATCAACACCATTTACGAGCTGGGTAAGAAAGCGCGTGGTCATGCGTTTAAGCTGCCGGCGG
AACTGATTGGTCCGGTTTGCGACATGTGCGCGGCGCTGACCGCGATTACCTACGCGGGTCTG
CTGGTGTATCGTGATGCGGTTATGAACATTCTGGGTGCGCCGGCGGGTTTCAGCCAGATGAT
GGCGACCGAGAGCCTGGAACAAATCACCGCGTATATGAAGAAAGTGGGTATTAAAAACCTGG
AGGAAAACCTGGACCCGGGTGTTTTCCTGGGCACCGCGGATAGCATGAACTTTGGCCCGATT
GCGGAGATTCTGCCGACCGTTCTGAAGAGCCTGGAAAAGCGTGCGAAATAAGTCGAC-3′.

3.3. Enzyme Production, Purification and Reconstitution

The apoenzymes of [Fe]-hydrogenase from *M. paynteri* were heterologously overproduced in *E. coli* BL21(DE3). The recombinant *E. coli* was cultivated in the tryptone–phosphate (TP) medium containing 50 μg/mL kanamycin at 37 °C [33]. When the optical density of *E. coli* at 600 nm became 0.6–0.8, 1-mM isopropyl β-D-thiogalactopyranoside (IPTG) was added to induce expression of the targeted gene. Cells were harvested by centrifugation using Avanti JXN-26 centrifuge with JLA-10.500 rotor (Beckman-Coulter, Krefeld, Germany) at 8000 rpm for 30 min at 4 °C. The wet cells (5–10 g) were suspended in 50-mM 3-(*N*-morpholino)propanesulfonic acid (MOPS)/KOH pH 7.0 containing 1-mM dithiothreitol (DTT). Cells were disrupted on ice by sonication for 10 min using SONOPULS GM200 (Bandelin, Berlin, Germany) with KE76 tip with 50 cycles. The cell debris and the unbroken cells were removed by centrifugation using an Avanti JXN-26 centrifuge with a JA-25.50 rotor (Beckman-Coulter) at 15,000 rpm for 30 min at 4 °C. Ammonium sulfate (2-M final concentration) was added to the supernatant. Precipitates were removed by centrifugation using an Avanti JXN-26 centrifuge with JA-25.50 rotor at 15,000 rpm for 30 min at 4 °C. The supernatant was loaded on a Phenyl Sepharose High Performance column (75 mL, GE Healthcare Life Sciences, Solingen, Germany) and eluted with a linear gradient of ammonium sulfate from 2 M to 0 M in 50-mM MOPS/KOH buffer pH 7.0 containing 1-mM DTT. Fractions containing the apoenzyme of pHmd were collected and concentrated by using Amicon Ultra-4 Centrifugation filters (30-kDa cut-off). To further purify the apoenzyme of pHmd, the concentrated apoenzyme sample was loaded to a HiPrep 16/60 Sephacryl S-200 HR gel filtration column (120 mL, GE Healthcare Life Sciences) using 25-mM Tris(hydroxymethyl)aminomethane (Tris)/HCl buffer pH 7.5 containing 150-mM NaCl, 5% glycerol and 2-mM DTT. To increase the purity of protein, the gel filtration repeated two times using the same conditions [20]. Finally, the purified apoenzyme was concentrated to 50–100 mg/mL.

Protein concentration was measured by Bradford method using bovine serum albumin as the standard. Reconstitution was performed under dark conditions in an anoxic tent (Coy) with a gas phase of 95%N_2/5%H_2 at 8 °C by mixing the 0.13-mM apoenzyme and the 0.18-mM FeGP cofactor (at a molecular ratio of 0.75:1), respectively, as previously described [14].

3.4. Enzyme Activity Assay

The enzyme activity was anaerobically measured as previously described using a 1-mL quartz cuvette containing 0.7 mL assay mixture [14]. For the reduction of methenyl-H_4MPT^+ with H_2, 120-mM potassium phosphate pH 7.5 containing 1-mM ethylenediaminetetraacetic acid (EDTA) was used as the assay buffer under 100% H_2. For the oxidation of methylene-H_4MPT under N_2, 120-mM potassium phosphate pH 6.0 containing 1-mM EDTA was used as assay buffer under 100% N_2. The assay was started by adding 10 μL of 0.01 mg/mL enzyme solution (final concentration in the assay was 0.14 μg/mL), and the decrease (reduction)/increase (oxidation) of absorbance at 336 nm was recorded. Its specific activity was calculated using the extinction coefficient of methenyl-H_4MPT^+ (ε_{336nm} = 21.6 $mM^{-1}\cdot cm^{-1}$) [14]. One unit (U) activity is the amount of the enzyme catalyzing the formation or consumption/formation of 1 μmol methenyl-H_4MPT per min.

The reconstitution rate of the holoenzyme was kinetically analyzed in a quartz cuvette. The assay solution for the reduction reaction contained 120-mM potassium phosphate buffer pH 7.5, 1-mM EDTA, 4-nM (0.15 μg/mL) pHmd apoenzyme and 20-μM methenyl-H_4MPT^+ under 100% H_2. In the oxidation reaction, the assay solution contained 120-mM potassium phosphate buffer pH 6.0, 1-mM EDTA, 4-nM (0.15 μg/mL) pHmd apoenzyme and 20-μM methylene-H_4MPT under 100% N_2 The reaction was started by the addition of 4–700-nM cofactor (final concentration). A change of the absorbance at 336 nm was recorded at 40 °C.

3.5. Simulation of the Enzyme Kinetics Data

Michalis–Menten kinetics-parameters were obtained by simulation of the substrate consumption and product formation by numerical integration of the equations derived from mass action kinetics (Figure S7). The reconstitution rate with the FeGP cofactor and simulated enzymatic activities were calculated by numerical integration using the equations shown in Figure S9. All simulations were coded in Python 3.7 using Spyder 4.1 development environment and the following libraries: SciPy [34], NumPy [35], Matplotlib [36] and pandas [37].

3.6. Crystallization

[Fe]-hydrogenase holoenzyme from *M. paynteri* (pHmd) was crystallized under 95%N_2/5%H_2 at 8 °C using 96-well two-drop MRC crystallization plates (sitting drop vapor diffusion method). For the crystallization of pHmd–GP and pHmd–FeGP complexes, 0.7 μL of 25 mg/mL reconstituted holoenzyme was mixed with 0.7 μL reservoir solution (from crystallization kits) under yellow light and incubated under dark conditions. The best diffracting crystal of pHmd–GP came out within two weeks in 25% *w/v* polyethylene glycol 1500 and 100-mM succinic acid/sodium dihydrogen phosphate/Glycine (SPG) buffer pH 8.5 (JBScreen Wizard 3&4 HTS, Jena Bioscience, Jena, Germany). For pHmd asymmetric dimer form, the crystals grew within two weeks in 30% *w/v* polyethylene glycol 4000, 200-mM lithium sulfate and 100-mM Tris pH 8.5 (JBScreen Wizard 3&4 HTS, Jena Bioscience).

The reconstituted holoenzyme with the Fe complex was prepared as previously described [28]. An Fe(II) complex [(2-CH_2CO-6-HOC_5H_3N)$Fe(CO)_3I$] (complex 5 in [26]) was dissolved in methanol containing 1% acetic acid [28]. For reconstitution, 0.4 mL of 10-mM Fe complex solution was mixed anaerobically with a 7.6 mL solution of 100-mM sodium acetate pH 5.6, 0.02 mM apoenzyme and 2-mM GMP (final concentrations) and incubated on ice for one hour. The buffer of the reconstituted enzyme was exchanged by three-time concentration/dilution cycles using a 30 kDa ultrafilter with 10 mM MOPS/KOH pH 7.0 and finally concentrated to 25 mg/mL. For crystallization, 0.7 μL of the reconstituted pHmd solution was mixed with 0.7 μL reservoir solution (from crystallization kits) under yellow light and incubated under dark conditions at 10 °C. The best crystal appeared within one month in 20% *w/v* polyethylene glycol 3350 and 200-mM magnesium formate reservoir solution (JBScreen Wizard 3&4 HTS, Jena Bioscience).

3.7. Data Collection and Refinement

The crystals of the asymmetric homodimer and the one containing GP were flash-frozen (3–5 s) in their crystallization reservoir solution supplemented with 10% *v/v* glycerol under 95%N_2/5%H_2. The crystal of pHmd bound with GMP was flash-frozen (3–5 s) in its crystallization reservoir solution containing 30% *v/v* glycerol under 95%N_2/5%H_2. All diffraction experiments were performed at 100 K on beamline BM30A (French Beamline for Investigation of Proteins) at the European Synchrotron Radiation Facility (ESRF) equipped with an ADSC Q315r charge-coupled device detector. The data were processed with XDS [38] and scaled with SCALA from the CCP4 suite [39]. The structure of pHmd–FeGP was determined by molecular replacement with PHASER [40] by decoupling the N- and C-terminal domain of the native Hmd from *M. marburgensis* in complex with 2-naphthylisocyanide (PDB: 4JJF) as templates. The structures of pHmd–GP and pHmd–GMP were solved with PHASER by using the monomer in the closed conformation of the asymmetric homodimer. The models were manually built with COOT [41] and refined with Phenix [42] and BUSTER (Bricogne G., Blanc E., Brandl M., Flensburg C., Keller P., Paciorek W., Roversi P, Sharff A., Smart O.S., Vonrhein C., Womack T.O. (2017). BUSTER version 2.10.3. Cambridge, United Kingdom: Global Phasing Ltd.). The final models were validated using the MolProbity server (http://molprobity.biochem.duke.edu) [40]. Data collection, refinement statistics and PDB code for the deposited model are listed in Table 1. The hydrogens were omitted in the final deposited model. The figures were generated and rendered with PyMOL (version 1.7, Schrödinger, Cambridge, UK). Alignments were performed by Clustal Omega [43]. The figures were made using ESPript 3.0 [44].

4. Conclusions

Based on the crystal structures of the pHmd apo- and holo-form—the latter bound with GMP, GP or the FeGP cofactor—we propose a trajectory of the isolated FeGP cofactor incorporation into the apoenzyme. First, binding of the GMP part guides correct positioning of the FeGP cofactor, which slightly opens the active-site cleft to engage the binding of the pyridinol part at the flexible Cys175 loop and induces small local conformational change at the loop containing Lys150. The iron site of the FeGP cofactor forms a covalent Cys175-S–Fe bonding upon exchange with the acetate/2-mercaptoethanol ligand bound in the free cofactor. Sequential binding of the GMP and pyridinol moieties might allow a correct covalent bonding between Cys175-thiolate and the Fe site. Lys150 might guide the binding of the pyridinol part to the specific position. The Lys150Ala mutation analysis supported this hypothesis. These results are of general interest for studying how nucleotide-containing cofactors and coenzymes are incorporated into the protein, and for developing semi-synthetic [Fe]-hydrogenase using mimic complexes. A plausible strategy of incorporation of the mimic complexes into the Hmd protein is to synthesize mimic compounds, which contains the GMP moiety at the position 4 of the pyridinol ring. Another strategy might be modifying the Hmd apoenzyme to enhance a smooth Fe–S bond formation of the mimic cofactor in the absence of the GMP moiety, in which the loop containing Lys150 might be the target of modification.

Supplementary Materials: The following are available online at http://www.mdpi.com/2304-6740/8/9/50/s1, Figure S1. Observation of subtle rearrangements of the asymmetric pHmd when the C-terminal domain of holo-jHmd and apo-jHmd are superposed, Figure S2. Superposition of the FeGP cofactor binding sites in pHmd (PDB: 6YKA), Hmd from *Methanococcus aeolicus* (aHmd) (PDB: 6HAC) and jHmd (PDB: 3F47), Figure S3. Comparison of the loop involved in the FeGP cofactor coordination in the jHmd and pHmd apo-forms. Figure S4. Alignments of Hmd amino acid sequences from different organisms, Figure S5. Comparison of the GP-binding states in the structure of Hmd from *M. marburgensis* obtained in an oxidized broken state with that in the structure of GP bound form of pHmd, Figure S6. Superposition of the active sites of apo- and holo-forms of Hmd. Figure S7. Equations used for the simulation of the modelled reaction and calculation of the kinetic parameters. Figure S8. Simulation of the progressive curves of the reactions. Figure S9. Equations used for the simulation of the binding constant of the FeGP cofactor to the pHmd apoenzyme, Figure S10. Simulation of the change of the activity in the reconstitution assay of pHmd.

Author Contributions: Conceptualization, S.S.; methodology, S.S.; software, T.W.; validation, T.W.; formal analysis, G.H., T.W., F.J.A.-G. and S.S.; investigation, G.H., T.W. and F.J.A.-G.; resources, S.S..; data curation, G.H., F.J.A.-G., T.W. and S.S.; writing—original draft preparation, S.S.; writing—review and editing, G.H., F.J.A.-G., T.W. and S.S.; visualization, G.H and T.W.; supervision, S.S.; project administration, S.S.; funding acquisition, S.S. and T.W. All authors have read and agreed to the published version of the manuscript.

Funding: This work was funded by the Max Planck Society (to T.W. and S.S.) and the Deutsche Forschungsgemeinschaft priority program (SPP 1927) (SH 87/1-1, to S.S.).

Acknowledgments: We thank Xile Hu (École polytechnique fédérale de Lausanne), who provided us the iron complex. The authors thank the staff from the BM30A (FIP) beamline at the European Synchrotron Radiation Facility (ESRF) for their availability and advice during data collection.

Conflicts of Interest: The authors declare no conflict of interest.

References

1. Zirngibl, C.; van Dongen, W.; Schwörer, B.; von Bünau, R.; Richter, M.; Klein, A.; Thauer, R.K. H_2-forming methylenetetrahydromethanopterin dehydrogenase, a novel type of hydrogenase without iron-sulfur clusters in methanogenic archaea. *Eur. J. Biochem.* **1992**, *208*, 511–520.
2. Huang, G.F.; Wagner, T.; Ermler, U.; Shima, S. Methanogenesis involves direct hydride transfer from H_2 to an organic substrate. *Nat. Rev. Chem.* **2020**, *4*, 213–221.
3. Shima, S.; Huang, G.; Wagner, T.; Ermler, U. Structural basis of hydrogenotrophic methanogenesis. *Annu. Rev. Microbiol.* **2020**, *74*, 713–733.
4. Pilak, O.; Mamat, B.; Vogt, S.; Hagemeier, C.H.; Thauer, R.K.; Shima, S.; Vonrhein, C.; Warkentin, E.; Ermler, U. The crystal structure of the apoenzyme of the iron-sulphur cluster-free hydrogenase. *J. Mol. Biol.* **2006**, *358*, 798–809.
5. Shima, S.; Pilak, O.; Vogt, S.; Schick, M.; Stagni, M.S.; Meyer-Klaucke, W.; Warkentin, E.; Thauer, R.K.; Ermler, U. The crystal structure of [Fe]-hydrogenase reveals the geometry of the active site. *Science* **2008**, *321*, 572–575.
6. Shima, S.; Ermler, U. Structure and function of [Fe]-hydrogenase and its iron-guanylylpyridinol (FeGP) cofactor. *Eur. J. Inorg. Chem.* **2011**, *2011*, 963–972.
7. Lyon, E.J.; Shima, S.; Boecher, R.; Thauer, R.K.; Grevels, F.W.; Bill, E.; Roseboom, W.; Albracht, S.P.J. Carbon monoxide as an intrinsic ligand to iron in the active site of the iron-sulfur-cluster-free hydrogenase H_2-forming methylenetetrahydromethanopterin dehydrogenase as revealed by infrared spectroscopy. *J. Am. Chem. Soc.* **2004**, *126*, 14239–14248.
8. Shima, S.; Lyon, E.J.; Sordel-Klippert, M.S.; Kauss, M.; Kahnt, J.; Thauer, R.K.; Steinbach, K.; Xie, X.L.; Verdier, L.; Griesinger, C. The cofactor of the iron-sulfur cluster free hydrogenase Hmd: Structure of the light-inactivation product. *Angew. Chem. Int. Ed.* **2004**, *43*, 2547–2551.
9. Shima, S.; Lyon, E.J.; Thauer, R.K.; Mienert, B.; Bill, E. Mössbauer studies of the iron-sulfur cluster-free hydrogenase: The electronic state of the mononuclear Fe active site. *J. Am. Chem. Soc.* **2005**, *127*, 10430–10435.
10. Hiromoto, T.; Ataka, K.; Pilak, O.; Vogt, S.; Stagni, M.S.; Meyer-Klaucke, W.; Warkentin, E.; Thauer, R.K.; Shima, S.; Ermler, U. The crystal structure of C176A mutated [Fe]-hydrogenase suggests an acyl- iron ligation in the active site iron complex. *FEBS Lett.* **2009**, *583*, 585–590.
11. Hiromoto, T.; Warkentin, E.; Moll, J.; Ermler, U.; Shima, S. The crystal structure of an [Fe]-hydrogenase-substrate complex reveals the framework for H_2 activation. *Angew. Chem. Int. Ed.* **2009**, *48*, 6457–6460.
12. Shima, S.; Schick, M.; Kahnt, J.; Ataka, K.; Steinbach, K.; Linne, U. Evidence for acyl–iron ligation in the active site of [Fe]-hydrogenase provided by mass spectrometry and infrared spectroscopy. *Dalton Trans.* **2012**, *41*, 767–771.
13. Korbas, M.; Vogt, S.; Meyer-Klaucke, W.; Bill, E.; Lyon, E.J.; Thauer, R.K.; Shima, S. The iron-sulfur cluster-free hydrogenase (Hmd) is a metalloenzyme with a novel iron binding motif. *J. Biol. Chem.* **2006**, *281*, 30804–30813.
14. Shima, S.; Schick, M.; Tamura, H. Preparation of [Fe]-hydrogenase from methanogenic archaea. *Methods Enzymol.* **2011**, *494*, 119–137.
15. Shima, S.; Vogt, S.; Göbels, A.; Bill, E. Iron-chromophore circular dichroism of [Fe]-hydrogenase: The conformational change required for H_2 activation. *Angew. Chem. Int. Ed.* **2010**, *49*, 9917–9921.

16. Lyon, E.J.; Shima, S.; Buurman, G.; Chowdhuri, S.; Batschauer, A.; Steinbach, K.; Thauer, R.K. UV-A/blue-light inactivation of the 'metal-free' hydrogenase (Hmd) from methanogenic archaea: The enzyme contains functional iron after all. *Eur. J. Biochem.* **2004**, *271*, 195–204.
17. Hidese, R.; Ataka, K.; Bill, E.; Shima, S. Cu^{I} and H_2O_2 Inactivate and Fe^{II} Inhibits [Fe]-hydrogenase at very low concentrations. *Chembiochem* **2015**, *16*, 1861–1865.
18. Huang, G.; Wagner, T.; Ermler, U.; Bill, E.; Ataka, K.; Shima, S. Dioxygen sensitivity of [Fe]-hydrogenase in the presence of reducing substrates. *Angew. Chem. Int. Ed.* **2018**, *57*, 4917–4920.
19. Buurman, G.; Shima, S.; Thauer, R.K. The metal-free hydrogenase from methanogenic archaea: Evidence for a bound cofactor. *FEBS Lett.* **2000**, *485*, 200–204.
20. Huang, G.F.; Wagner, T.; Wodrich, M.D.; Ataka, K.; Bill, E.; Ermler, U.; Hu, X.L.; Shima, S. The atomic-resolution crystal structure of activated [Fe]-hydrogenase. *Nat. Catal.* **2019**, *2*, 537–543.
21. Song, L.C.; Zhu, L.; Liu, B.B. A biomimetic model for the active site of [Fe]-H_2ase featuring a 2-methoxy-3,5-dimethyl-4-phosphato-6-acylmethylpyridine ligand. *Organometallics* **2019**, *38*, 4071–4075.
22. Seo, J.; Manes, T.A.; Rose, M.J. Structural and functional synthetic model of mono-iron hydrogenase featuring an anthracene scaffold. *Nat. Chem.* **2017**, *9*, 552–557.
23. Kerns, S.A.; Magtaan, A.C.; Vong, P.R.; Rose, M.J. Functional hydride transfer by a thiolate-containing model of mono-iron hydrogenase featuring an anthracene scaffold. *Angew. Chem. Int. Ed.* **2018**, *57*, 2855–2858.
24. Xu, T.; Yin, C.J.M.; Wodrich, M.D.; Mazza, S.; Schultz, K.M.; Scopelliti, R.; Hu, X. A functional model of [Fe]-hydrogenase. *J. Am. Chem. Soc.* **2016**, *138*, 3270–3273.
25. Pan, H.J.; Huang, G.F.; Wodrich, M.D.; Tirani, F.F.; Ataka, K.; Shima, S.; Hu, X.L. A catalytically active [Mn]-hydrogenase incorporating a non-native metal cofactor. *Nat. Chem.* **2019**, *11*, 669–675.
26. Hu, B.; Chen, D.; Hu, X.L. Synthesis and reactivity of mononuclear iron models of [Fe]-hydrogenase that contain an acylmethylpyridinol ligand. *Chem. Eur. J.* **2014**, *20*, 1677–1682.
27. Chen, D.; Scopelliti, R.; Hu, X. A five-coordinate iron center in the active site of [Fe]-hydrogenase: Hints from a model study. *Angew. Chem. Int. Ed.* **2011**, *50*, 5671–5673.
28. Shima, S.; Chen, D.; Xu, T.; Wodrich, M.D.; Fujishiro, T.; Schultz, K.M.; Kahnt, J.; Ataka, K.; Hu, X.L. Reconstitution of [Fe]-hydrogenase using model complexes. *Nat. Chem.* **2015**, *7*, 995–1002.
29. Schick, M.; Xie, X.L.; Ataka, K.; Kahnt, J.; Linne, U.; Shima, S. Biosynthesis of the iron-guanylylpyridinol cofactor of [Fe]-hydrogenase in methanogenic archaea as elucidated by stable-isotope labeling. *J. Am. Chem. Soc.* **2012**, *134*, 3271–3280.
30. Schwörer, B.; Thauer, R.K. Activities of formylmethanofuran dehydrogenase, methylenetetrahydromethanopterin dehydrogenase, methylenetetrahydromethanopterin reductase, and heterodisulfide reductase in methanogenic bacteria. *Arch. Microbiol.* **1991**, *155*, 459–465.
31. de Poorter, L.M.I.; Keltjens, J.T. Convenient fluorescence-based methods to measure membrane potential and intracellular pH in the Archaeon *Methanobacterium thermoautotrophicum*. *J. Microbiol. Methods* **2001**, *47*, 233–241.
32. Shima, S.; Thauer, R.K. Tetrahydromethanopterin-specific enzymes from *Methanopyrus kandleri*. *Methods Enzymol.* **2001**, *331*, 317–353.
33. Moore, J.T.; Uppal, A.; Maley, F.; Maley, G.F. Overcoming inclusion body formation in a high-level expression system. *Protein Expres. Purif.* **1993**, *4*, 160–163.
34. Virtanen, P.; Gommers, R.; Oliphant, T.E.; Haberland, M.; Reddy, T.; Cournapeau, D.; Burovski, E.; Peterson, P.; Weckesser, W.; Bright, J.; et al. SciPy 1.0: Fundamental algorithms for scientific computing in Python. *Nat. Methods* **2020**, *17*, 261–272.
35. Van der Walt, S.; Colbert, S.C.; Varoquaux, G. The NumPy Array: A structure for efficient numerical computation. *Comput. Sci. Eng.* **2011**, *13*, 22–30.
36. Hunter, J.D. Matplotlib: A 2D graphics environment. *Comput. Sci. Eng.* **2007**, *9*, 90–95.
37. McKinney, W. Data structures for statistical computing in Python. In Proceedings of the 9th Python in Science Conference, Austin, TX, USA, 28–30 June 2010; pp. 51–56.
38. Kabsch, W. Xds. *Acta Crystallogr. D* **2010**, *66*, 125–132.
39. Winn, M.D.; Ballard, C.C.; Cowtan, K.D.; Dodson, E.J.; Emsley, P.; Evans, P.R.; Keegan, R.M.; Krissinel, E.B.; Leslie, A.G.; McCoy, A.; et al. Overview of the *CCP4* suite and current developments. *Acta Crystallogr. D* **2011**, *67*, 235–242.

40. McCoy, A.J.; Grosse-Kunstleve, R.W.; Adams, P.D.; Winn, M.D.; Storoni, L.C.; Read, R.J. Phaser crystallographic software. *J. Appl. Crystallogr.* **2007**, *40*, 658–674.
41. Emsley, P.; Lohkamp, B.; Scott, W.G.; Cowtan, K. Features and development of Coot. *Acta Crystallog. D* **2010**, *66*, 486–501.
42. Liebschner, D.; Afonine, P.V.; Baker, M.L.; Bunkóczi, G.; Chen, V.B.; Croll, T.I.; Hintze, B.; Hung, L.W.; Jain, S.; McCoy, A.J.; et al. Macromolecular structure determination using X-rays, neutrons and electrons: Recent developments in Phenix. *Acta Crystallogr. D* **2019**, *75*, 861–877.
43. Madeira, F.; Park, Y.M.; Lee, J.; Buso, N.; Gur, T.; Madhusoodanan, N.; Basutkar, P.; Tivey, A.R.N.; Potter, S.C.; Finn, R.D.; et al. The EMBL-EBI search and sequence analysis tools APIs in 2019. *Nucleic Acids Res.* **2019**, *47*, W636–W641.
44. Robert, X.; Gouet, P. Deciphering key features in protein structures with the new ENDscript server. *Nucleic Acids Res.* **2014**, *42*, W320–W324.

Article

Theoretical Insights into the Aerobic Hydrogenase Activity of Molybdenum–Copper CO Dehydrogenase

Anna Rovaletti [1], Maurizio Bruschi [1], Giorgio Moro [2], Ugo Cosentino [1] and Claudio Greco [1,*] and Ulf Ryde [3,*]

1 Department of Earth and Environmental Sciences, University of Milano-Bicocca, Piazza della Scienza 1, 20126 Milan, Italy; a.rovaletti@campus.unimib.it (A.R.); maurizio.bruschi@unimib.it (M.B.); ugo.cosentino@unimib.it (U.C.)

2 Department of Biotechnology and Biosciences, University of Milano-Bicocca, Piazza della Scienza 2, 20126 Milan, Italy; giorgio.moro@unimib.it

3 Department of Theoretical Chemistry, Lund University, Chemical Centre, P.O. Box 124, SE-221 00 Lund, Sweden

* Correspondence: claudio.greco@unimib.it (C.G.); ulf.ryde@teokem.lu.se (U.R.)

Received: 14 September 2019; Accepted: 4 November 2019; Published: 9 November 2019

Abstract: The Mo/Cu-dependent CO dehydrogenase from *O. carboxidovorans* is an enzyme that is able to catalyse CO oxidation to CO_2; moreover, it also expresses hydrogenase activity, as it is able to oxidize H_2. Here, we have studied the dihydrogen oxidation catalysis by this enzyme using QM/MM calculations. Our results indicate that the equatorial oxo ligand of Mo is the best suited base for catalysis. Moreover, extraction of the first proton from H_2 by means of this basic centre leads to the formation of a Mo–OH–Cu^{I}H hydride that allows for the stabilization of the copper hydride, otherwise known to be very unstable. In light of our results, two mechanisms for the hydrogenase activity of the enzyme are proposed. The first reactive channel depends on protonation of the sulphur atom of a Cu-bound cysteine residues, which appears to favour the binding and activation of the substrate. The second reactive channel involves a frustrated Lewis pair, formed by the equatorial oxo group bound to Mo and by the copper centre. In this case, no binding of the hydrogen molecule to the Cu center is observed but once H_2 enters into the active site, it can be split following a low-energy path.

Keywords: CO dehydrogenase; dihydrogen; hydrogenase; quantum/classical modeling; density functional theory

---

1. Introduction

Carbon monoxide dehydrogenases (CODHs) have proved to be an essential component for the biogeochemical carbon monoxide (CO) consumption. They contribute to maintenance of sub-toxic concentration of CO in the lower atmosphere by processing approximately 2×10^8 tons of it annually [1]. To date, only two enzymes have been found to be able to use CO as carbon and energy source [2]. The first type is represented by the oxygen-sensitive Ni,Fe-dependent CO dehydrogenase enzyme (NiFe-CODH), found in anaerobic bacteria and archea, whereas in aerobic carboxido-bacteria, such as *Oligotropha carboxidovorans*, the same function is performed by the oxygen-tolerant MoCu-dependent CO dehydrogenase (MoCu-CODH) enzyme [3]. These bacterial enzymes catalyze the oxidation of CO to CO_2 following the reaction:

$$CO + H_2O \longrightarrow CO_2 + 2\,H^+ + 2\,e^-$$

In addition to this reaction, both enzymes show the ability to catalyze other reactions, although at lower activity. Reduction of CO_2, following the reverse mechanism, is explicated by NiFe-CODH, while the ability of oxidizing H_2 to protons has been reported for MoCu-CODH [3,4].

The latter belongs to the xanthine oxidase family of enzymes, whose members are usually characterised by a mononuclear redox-active site, $LMo^{VI}OS(OH)$, with a square-pyramidal coordination geometry [5]. The oxo (O^{2-}) group is an apical ligand and the equatorial plane has a dithiolene ligand from a molybdopterin–cytosine dinucleotide (MCD) cofactor, a catalytically labile hydroxyl group (OH^-) and a sulphide ion (S^{2-}) [6]. MoCu-CODH presents a noncanonical binuclear active site, in which the sulphido ligand bridges to a second metal, –S–Cu–S-Cys and the equatorial hydroxyl group is deprotonated to form another oxo group. The copper ion is found in a +1 oxidation state and it is also coordinated by a weakly-bound water molecule [7].

The protein is a $(\alpha\beta\gamma)_2$ heterodimer. The active site is located in the large subunit (CoxL; 89 kDa), the medium subunit (CoxM; 30 kDa) contains a FAD cofactor, whereas the small one (CoxS; 18 kDa) possesses two [2Fe–2S] iron–sulfur clusters [7].

The protein resting state is characterised by a $Mo^{VI}(=O)SCu^{I}$ core. Binding of either CO or H_2 is believed to occur at the Cu^I centre, allowing the proper placement and activation of the substrate [8,9]. The subsequent steps lead to the transfer of two electrons to the Mo ion. This process is favoured by the presence of a highly delocalised redox-active orbital over the Mo–μS–Cu moiety [10]. The catalytic activity of MoCu-CODH is also promoted by the presence of highly conserved amino acids in the metal second coordination sphere [11]. A glutamate residue (Glu763) positioned in proximity of the equatorial oxygen ligand of molybdenum is believed to promote deprotonation events [9,12]. In addition, a phenylalanine (Phe390), situated in a flexible loop in front of the copper centre, has been proposed to contribute to the proper orientation of the Cu-bound ligand [13].

Notwithstanding the fact that the general features of the catalytic reaction cycle of the enzyme has been clarified thanks to previous experimental and theoretical efforts, a detailed understanding of the reaction mechanisms for the oxidation of both CO and H_2 has not yet been reached. As for H_2, a mechanism for the dihydrogen molecule splitting was proposed by Wilcoxen and Hille (see Figure 1) [9]. According to electron paramagnetic resonance (EPR) spectroscopy studies, H_2 is believed to coordinate Cu^I in a side-on fashion. Once the hydrogen is bound, it would be activated by the metal and could easily be split by extraction of a proton thanks to a nearby base. Subsequently, a second proton would leave the active site and the two electrons would be transferred to the Mo centre. To date, no theoretical data have been reported in support of the experimental proposal. Only one previous theoretical work investigating the hydrogenase activity has been published so far [14]. In this study, it was unexpectedly found that the copper ion could not bind the H_2 substrate. Therefore, the authors suggested that changes in the active site are needed, for example, by protonation events, before the catalytic cycle can start. In such context, it is interesting to notice that, to the best of our knowledge, side-on binding of H_2 to molecular species containing a Cu(I) ion has been previously reported only in a couple of gas-phase case studies, that is, H_2–CuCl [15] and H_2–CuF [16]. Somehow, the interaction between H_2 and the enzyme appears to find a closer parallel with the case of porous materials containing electron-rich copper ions, such as certain zeolites proposed for H_2 storage applications [17–19] and Prussian-blue analogues [20].

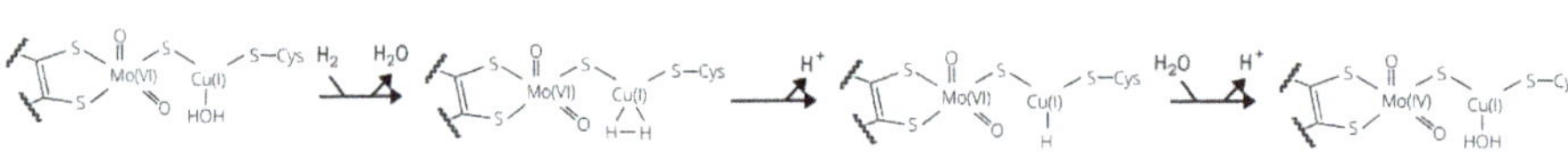

Figure 1. Reaction mechanism for H_2 activation proposed by Wilcoxen and Hille on the basis of experimental results [9].

In the present work we present a plausible mechanistic picture for the MoCu-CODH hydrogenase activity, taking into account the effects of the protein matrix by means of hybrid quantum mechanical/molecular mechanical (QM/MM) enzyme models.

2. Results and Discussion

2.1. Study of H_2 Binding Modes to the Copper Centre

We started with an investigation of the most plausible binding sites for the H_2 substrate. Considering the resting state of the protein, two types of coordination to the copper centre were considered (**1HR** and **2HR** in Figure 2). We find that H_2 can bind to the Cu ion both when the latter is bi-coordinated (μS–Cu–S-Cys388) and also when the Cu coordination sphere includes a weakly bound water molecule, as proposed in previous investigations of the enzyme resting state [7,13]. The hydrogen molecule is found to be slightly activated, with a H–H distance of 0.77 Å in both cases (the equilibrium distance of H_2 at BP86/def2-TZVP level is 0.75 Å).

For both binding modes, we estimated the binding energy of H_2 to Cu^I by comparing the energy of the coordinated structure to the corresponding one in which H_2 is not linked to copper but it is already present in the active site (**1HNoB** and **2HNoB** in Figure 3). The resulting QM/MM total energy differences were found to be +46.9 and +46.4 kJ/mol, in the absence (**1H**) or presence of the coordinated water molecule (**2H**), respectively (see Table 1).

Table 1. Relative energies (kJ/mol) calculated using the quantum mechanical/molecular mechanical (QM/MM) approach at B3LYP-D3(BJ)/def2-TZVPD level.

	NoB	R	P1	TS	P	TS2	P2
1H	−46.9	0.0	57.7		−2.9	31.8	−130.1
2H	−46.4	0.0	83.3				
3H	−10.9	0.0		29.3	28.0	56.4	−90.8
1H-FLP	0.8	0.0		45.2	44.4	79.1	−82.8

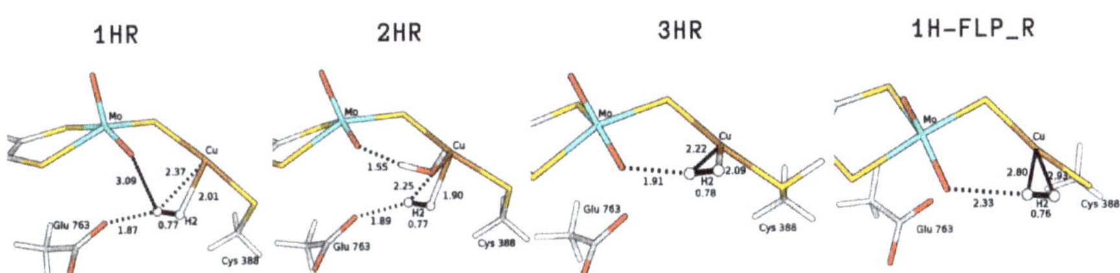

Figure 2. Binding modes of H_2 to Cu^I characterised by means of QM/MM calculations considering different copper coordination modes (**1HR**, bi-coordinated; **2HR**, tri-coordinated), active site protonation states (**3HR**, protonated Cys388) and the stable intermediate **1H-FLP_R**. All distances are in Å.

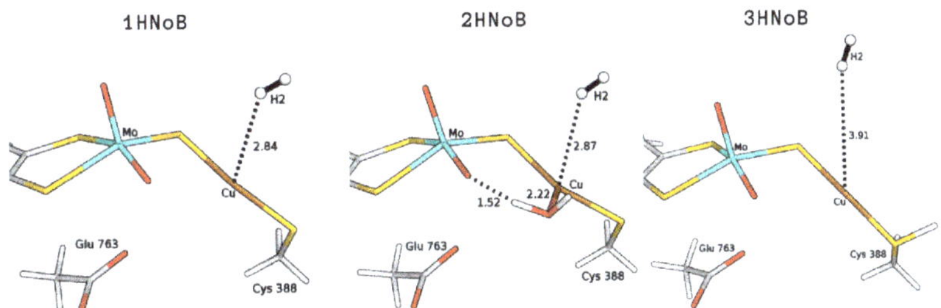

Figure 3. QM/MM optimised structures of intermediates where the H_2 molecule is present in the active site but not bound to the Cu ion, considering different copper coordination modes (**1HNoB**, bi-coordinated; **2HNoB**, tri-coordinated), active-site protonation states (**3HNoB**, protonated Cys388). All distances are in Å.

Thus, formation of these reactants was found to be strongly unfavourable from an energetic point of view. Therefore, we contemplated the possibility that a sterically hindered Lewis acid–Lewis base pair (a so-called frustrated Lewis pair, FLP), may form in the active site, which would eventually lead to heterolytic H_2 cleavage. Indeed, combination of Lewis acid and bases have been demonstrated to activate a wide range of small molecules [21]. Moreover, the plausible involvement of an FLPs in enzymatic catalysis has been recently proposed for the H_2 splitting by [NiFe]-hydrogenase [22] and an FLP might actually be active also in the case of [FeFe]-hydrogenases [23–25].

We found a variant of **1HR** (**1H-FLP_R** in Figure 2), in which H_2 assumes a different position in the active site, such that the Mo–O_{eq} atom and the copper centre appear to be able to act as a FLP. The **1H-FLP_R** reactant is characterised by a H_2 molecule positioned in front of the copper atom, with long Cu–H_{H_2} distances (2.80 and 2.93 Å) and a O_{eq}–H_{H_2} distance of 2.33 Å. **1H-FLP_R** was found to be 0.8 kJ/mol lower in energy than the **1HNoB** species in which the hydrogen molecule is separated from the bimetallic centre.

Based on previous literature, which proposed that the active site could be protonated prior to substrate binding [14,26], we also evaluated the influence of protonation on the coordination of H_2 to the metal. Protonation of the bridging sulphide ion turned out not to favour dihydrogen binding, as no H_2-bound geometry was found on the QM/MM potential energy surface. On the other hand, protonation of the sulphur atom of Cys388 favours coordination of H_2 to Cu. The latter process would lead to the formation of a H_2-coordinated geometry (**3HR** in Figure 2), in which dihydrogen is activated to a H–H bond of 0.78 Å and the computed binding energy is +10.9 kJ/mol.

It is noticeable that energy minimizations of the one-electron reduced counterparts of **1HR**, **2HR** and **3HR** lead to the detachment of the hydrogen molecule from the copper centre. This result is not compatible with the experimental outcomes that identified a H_2-bound $Mo^{V}Cu^{I}$ species by means of EPR experiments [9]. However, since one-electron reduced counterparts do not represent plausible intermediates in the catalytic cycle—there are no evidences that H_2 binding is followed by Mo reduction, prior to substrate oxidation—we have not investigated this aspect any further. The aim of the work is focused on the mechanistic characteristics of the H_2 oxidation reaction, the details of which are described in the subsequent sections.

2.2. Exploring Basic Residues in the Active Site

Next, we investigated which functional group may represent the base for the abstraction of the first proton from the activated hydrogen molecule. We considered the Glu763 residue, which has been suggested to take part into deprotonation events (*vide infra*) and the equatorial oxo ligand of molybdenum. As for the reactants, we considered not only **1H-FLP_R** and **3HR** but also **1HR** and **2HR**. In fact, even though H_2 binding in the latter two cases was computed to be disfavoured, one cannot completely rule out the hypothesis that such adducts have some minor role in dihydrogen oxidation catalysis, a possibility that would be enabled in case the subsequent reaction steps are associated with an overall smooth energy profile.

The two bases would lead to the formation of two copper hydrides (see Figure 4), namely either GluOOH–Cu^{I}H (**P1**) or Mo–OH–Cu^{I}H (**P**). For the **1HR** and **2HR** states (see Figure 2), the oxygen of the glutamate residue is closest (the O_{Glu}–H_{H_2} distance is 1.87 Å and 1.89 Å for **1HR** and **2HR**), while the Mo–oxo ligand is 3.09 and 3.28 Å from the closest H atom of H_2, respectively. However, for **3HR**, O_{eq} is the closest base (the O–H distance is 1.91 Å) while the oxygen of Glu763 is 3.83 Å away.

The search for the GluOOH–$Cu^{I}H$ products was found to be challenging, as the glutamic acid sidechain in the guess geometries turned out to release a proton to reform a Cu-bound H_2 along optimizations. Therefore, to optimize the intermediates **1HP1** and **2HP1**, (see Figure 5) we had to fix the O_{Glu}–H distance; this was done in the hypothesis of the existence of a barrier for proton transfer, although so small that convergence on the desired minimum could be hampered by the choices made on the starting input structures to be optimized. The total energy differences of these structures, with respect to the corresponding reagents, were found to be 57.7 and 83.3 kJ/mol for **1H** and **2H**, respectively. In the case of **3H**, no GluOOH–$Cu^{I}H$ (**3HP1**) product was found.

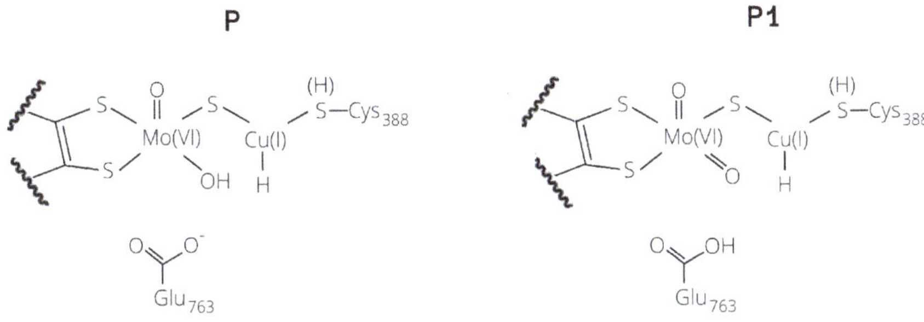

Figure 4. Copper–hydride complexes formed by proton extraction by means of Glu763 (**P1**) or Mo–O_{eq} (**P**).

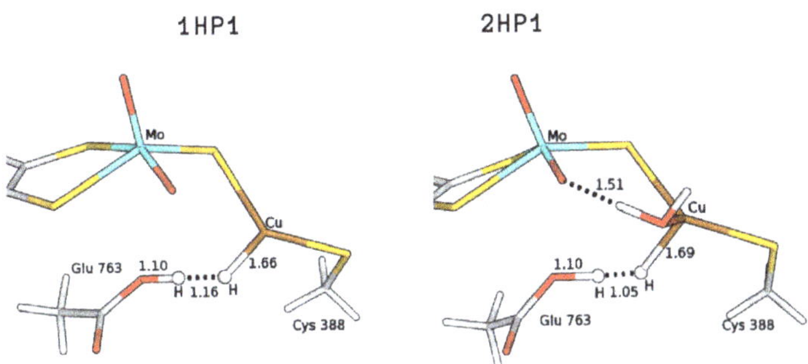

Figure 5. QM/MM optimised structures of the copper hydride complexes formed by proton abstraction by means of Glu763 (**P1**). All distances are in Å.

The product formed by extraction of the proton by the equatorial ligand of Mo (**P**) was not found in the case of **2H**, whereas it corresponds to structures with a total energy difference of −2.9 and 28.0 kJ/mol with respect to the reagents for **1H** and **3H**, respectively (**1H-FLP_P** in Figure 6 and **3HP** in Figure 7). **1H-FLP_P** can also represent the product for the H_2 splitting reaction starting from the **1H-FLP_R** structure. In this case, the resulting Cu–hydride was found to be 44.4 kJ/mol less stable than the reagent.

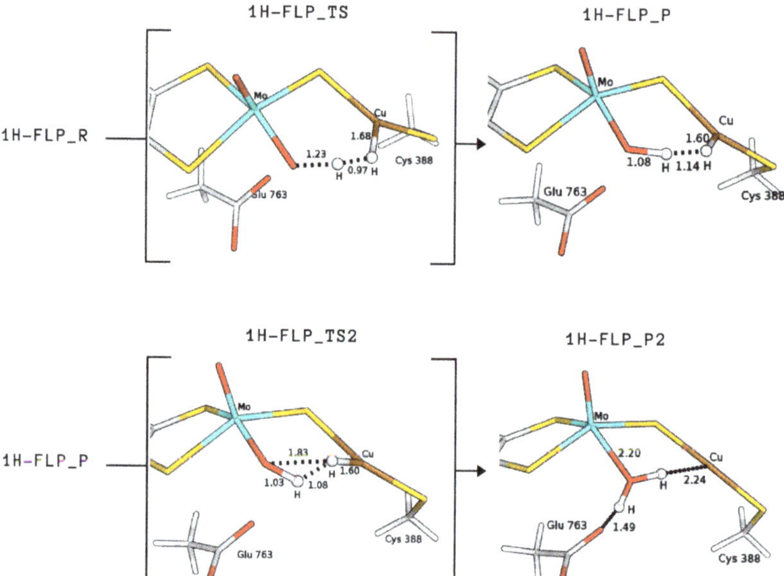

Figure 6. QM/MM structures of intermediates and transition states (in square brackets) of species involved in the catalytic cycle **B** (see Figure 8). All distances are in Å.

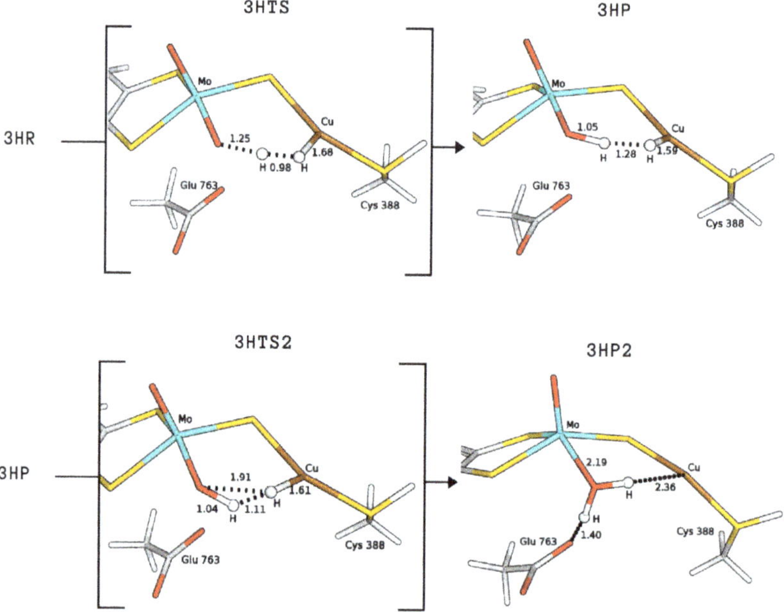

Figure 7. QM/MM structures of intermediates and transition states (in square brackets) of species involved in the catalytic cycle **A** (see Figure 8). All distances are in Å.

2.3. Plausible Activation Mechanisms for H_2 Splitting

Based on the above results on the relative energies of the intermediates potentially involved in the catalysis, we proceeded with the computation of the activation barriers in order to estimate which reaction mechanism would represent the most energetically favourable path for the hydrogenase activity of MoCu-CODH.

Starting from the H_2-bound conformation **1HR**, we have just shown that the equatorial Mo–oxo ligand represents the most favourable basic group for the first proton abstraction, leading to a rather stable product (**1H-FLP_P**). However, despite extensive efforts, we were not able to locate any transition state for the reaction **1HR** → **1H-FLP_P**. This issue is probably due to the large distance between the labile proton and the base. We also considered the **1HR** → **1HP1** reaction because, even if the GluOOH–Cu^{I}H product is not as stable as the Mo–OH–Cu^{I}H product, the whole catalysis may be driven by subsequent exoenergetic reactions. However, again, no transition state was found.

On the other hand, the **1H-FLP_R** species can be linked to the **1H-FLP_P** product through a low-lying transition state (**1H-FLP_TS**), reported in Figure 6. The reaction **1H-FLP_R** → **1H-FLP_P** was found to be endergonic (ΔE = 44.4 kJ/mol) with an associated activation barrier of 45.2 kJ/mol. The latter results support the hypothesis that Cu and the totally oxidised equatorial ligand of molybdenum could play the role of a FLP.

For the complex with a water molecule at the copper centre (**2HR**), we could not locate any transition state geometry that would allow the deprotonation of the hydrogen molecule by the anionic Glu763 residue.

Finally, we considered protonation of Cys388 and found that such protonation can be functional for the hydrogenase activity. In fact, in this case the activation energy for H_2 cleavage involving the equatorial oxo-ligand is only 29.3 kJ/mol (**3HTS** in Figure 7), that is, 15.9 kJ/mol lower than for **1H-FLP**.

Independently of the protonation state of Cys388, our results support a mechanism in which the equatorial oxo group of Mo extracts the first proton. Such process is characterised by a rather low activation barrier that may allow a relatively rapid proton exchange, in accordance with experimental evidence [9]. However, differently from what was hypothesised in the literature [9], we propose that this rapid proton exchange is promoted by the oxo ligand on Mo, rather than by Glu763. Moreover, the equilibrium of this step favours the H_2 reactant complex over copper hydride in both cases, as discussed by Wilcoxen and Hille in light of the pH independence of the reaction. Finally, the intermediates **3HP** or **1H-FLP_P** should be converted to the fully reduced form of the enzyme by the transfer of the Cu-bound hydride to the (now protonated) O_{eq}, giving $Mo^{IV}O(OH_2)Cu^{I}$ [8,13,27] (**P2**; see Figure 8). In both cases, the resulting reduced intermediates were found to be very stable, −90.8 and −82.8 kJ/mol for **3HP2** and **1H-FLP_P2**, respectively (see Table 1 and Figures 6 and 7). Moreover, the reaction barriers for the above two reactions are relatively low (28.4 and 34.7 kJ/mol, respectively). Finally, the reverse reactions are energetically impeded since the reverse activation barriers exceed 140 kJ/mol in both cases (TSs energies are and 147.2 and 161.9 kJ/mol respectively). Again, this is in good agreement with experiments, showing no production of H_2 by the enzyme [9].

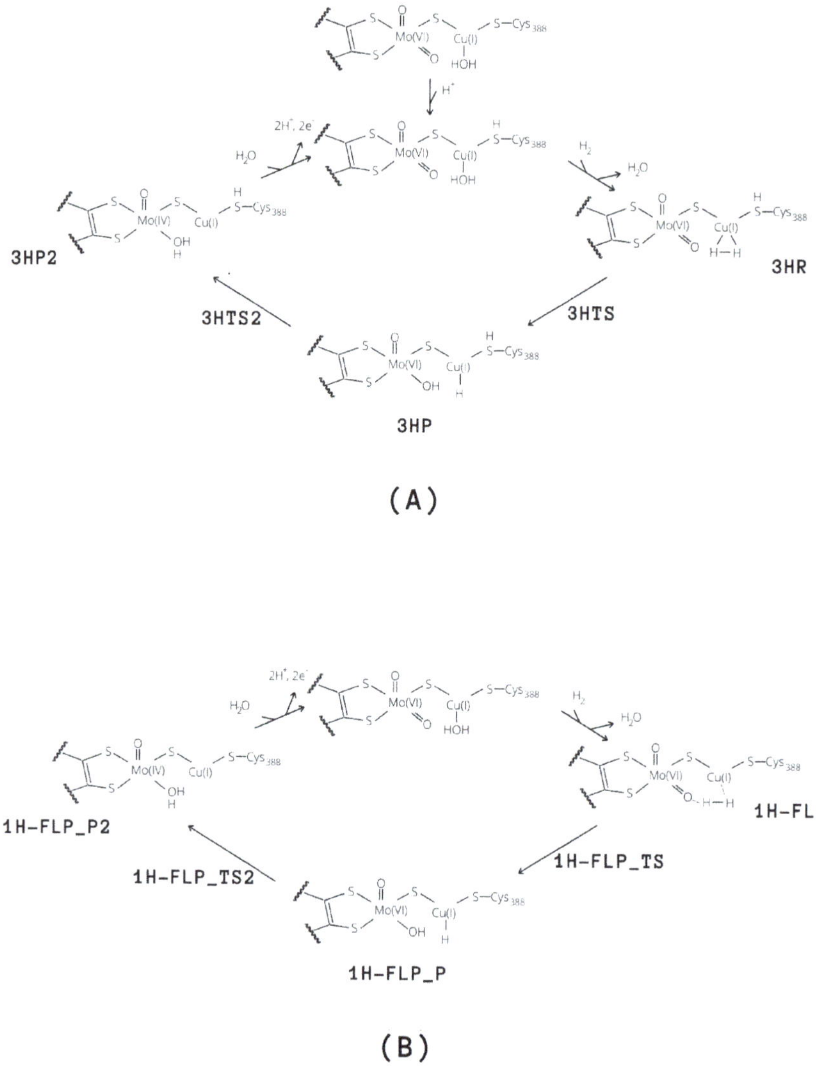

(A)

(B)

Figure 8. Proposed catalytic cycles for the oxidation of H_2 by means of MoCu-CODH considering either the presence of protonated (A) and deprotonated (B) Cys388.

3. Methods

3.1. The Protein

All calculations were based on the crystal structure of MoCu-CODH hydrogenase in its oxidised form (PDB ID: 1N5W) [7]. Only the large subunit (CoxL) of one monomer, containing the active site, was considered in this study. The enzyme was set up in the same way as in our previous study [28]. The protonation state of all the residues was determined based on calculations with PROPKA [29] and on studies of the hydrogen-bond pattern, of the solvent accessibility and of the possible formation of ionic pairs. All Arg, Lys, Asp and Glu residues were assumed to be charged, with exception of Glu29 and Glu488 that were protonated on OE2, whereas Asp684 was protonated on OD1. Cysteine ligands coordinating to metals were deprotonated. Among the His residues, His61, 339, 766 and 793 were protonated on the ND1 atom, His177, 178, 210, 213, 243, 700, 753, 754 and 788 were assumed to be protonated on NE2 atom, whereas the other His residues were assumed to be doubly protonated. The protein was solvated with water molecules, forming a sphere with a radius of 60 Å around the geometric centre of the protein. The added protons and water molecules were then optimised by a 1-ns simulated-annealing molecular-dynamics simulation, followed by energy minimization [28] (for details on the adopted force field, see the next subsection).

3.2. *QM/MM Calculations*

The QM/MM calculations were performed with the ComQum software [30,31]. According to this approach, the protein and the solvent are split into three subsystems—System 1 corresponds to the QM region and it was relaxed by QM methods. It consisted of the molybdenum ion and its first coordination sphere (two O^{2-}, the bridging sulphide and the MCD cofactor, truncated to exclude the phosphate and cytosine moieties), the copper ion, the ligand molecules H_2, H_2O, as well as the sidechains of residues Cys388 and Glu763 (see Figure 9).

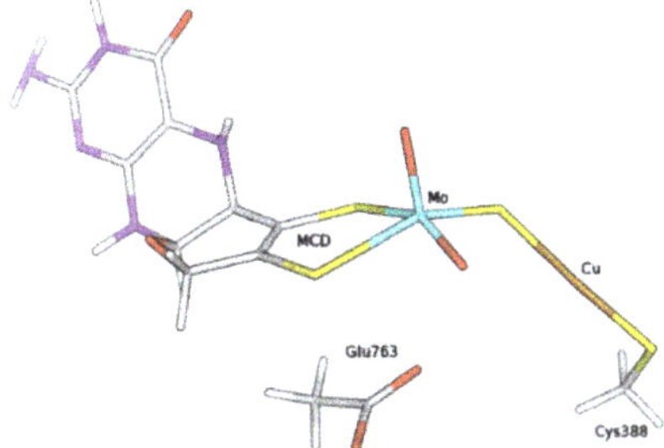

Figure 9. Composition of the QM system within the hybrid QM/MM model of the enzyme, in the form free from H_2/H_2O ligands at the Cu centre.

System 2 consists of all residues with any atom within 6 Å of any atom in System 1, whereas System 3 contains the remaining part of the protein and the water molecules. The latter two systems were kept fixed at the crystallographic coordinates in the QM/MM calculations.

In the QM calculations, System 1 was represented by a wavefunction whereas all the other atoms were represented by an array of partial point charges.

When there is a covalent bond between the QM and MM systems, the QM system was truncated using the hydrogen link-atom approach [32]. The QM system was capped with hydrogen atoms (hydrogen link atoms, HL), the position of which are linearly related to the corresponding carbon atoms (carbon link atoms, CL) in the full system [30]. All atoms were included in the point-charge model, except the CL atoms. ComQum employs a subtractive scheme with electrostatic embedding and van der Waals link-atom corrections [33]. The total QM/MM energy is calculated as

$$E_{\mathrm{QM/MM}} = E^{\mathrm{HL}}_{\mathrm{QM1+ptch23}} + E^{\mathrm{CL}}_{\mathrm{MM123},q_1=0} - E^{\mathrm{HL}}_{\mathrm{MM1},q_1=0} \tag{1}$$

where $\mathrm{E}^{\mathrm{HL}}_{\mathrm{QM1+ptch23}}$ is the QM energy of System 1 truncated by HL atoms and embedded in the set of point charges modeling Systems 2 and 3. $\mathrm{E}^{\mathrm{HL}}_{\mathrm{MM1},q_1=0}$ is the MM energy of System 1, truncated by HL atoms, without any electrostatic interactions. $\mathrm{E}^{\mathrm{CL}}_{\mathrm{MM123},q_1=0}$ is the classical energy of the whole system, with CL atoms and with the charges in System 1 set to zero, to avoid double counting of the electrostatic interactions. The QM calculations were carried out using TURBOMOLE 7.1 software [34]. Geometry optimizations and TS searches (the latter based on minimizations with geometric constraints imposed step-wise to selected atoms, along the putative reactive path) were performed using the BP86 functional [35,36] in combination with def2-TZVP basis set [37]. All calculations included Grimme's dispersion correction with Becke–Johnson damping (D3(BJ)) [38]. The resolution-of-identity technique was employed to accelerate the calculations [39].

The MM calculations were performed with the Amber software [40], using the Amber ff14SB force field for the protein [41] and the general Amber force field [42] with restrained electrostatic potential (RESP) charges [43] for H_2 and MCD. The two Fe_2S_2 clusters were described with RESP charges and a non-bonded model (they are kept fixed in the calculations).

4. Conclusions

We have studied dihydrogen oxidation by MoCu-CODH using QM/MM calculations. Our results indicate that the equatorial oxo ligand of Mo is a better base than Glu763 during catalysis. Moreover, extraction of the first proton from H_2 by means of this basic centre leads to the formation of a Mo-OH–Cu^{I}H hydride that allows for the stabilization of the copper hydride, otherwise known to be very unstable [44]. It is intriguing to notice that our proposal of a direct involvement of the Mo-bound oxo group in H_2 splitting finds a conceptually similar case in a recently published mechanistic study on the [Fe]-hydrogenase. In the latter enzyme, a deprotonared OH group that belongs to the iron–guanylylpyridinol cofactor and that is in γ-position with respect to the metal-activated H_2, was suggested to be involved in substrate splitting [45]. Such mechanistic picture on the [Fe]-hydrogenase was the result of crystallographic studies that took advantage of the fact that the latter enzyme exists in two forms—an open substrate-accessible form and a closed catalytically active form. This is not the case for MoCu-CODH; however, key details on the proton transfer events following H_2 binding might still be gained at experimental level, most likely by means of application of high-sensitivity infrared spectroscopy techniques (see for example, Reference [46]).

In light of our results, two plausible mechanisms for the hydrogenase activity of the enzyme can be proposed, as reported in Figure 8. The first reactive channel (cycle **A** in Figure 8) depends on protonation of the sulphur atom of Cys388, which appears to favour the binding and activation of the hydrogen molecule. The second reactive channel involves a frustrated Lewis pair, formed by the Mo–O_{eq} oxo group and the copper centre (cycle **B** in Figure 8). In this case, no binding of the hydrogen molecule to the Cu center is observed but once H_2 enters into the active site, it can be split following a low-energy path. All in all, we cannot exclude that both pathways may be active in MoCu-CODH, thus enhancing the overall catalytic turnover. The two mechanisms for H_2 oxidation are characterized by energy profiles that were found to be quite similar (see Figure 10) and although the FLP-based mechanism features a ~20 kJ/mol higher rate-determining barrier, it does not require any preliminary active-site protonation. However, the H^+ source for protonation of Cys388 might be the H_2 molecule itself, as a result of a possible proton transfer following initial H_2 splitting at the active site by means of the FLP mechanism. Alternatively, protonation of Cys388 may occur as a parallel, concerted process at the onset of H_2 oxidation catalysis. Notably, the existence of a pair of distinct proton pathways has been recently suggested for [FeFe]-hydrogenases. There, a "regulatory" protonation of the [4Fe–4S] portion of the hexanuclear active site was proposed to occur in concomitance with a "catalytic" protonation step occurring at the di-iron portion of the same active site [47]. The results presented in the latter study and—hopefully—those discussed in the present paper are likely to provide useful information for future efforts to deepen insights into proton transfer mechanisms from and to the MoCu-CODH active site.

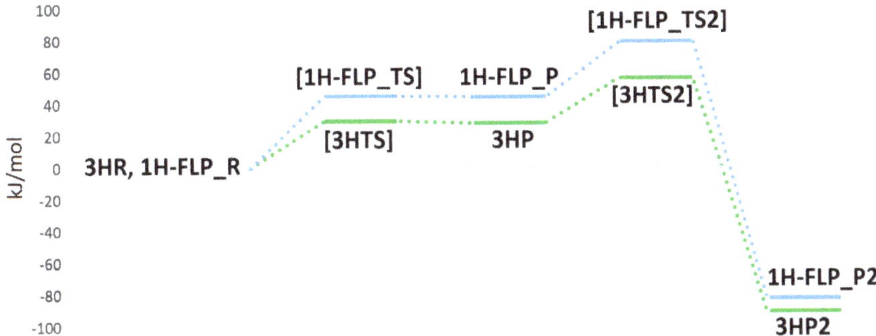

Figure 10. QM/MM energy profiles (in kJ/mol) for the oxidation of H_2 in MoCu-CODH, considering either the presence of protonated (profile traced in green) or deprotonated (profile traced in blue) Cys388.

Author Contributions: Conceptualization, C.G. and U.R.; investigation and formal analysis, A.R., C.G. and U.R.; data curation, A.R.; writing—original draft preparation, A.R.; writing—review and editing, U.C., M.B., G.M., C.G. and U.R.

Funding: This investigation has been supported by grants from the Swedish research council (project 2018-05003). The computations were performed on computer resources provided by the Swedish National Infrastructure for Computing (SNIC) at Lunarc at Lund University.

Conflicts of Interest: The authors declare no conflict of interest.

References

1. Mörsdorf, G.; Frunzke, K.; Gadkari, D.; Meyer, O. Microbial growth on carbon monoxide. *Biodegradation* **1992**, *3*, 61–82. [CrossRef]
2. Xavier, J.C.; Preiner, M.; Martin, W.F. Something special about CO-dependent CO_2 fixation. *FEBS J.* **2018**, *285*, 4181–4195. [CrossRef] [PubMed]
3. Jeoung, J.-H.; Martins, B.M.; Dobbek, H. X-ray Crystallography of Carbon Monoxide Dehydrogenases. In *Metalloproteins*; Hu, Y., Ed.; Humana Press: New York, NY, USA, 2019; pp. 167–178.
4. Santiago, B.; Meyer, O. Characterization of hydrogenase activities associated with the molybdenum CO dehydrogenase from *Oligotropha carboxidovorans*. *FEMS Microbiol. Lett.* **1996**, *136*, 157–162. [CrossRef]
5. Nishino, T.; Okamoto, K.; Leimkühler, S. Enzymes of the Xanthine Oxidase Family. In *Molybdenum and Tungsten Enzymes: Biochemistry*; Hille, R., Schulzke, C., Kirk, M.L., Eds.; The Royal Society of Chemistry: London, UK, 2016; pp. 192–239.
6. Hille, R.; Hall, J.; Basu, P. The mononuclear molybdenum enzymes. *Chem. Rev.* **2014**, *114*, 3963–4038. [CrossRef] [PubMed]
7. Dobbek, H.; Gremer, L.; Kiefersauer, R.; Huber, R.; Meyer, O. Catalysis at a dinuclear [CuSMo(=O)OH] cluster in a CO dehydrogenase resolved at 1.1-Å resolution. *Proc. Natl. Acad. Sci. USA* **2002**, *99*, 15971–15976. [CrossRef] [PubMed]
8. Zhang, B.; Hemann, C.F.; Hille, R. Kinetic and spectroscopic studies of the molybdenum-copper CO dehydrogenase from Oligotropha carboxidovorans. *J. Biol. Chem.* **2010**, *285*, 12571–12578. [CrossRef] [PubMed]
9. Wilcoxen, J.; Hille, R. The Hydrogenase Activity of the Molybdenum/Copper-containing Carbon Monoxide Dehydrogenase of *Oligotropha carboxidovorans*. *J. Biol. Chem.* **2013**, *288*, 36052–36060. [CrossRef] [PubMed]
10. Gourlay, C; Nielsen, D.J.; White, J.M.; Knottenbelt, S.Z.; Kirk, M.L.; Young, C.G. Paramagnetic active site models for the molybdenum–copper carbon monoxide dehydrogenase. *J. Am. Chem. Soc.* **2006**, *128*, 2164–2165. [CrossRef] [PubMed]
11. Kaufmann, P.; Duffus, B.R.; Teutloff, C.; Leimkühler, S. Functional Studies on *Oligotropha carboxidovorans* Molybdenum-Copper CO Dehydrogenase Produced in *Escherichia coli*. *Biochemistry* **2018**, *57*, 2889–2901. [CrossRef] [PubMed]

12. Hille, R.; Dingwall, S.; Wilcoxen, J. The aerobic CO dehydrogenase from *Oligotropha carboxidovorans*. *J. Biol. Inorg. Chem. JBIC* **2015**, *20*, 243–251. [CrossRef] [PubMed]
13. Rokhsana, D.; Large, T.A.G.; Dienst, M.C.; Retegan, M.; Neese, F. A realistic in silico model for structure/function studies of molybdenum–copper CO dehydrogenase. *J. Biol. Inorg. Chem. JBIC* **2016**, *21*, 491–499. [CrossRef] [PubMed]
14. Breglia, R.; Bruschi, M.; Cosentino, U.; De Gioia, L.; Greco, C.; Miyake, T.; Moro, G. A theoretical study on the reactivity of the Mo/Cu-containing carbon monoxide dehydrogenase with dihydrogen. *Protein Eng. Des. Sel.* **2017**, *30*, 169–174. [CrossRef] [PubMed]
15. Plitt, H.S.; Bär, M.R.; Ahlrichs, R.; Schnöckel, H. [Cu(η^2-H_2)Cl], a model compound for H_2 complexes. Ab initio calculations and identification by IR spectroscopy. *Angew. Chem. Int. Ed.* **1991**, *30*, 832–834. [CrossRef]
16. Frohman, D.J; Grubbs, G.S., II; Yu, Z.; Novick, S.E. Probing the Chemical Nature of Dihydrogen Complexation to Transition Metals, a Gas Phase Case Study: H_2–CuF. *Inorg. Chem.* **2013**, *52*, 816–822. [CrossRef] [PubMed]
17. Serykh, A.I.; Kazansky, V.B. Unusually strong adsorption of molecular hydrogen on Cu^+ sites in copper-modified ZSM-5. *Phys. Chem. Chem. Phys.* **2004**, *6*, 5250–5255. [CrossRef]
18. Spoto, G.; Gribov, E.; Bordiga, S.; Lamberti, C.; Ricchiardi, G.; Scarano, D.; Zecchina, A. $Cu^+(H_2)$ and $Na^+(H_2)$ adducts in exchanged ZSM-5 zeolites. *Chem. Commun.* **2004**, 2768–2769. [CrossRef] [PubMed]
19. Georgiev, P.A.; Albinati, A.; Mojet, B.L.; Ollivier, J.; Eckert, J. Observation of Exceptionally Strong Binding of Molecular Hydrogen in a Porous Material: Formation of an η^2-H_2 Complex in a Cu-Exchanged ZSM-5 Zeolite. *J. Am. Chem. Soc.* **2007**, *129*, 8086–8087. [CrossRef] [PubMed]
20. Reguera, L.; Krap, C.P.; Balmaseda, J.; Reguera, E. Hydrogen storage in copper prussian blue analogues: Evidence of H_2 coordination to the copper atom. *J. Phys. Chem. C* **2008**, *112*, 15893–15899. [CrossRef]
21. Stephan, D.W. "Frustrated Lewis pairs": A concept for new reactivity and catalysis. *Org. Biomol. Chem.* **2008**, *6*, 1535–1539. [CrossRef] [PubMed]
22. Carr, S.B.; Evans, R.M.; Brooke, E.J.; Wehlin, S.A.M.; Nomerotskaia, E.; Sargent, F.; Armstrong, F.A.; Phillips, S.E.V. Hydrogen activation by [NiFe]-hydrogenases. *Biochem. Soc. Trans.* **2016**, *44*, 863–868. [CrossRef] [PubMed]
23. Miyake, T.; Bruschi, M.; Cosentino, U.; Baffert, C.; Fourmond, V.; Legér, C.; Moro, G.; De Gioia, L.; Greco, C. Does the environment around the H-cluster allow coordination of the pendant amine to the catalytic iron center in [FeFe] hydrogenases? Answers from theory. *J. Biol. Inorg. Chem. JBIC* **2013**, *18*, 693–700. [CrossRef] [PubMed]
24. Greco, C.; Silakov, A.; Bruschi, M.; Ryde, U.; De Gioia, L.; Lubitz, W. Magnetic Properties of [FeFe]-Hydrogenases: A Theoretical Investigation Based on Extended QM and QM/MM Models of the H-Cluster and Its Surroundings. *Eur. J. Inorg. Chem.* **2011**, 1043–1049. [CrossRef]
25. Greco, C. H_2 Binding and Splitting on a New-Generation [FeFe]-Hydrogenase Model Featuring a Redox-Active Decamethylferrocenyl Phosphine Ligand: A Theoretical Investigation. *Inorg. Chem.* **2013**, *52*, 1901–1908. [CrossRef] [PubMed]
26. Siegbahn, P.E.M.; Shestakov, A.F. Quantum chemical modeling of CO oxidation by the active site of molybdenum CO dehydrogenase. *J. Comput. Chem.* **2005**, *26*, 888–898. [CrossRef] [PubMed]
27. Gnida, M.; Ferner, R.; Gremer, L.; Meyer, O.; Meyer-Klaucke, W. A novel binuclear [CuSMo] cluster at the active site of carbon monoxide dehydrogenase: Characterization by X-ray absorption spectroscopy. *Biochemistry* **2003**, *42*, 222–230. [CrossRef] [PubMed]
28. Rovaletti, A.; Bruschi, M.; Moro, G.; Cosentino, U.; Ryde, U.; Greco, C. A thiocarbonate sink on the enzymatic energy landscape of aerobic CO oxidation? Answers from DFT and QM/MM models of MoCu CO-dehydrogenases. *J. Catal.* **2019**, *372*, 201–205. [CrossRef]
29. Olsson, M.H.M.; Søndergaard, C.R.; Rostkowski, M.; Jensen, J.H. PROPKA3: Consistent treatment of internal and surface residues in empirical pK_a Predictions. *J. Chem. Theory Comput.* **2011**, *7*, 525–537. [CrossRef] [PubMed]
30. Ryde, U. The coordination of the catalytic zinc ion in alcohol dehydrogenase studied by combined quantum-chemical and molecular mechanics calculations. *J. Comput.-Aided Mol. Des.* **1996**, *10*, 153–164. [CrossRef] [PubMed]
31. Ryde, U.; Olsson, M.H.M. Structure, strain, and reorganization energy of blue copper models in the protein. *Int. J. Quantum Chem.* **2001**, *81*, 335–347. [CrossRef]

32. Reuter, N.; Dejaegere, A.; Maigret, B.; Karplus, M. Frontier bonds in QM/MM methods: A comparison of different approaches. *J. Phys. Chem. A* **2000**, *104*, 1720–1735. [CrossRef]
33. Cao, L.; Ryde, U. On the difference between additive and subtractive QM/MM calculations. *Front. Chem.* **2018**, *6*, 89. [CrossRef] [PubMed]
34. TURBOMOLE V6.3 2011, a Development of University of Karlsruhe and Forschungszentrum Karlsruhe GmbH, 1989–2007; TURBOMOLE GmbH, Since 2007. Available online: http://www.turbomole.com (accessed on 8 November 2019)
35. Becke, A. Density-functional exchange-energy approximation with correct asymptotic behavior. *Phys. Rev. A* **1988**, *38*, 3098–3100. [CrossRef] [PubMed]
36. Perdew, J.P. Density-functional approximation for the correlation energy of the inhomogeneous electron gas. *Phys. Rev. B* **1986**, *33*, 8822–8824. [CrossRef] [PubMed]
37. Weigend, F.; Ahlrichs, R. Balanced basis sets of split valence, triple zeta valence and quadruple zeta valence quality for H to Rn: Design and assessment of accuracy. *Phys. Chem. Chem. Phys.* **2005**, *7*, 3297–3305. [CrossRef] [PubMed]
38. Grimme, S.; Ehrlich, S.; Goerigk, L. Effect of the damping function in dispersion corrected density functional theory. *J. Comput. Chem.* **2011**, *32*, 1456–1465. [CrossRef] [PubMed]
39. Eichkorn, K.; Weigend, F.; Treutler, O.; Ahlrichs, R. Auxiliary basis sets for main row atoms and transition metals and their use to approximate Coulomb potentials. *Theor. Chem. Acc.* **1997**, *97*, 119–124. [CrossRef]
40. Case, D.A.; Babin, V.; Berryman, J.; Betz, R.M.; Cai, Q.; Cerutti, D.S.; Cheatham, T.E., III; Darden, T.A.; Duke, R.E.; Gohlke, H.; et al. *AMBER 14*; University of California: San Francisco, CA, USA, 2014.
41. Maier, J.A.; Martinez, C.; Kasavajhala, K.; Wickstrom, L.; Hauser, K.E.; Simmerling, C. ff14SB: Improving the accuracy of protein side chain and backbone parameters from ff99SB. *J. Chem. Theory Comput.* **2015**, *11*, 3696–3713. [CrossRef] [PubMed]
42. Wang, J.; Wolf, R.M.; Caldwell, J.W.; Kollman, P.A.; Case, D.A. Development and testing of a general Amber force field. *J. Comput. Chem.* **2004**, *25*, 1157–1174. [CrossRef] [PubMed]
43. Bayly, C.I.; Cieplak, P.; Cornell, W.D.; Kollman, P.A. A Well-Behaved Electrostatic Potential Based Method Using Charge Restraints for Deriving Atomic Charges: The RESP Model. *J. Phys. Chem.* **1993**, *97*, 10269–10280. [CrossRef]
44. Romero, E.A.; Olsen, P.M.; Jazzar, R.; Soleilhavoup, M.; Gembicky, M.; Bertrand, G. Spectroscopic Evidence for a Monomeric Copper(I) Hydride and Crystallographic Characterization of a Monomeric Silver(I) Hydride. *Angew. Chem. Int. Ed.* **2017**, *56*, 4024–4027. [CrossRef] [PubMed]
45. Huang, G.; Wagner, T.; Wodrich, M. D.; Ataka, K.; Bill, E.; Ermler, U.; Hu, X.; Shima, S. The atomic-resolution crystal structure of activated [Fe]-hydrogenase. *Nat. Catal.* **2019**, *2*, 537–543 [CrossRef]
46. Tai, H.; Nishikawa, K.; Higuchi, Y.; Mao, Z.-W.; Hirota, S. Cysteine SH and Glutamate COOH Contributions to [NiFe] Hydrogenase Proton Transfer Revealed by Highly Sensitive FTIR Spectroscopy. *Angew. Chem. Int. Ed.* **2019**, *58*, 13285–13290. [CrossRef] [PubMed]
47. Senger, M.; Mebs, S.; Duan, J.; Shulenina, O.; Laun, K.; Kertess, L.; Wittkamp, F.; Apfel, U.P.; Happe, T.; Winkler, M. Protonation/reduction dynamics at the [4Fe–4S] cluster of the hydrogen-forming cofactor in [FeFe]-hydrogenases. *Phys. Chem. Chem. Phys.* **2018**, *20*, 3128–3140. [CrossRef] [PubMed]

Article

A Mixed-Valence Tetra-Nuclear Nickel Dithiolene Complex: Synthesis, Crystal Structure, and the Lability of Its Nickel Sulfur Bonds

Mohsen Ahmadi, Jevy Correia, Nicolas Chrysochos and Carola Schulzke *

Institut für Biochemie, Universität Greifswald, Felix-Hausdorff-Straße 4, 17489 Greifswald, Germany; ahmadim.ug@gmail.com (M.A.); correiaj@uni-greifswald.de (J.C.); nc093759@uni-greifswald.de (N.C.)

* Correspondence: carola.schulzke@uni-greifswald.de; Tel.: +49-3834-420-4321

Received: 9 March 2020; Accepted: 3 April 2020; Published: 9 April 2020

Abstract: In this study, by employing a common synthetic protocol, an unusual and unexpected tetra-nuclear nickel dithiolene complex was obtained. The synthesis of the $[Ni_4(ecpdt)_6]^{2-}$ dianion (ecpdt = (Z)-3-ethoxy-3-oxo-1-phenylprop-1-ene-1,2-bis-thiolate) with two K^+ as counter ions was then intentionally reproduced. The formation of this specific complex is attributed to the distinct dithiolene precursor used and the combination with the then coordinated counter ion in the molecular solid-state structure, as determined by X-ray diffraction. $K_2[Ni_4(ecpdt)_6]$ was further characterized by ESI-MS, FT-IR, UV-Vis, and cyclic voltammetry. The tetra-nuclear complex was found to have an uncommon geometry arising from the combination of four nickel centers and six dithiolene ligands. In the center of the arrangement, suspiciously long Ni–S distances were found, suggesting that the tetrameric structure can be easily split into two identical dimeric fragments or two distinct groups of monomeric fragments, for instance, upon dissolving. A proposed variable magnetism in the solid-state and in solution due to the postulated dissociation was confirmed. The Ni–S bonds of the "inner" and "outer" nickel centers differed concurrently with their coordination geometries. This observation also correlates with the fact that the complex bears two anionic charges requiring the four nickel centers to be present in two distinct oxidation states (2 × +2 and 2 × +3), i.e., to be hetero-valent. The different coordination geometries observed, together with the magnetic investigation, allowed the square planar "outer" geometry to be assigned to d^8 centers, i.e., Ni^{2+}, while the Ni^{3+} centers (d^7) were in a square pyramidal geometry with longer Ni–S distances due to the increased number of donor atoms and interactions.

Keywords: mixed-valence complex; dithiolene ligand; tetra-nuclear nickel complex; X-ray structure; magnetic moment

---

1. Introduction

Ene-dithiolate or dithiolene ligands have been compounds of interest in the scientific community since the 1960s [1–5]. At first, they were mostly investigated for their complexes' electronic structures and reactivity, and then in the context of materials chemistry, e.g., regarding electronic and non-linear optical properties [6,7]. They have since also attracted much attention as model ligands for molybdopterin (mpt—a natural ligand found in the cofactors of molybdenum and tungsten-dependent oxidoreductases) [8–11]. A specific characteristic of dithiolene ligands is their non-innocence, i.e., their ability to donate electrons to the coordinated metal ion. Such behavior was also proposed and discussed for the molybdopterin ligand coordinated to molybdenum or tungsten concurrent with enzymatic turnover processes [12]. More recently, dithiolene ligands were employed in bioinorganic model chemistry targeting hydrogenase enzymes' active sites [13].

Nickel, as a bio-metal, is most often found in combination with sulfur donor atoms, including Ni–Fe hydrogenases, in which the nickel ion and the iron ion of the active site are bridged by cysteinate sulfur atoms. This has led to many related bio-inspired model compounds, including those in which two sulfur bridged nickel ions were used for such species [14–20].

From the very beginning of dithiolene chemistry, nickel has always been an integral part of many respective investigations, either in the course of the synthesis of dithiolene ligands [21,22] as a dithiolene ligand transmitter to other metal ions (e.g., molybdenum or tungsten) [21,23–27] or as a central metal of interest for potential applications [7,28–33]. Such applications of dithiolene-bearing compounds include molecular materials with conducting [34], magnetic [35,36], and optical [37] properties on account of their unique electronic structure.

A combination of sulfur donor dithiolene ligands and coordinated nickel ions is therefore relevant for problems related to bioinorganic and inorganic/materials chemistry and with respect to a focus on either ligands, metals, or both (i.e., their specific combination).

Homoleptic bis-dithiolene complexes, in particular those of group 10 and group 11 metals, $[M(dithiolene)_2]^{n-}$ (n = 0–2), most often exhibit a square planar geometry, which is occasionally slightly distorted [38–42]. Depending on the electron configuration, a tetrahedral geometry might also be accessible, for instance with Ag^+ [43]. Typically, when homoleptic nickel bis-dithiolene complexes oligomerize, sulfur donor atoms become bridging ligands occupying the apical position of one of the coordinated nickel centers in the square pyramidal geometry [31,44].

Metal dithiolene complexes are generally able to form different types of networks by covalent [45] and non-covalent bonds [46], while extended networks with nickel centers are comparatively rare. In cases where two or more nickel ions are bridged by dithiolene sulfur donor atoms, the metal centers might adopt different or partial formal oxidation states. These species can be stabilized by a significant contribution of the extended π-ligands to their frontier orbitals [35]. Respective observations were made with oligo-nuclear nickel complexes, such as $[Ni_2(edt)_3]^{2-}$, $[Ni_3(edt)_4]^{2-}$ (edt = ethane-1,2-di-thiolate) [47], $[Ni_2(dmit)_3]$ (*dmit* = 2-thione-1,3-dithiole-4,5-dithiolate) [48], $[AsPh_4]_2[Ni_3(pdt)_2(dmit)_2]$, and $[NEt_4]_2[Ni_5(edt)_4(dmit)_2]^{2-}$ (pdt = propane-1,2-dithiol, edt = ethane-1,2-dithiol, dmit = 2-thione-1,3-dithiole-4,5-dithiolate) [49]. In 2011, Neves et al. [35] reported an interesting mixed-valence tetra-nuclear nickel dithiolene complex $[K(18\text{-crown-}6)]_2[Ni_4(\alpha\text{-tpdt})_4]$ (α-tpdt = 2,3-thiophenedithiolate) with a structure that was unique at that time and constitutes the only known analog of the novel tetrameric complex which is presented here. The originally targeted nickel dithiolene complex in the present investigation was initially merely of interest as a dithiolene ligand transmitter to a molybdenum center in a procedure, which typically works rather well and is commonly applied in our group [21,23–25,50,51]. With the specific ligand system used (a dithiolene with one phenyl substituent on one carbon of the ene function and one carboxylic acid ethyl ester group on the other), the reaction did not turn out as expected. Instead of the envisaged mono-nuclear nickel bis-dithiolene complex, a mixed valence tetra-nuclear nickel complex was obtained, in which half of all dithiolene sulfur donor atoms were in bridging positions between the two nickel centers. As the obtained complex is a very uncommon type (only one precedent, [35]), nickel centers bridged to other cations (whether nickel or any other metal) by sulfur donor atoms are biologically relevant with regard to hydrogenases, and nickel dithiolene chemistry is generally important (as evidenced by significant ongoing publication activities—vide supra), the complex obtained was investigated comprehensively and in detail, and the results of this study are presented here.

2. Results and Discussion

One of the most reliable and comparatively convenient procedures for the synthesis of dithiolene ligands (originally developed by Garner's group in particular for those with unsymmetrical substitutions on the ene) employs the xanthogenation of α-bromoketones, followed by an acidic cyclization reaction [9,52]. Accordingly, the starting material in this study, ethyl benzoyl acetate, was treated with *N*-bromosuccinimide (NBS) in solvent-free conditions (Scheme 1) and purified by Kugelrohr distillation. The colorless oily product, **1**, was subsequently reacted with the potassium salt of *o*-isopropyl xanthate to replace the bromine substituent by the R-CS_2^- moiety, affording xanthogenated compound **2** in 84% yield. The targeted dithiolene precursor 4-ethylcarboxylate-5-phenyl-1,3-dithiole-2-one (ecpdt = CO, **3**) was obtained as a colorless crystalline solid through an acidic cyclization reaction with H_2SO_4 in a diethyl ether/dichloromethane solvent system and purified by column chromatography. Ligand precursor **3** was characterized inter alia by single-crystal X-ray diffraction (Figure 1).

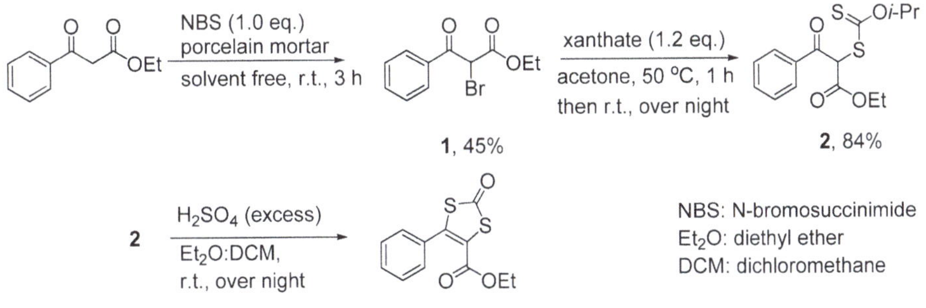

NBS (1.0 eq.)
porcelain mortar
solvent free, r.t., 3 h

xanthate (1.2 eq.)
acetone, 50 °C, 1 h
then r.t., over night

1, 45%

2, 84%

2

H_2SO_4 (excess)
Et_2O:DCM,
r.t., over night

3, 66%

NBS: N-bromosuccinimide
Et_2O: diethyl ether
DCM: dichloromethane

Scheme 1. The three-step preparation of ligand precursor ecpdt = CO (**3**) from commercially available ethyl benzoyl acetate.

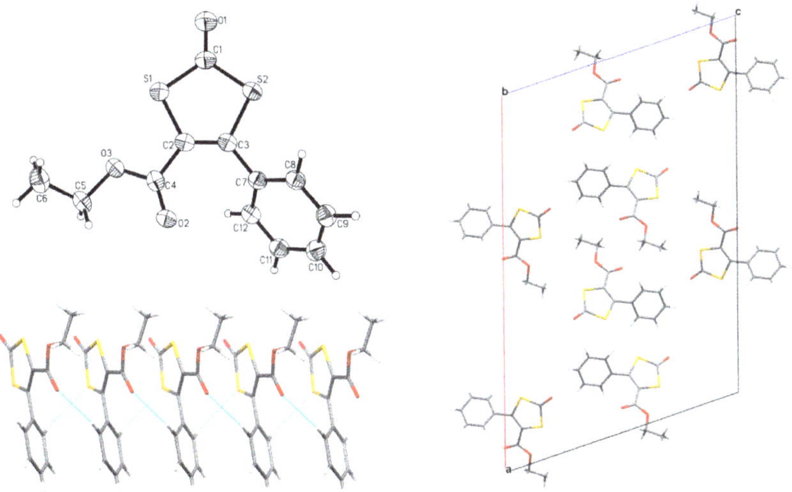

Figure 1. Top left: Molecular structure of **3** shown with 50% probability thermal ellipsoids. Selected bond lengths (Å) and angels (°): [C1–O1: 1.221(5), S1–C1: 1.761(4), S1–C2: 1.756(4), S2–C1: 1.764(4), S2–C3: 1.747(4), C2–C3, 1.353(5)]; [C2–C3–S2: 96.10(19), C3–C2–S1: 97.27(19), S1–C1–S2: 123.8(3)]. Bottom left: Non-classical intermolecular hydrogen bonding interactions (shown in light blue) forming chains, which protrude along the crystallographic *b*-axis (carbon: gray, hydrogen: white, oxygen: red, and sulfur: yellow). Right: Projection view of the crystal packing along the crystallographic *b*-axis.

The nickel cluster as part of **4** was first synthesized unintentionally, and then on purpose, by reacting nickel(II) chloride with the de-protected form of **3**, i.e., the ecpdt^{2-} ligand, which was generated in situ by the addition of KOH in dry and degassed MeOH (Scheme 2). This reaction with the typical 2:1 stoichiometry, using molecular oxygen in air as the oxidant, usually leads to a nickel(III) bis-dithiolene complex, which had been the original target and was planned to be used in a dithiolene *trans*-metalation reaction. Here, however, an unexpected transformation took place. The raw product was initially obtained upon drying as a brownish powder and then recrystallized by the slow diffusion of diethyl ether into an acetonitrile solution of **4** at room temperature while exposed to air, yielding black cube-shaped crystals with a reddish shine. The molecular structure of the Ni-tetrameric anion of **4** is shown in Figure 2, while all components of the crystal structure are displayed to the left in Figure 3.

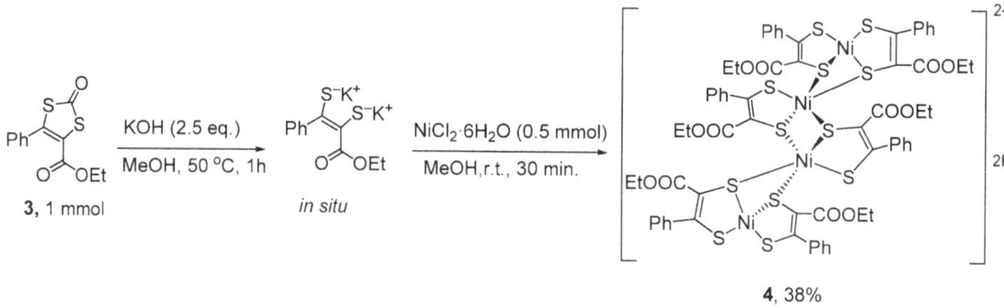

KOH (2.5 eq.)
MeOH, 50 °C, 1h
3, 1 mmol
in situ
NiCl2·6H2O (0.5 mmol)
MeOH, r.t., 30 min.
4, 38%
2K+

Scheme 2. Preparation of the nickel-tetramer salt $K_2[Ni_4(ecpdt)_6]$ (**4**) (Et = ethyl, Ph = phenyl).

S4A
O4A
S2A
Ni1A
O3A
S3A
S1A
O5
O6
Ni2A
O2A
S5
O1A
S6
S6A
S5A
O5A
O6A
O2
O1
Ni2
O3
S1
S3
S2
Ni1
O4
S4

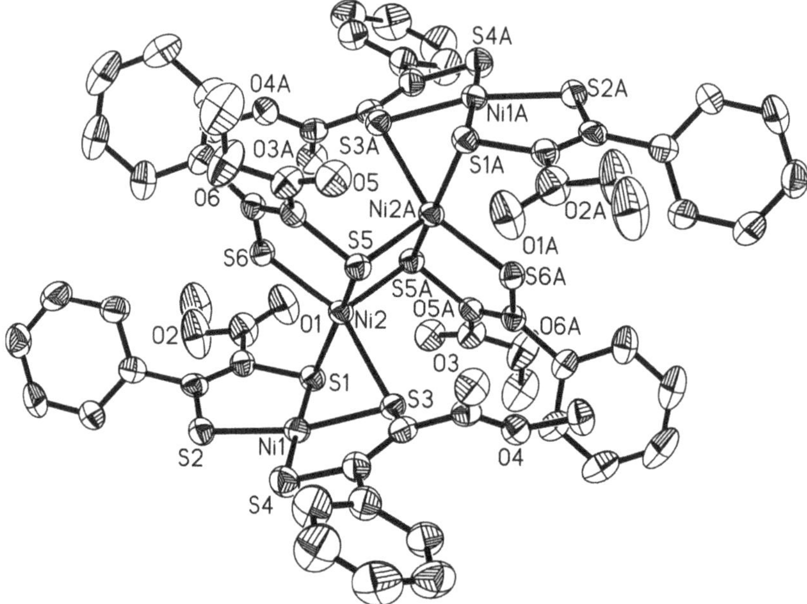

Figure 2. Molecular structure of the di-anion of **4·2CH$_3$CN·2Et$_2$O** shown with 50% probability thermal ellipsoids (the two potassium counter cations, disordered minor occupancies, H atoms, and lattice solvents are omitted and only heteroatom labels are shown for clarity reasons; see the Supplementary Material, Figure S16, for a fully labeled version; label extension A refers to symmetry (1−x, 1−y, 1−z) generated atoms).

4·2CH$_3$CN·2Et$_2$O crystallizes in monoclinic space group $P2_1/n$. The asymmetric unit consists of half the tetrameric di-anion, one potassium cation, one acetonitrile, and one diethyl ether. The complete molecular structure is generated by a symmetry operation (i.e., inversion in the center of the unit cell; 1−x, 1−y, 1−z). The unit cell of **4·2CH$_3$CN·2Et$_2$O** contains a total of two di-anions and four potassium ions as counter cations, although one anion is located in the very center. Each of the potassium ions is coordinated by one dithiolene ligand's sulfur donor atom plus three carbonyl oxygen donor atoms of the ester substituents (all belong to the Ni-tetrameric di-anion), as well as by one nitrogen donor atom and another oxygen donor atom from the two coordinated solvent molecules (acetonitrile and diethyl ether), thereby resulting in a coordination number of six for the potassium cations (Figure 3). The four nickel centers of the anion are arranged in-line and the whole complex and entire structure are centrosymmetric. Selected bond lengths and angles of the anionic tetrameric nickel complex of **4** are listed in Table 1.

Table 1. Selected bond lengths [Å] and angles [°] for **4·2CH$_3$CN·2Et$_2$O** (label extension A refers to symmetry (1−x, 1−y, 1−z) generated atoms; all data are presented in the Supplementary Material, Table S2).

Bond Lengths		Bond Angles	
Ni(1)–Ni(2)	2.6747(11)	S(1)–Ni(1)–S(2)	91.82(6)
Ni(2)–Ni(2A)	2.9187(16)	S(1)–Ni(1)–S(3)	84.52(6)
Ni(1)–S(1)	2.1487(17)	S(1)–Ni(1)–S(4)	176.32(7)
Ni(1)–S(2)	2.1482(17)	S(1)–Ni(2)–S(3)	77.09(6)
Ni(1)–S(3)	2.1725(17)	S(1)–Ni(2)–S(5)	164.00(6)
Ni(1)–S(4)	2.1464(18)	S(1)–Ni(2)–S(5A)	95.06(6)
Ni(2)–S(1)	2.3125(16)	S(1)–Ni(2)–S(6)	92.80(6)
Ni(2)–S(3)	2.3509(16)	S(3)–Ni(2)–S(5)	94.06(6)
Ni(2)–S(5)	2.1758(16)	S(3)–Ni(2)–S(5A)	102.43(5)
Ni(2)–S(5A)	2.3554(16)	S(3)–Ni(2)–S(6)	156.26(6)
Ni(2)–S(6)	2.1793(16)	S(5)–Ni(2)–S(5A)	99.91(6)
C(1)–C(4)	1.366(8)	S(5)–Ni(2)–S(1)	90.24(6)
C(11)–C(14)	1.364(8)	Ni(1)–Ni(2)–S(5)	113.74(5)
C(21)–C(24)	1.369(8)	Ni(1)–Ni(2)–S(5A)	136.58(5)
		Ni(1)–Ni(2)–Ni(2A)	152.41(4)

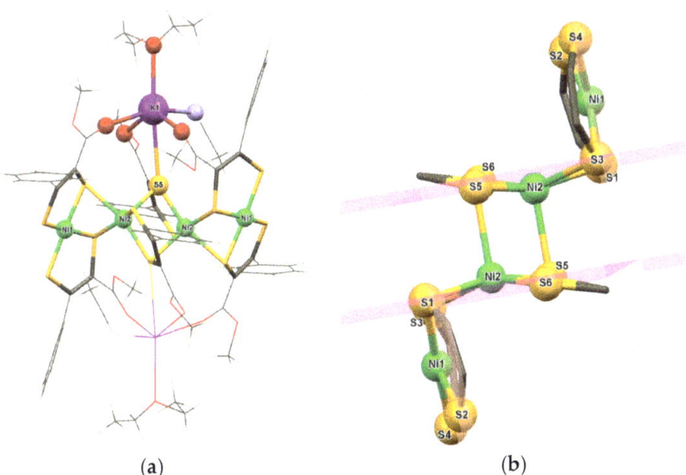

Figure 3. (**a**) The entire molecular structure of crystalline K$_2$[Ni$_4$(ecpdt)$_6$]·2CH$_3$CN·2Et$_2$O (**4·2CH$_3$CN·2Et$_2$O**) highlighting all contacts around the three distinct metal centers (Ni1, Ni2, and K1); disordered minor occupancies not shown; carbon: gray, hydrogen: white, oxygen: red, sulfur: yellow, nickel: green, and potassium: purple. Selected bond lengths in Å: K1–O1: 2.591(5), K1–O3: 2.651(5), K1–O5: 2.676(5), K1–O7: 2.744(8), K1–S5: 3.378(2), and K1–N1: 2.831(8); (**b**) arrangement of nickel centers and coordinated dithiolene ligands (the ligand substituents are omitted for clarity); notably, the Ni2 centers are slightly dislocated above and below the S1, S3, S5, and S6 planes (darker pink), while Ni1 resides more firmly in the plane (brighter pink) of its four surrounding sulfur donor atoms (S1, S2, S3, and S4).

The average Ni–S distance for Ni1 (the outer square-planar nickel center) is 2.154 Å, while the three distances range from 2.1464 Å to 2.1487 Å and one is notably longer, with a value of 2.1725 Å (S3). For Ni2 (the inner square-pyramidal nickel center), the average bond length is 2.275 Å and the distances can be divided into two sets. One dithiolene ligand is only coordinated to Ni2 and has no interaction with Ni1. The respective Ni–S distances are 2.1758 Å (S5) and 2.1793 Å (S6). S5 binds to both Ni2; once in a basal position as part of the ene-dithiolate chelate and secondly in a bridging fashion (Ni2–S5–Ni2) occupying the apex of the second nickel's pyramidal coordination polyhedron. The distance to this second Ni2 is 2.3554 Å. The other two Ni2–S distances each involve one sulfur donor atom of the two dithiolene ligands forming chelates with Ni1 (Ni2–S1: 2.3125 Å and Ni2–S3: 2.3509 Å). Conspicuously, both S1 and S3 belong to the ester sides of the dithiolene ligands (not the phenyl side). All individual and average Ni–S distances are longer for the inner five-coordinated nickel centers than they are for the outer four-coordinated nickel centers. This points to Ni1 being smaller, with a more oxidized center, and Ni2 being larger and more reduced. This contrasts with the nearly perfectly square-planar geometry around Ni1 being most often associated with a d^8 metal center (i.e., corresponding to more reduced Ni^{2+}). At the same time, the coordination shell around Ni2 is more crowded, engaging five instead of only four ligands. This might contribute to elongation of the Ni–S bonds, despite a higher oxidation state. Notably, the longer ones of the Ni–S bonds, i.e., those which surpass 2.3 Å, mean that the Ni and S atoms have distances which are longer than the sum of their covalent radii (2.29 Å [53]), suggesting only comparably weak attraction/interaction. While Ni1 is clearly strongly coordinated by all four sulfur donor atoms of its two dithiolene ligands, three out of the five Ni–S interactions of Ni2 are significantly weakened beyond a typical Ni–S single bond. The longest, and thus weakest, Ni–S distance is the one which holds the two identical halves of the tetramer together, i.e., the apical coordination of S5 to the second Ni2 in the complex's center. Notably, S5 not only bridges the two Ni2 centers, but also interacts with the potassium counter ion (S5–K1: 3.378(2) Å). This distance is very close to the sum of their covalent radii (3.39 Å [53]) and it will take away at least some electron density from the Ni2–S5 bonds, thereby weakening them. The longer nickel–sulfur bonds are, in fact, consistent with a dissociation (vide infra) of the tetrameric complex in solution, as has been observed previously for dimeric species with different central metals, which are closely related to the central structural Ni2/Ni2 motif of **4·2CH$_3$CN·2Et$_2$O** [54,55]. The central Ni2–Ni2A distance across the perfectly planar, lozenge-shaped four-membered ring (formed with two S5 sulfur bridges) is with 2.9178(15) Å longer than the Ni1–Ni2 distance of 2.6750(10) Å. The observed Ni–Ni distances are both in the range, yet slightly shorter compared to the respective distances in the previously described related $[Ni_4(\alpha\text{-tpdt})_6]^{2-}$ complex (3.142(3) Å and 2.756(2) Å) [35]. Based on a detailed DFT computational study and due to the absence of bond critical points and the occupation of antibonding molecular orbitals, Neves et al. dismissed the presence of relevant Ni–Ni bonding in their Ni_4 species [35] and due to the very close similarity of the two tetrameric complexes, it can be assumed that there are no actual metal–metal bonds in **4·2CH$_3$CN·2Et$_2$O** either.

With regard to assigning the two oxidation states (+2 and +3) to the inner and outer nickel centers, metrical structural parameters of the C–S and C=C bonds in the three dithiolene ligands of the asymmetric unit are also of interest. Dithiolene ligands are non-innocent and can push more electron density than usual towards a coordinated metal ion using their π system, including the lone pairs in the sulfur *p*-orbitals (see the review by Yang et al. in Special Issue "Transition Metals in Catalysis: The Functional Relationship of Fe–S Clusters and Molybdenum or Tungsten Cofactor-Containing Enzyme Systems" of *Inorganics*) [54,56]. The ligand's extra electron density donation beyond a single coordinative σ-bond comes at the expense of the C=C double bond strength and is concurrent with the formation of a partial C=S bond character. Non-innocent behavior, therefore, results in slightly elongated C–C and slightly shortened C–S distances (with the latter quite often only being on one side of the ene-dithiolate ligand). Typically, the resultant bond lengths suggest a mixed single/double bond character for both interactions; most often residing more on the side of the ene-dithiolate forms.

In the protected ligand precursor **3**, the C=C double bond is 1.353(5) Å long and the C–S distances are 1.747(4) Å (phenyl side) and 1.756(4) Å (ester side) (see Table 2). In the ligand, which is only coordinated to Ni2 in **4·2CH$_3$CN·2Et$_2$O**, the double bond is elongated (C21–C24: 1.369(8) Å), the ester side C–S distance is slightly shortened (C21–S5: 1.748(6) Å), and the phenyl side C–S distance is significantly shortened (C24–S6: 1.712(6) Å). This supports the presence of a moderate non-innocent effect and a sulfur-mediated push of electron density out of the C=C double bond towards the coordinated nickel center.

Table 2. Atom to atom distances [Å] of C=C and C–S bonds in ligand precursor **3** and complex salt **4·2CH$_3$CN·2Et$_2$O**.

3		4·2CH$_3$CN·2Et$_2$O	
C(2)–C(3)	1.353(5)	C(1)–C(4)	1.366(8)
C(2)–S(1) (ester side)	1.756(4)	C(1)–S(1) (ester side)	1.745(6)
C(3)–S(2) (phenyl side)	1.747(4)	C(4)–S(2) (phenyl side)	1.711(6)
		C(11)–C(14)	1.364(8)
		C(11)–S(3) (ester side)	1.749(6)
		C(14)–S(4) (phenyl side)	1.729(6)
		C(21)–C(24)	1.369(8)
		C(21)–S(5) (ester side)	1.748(6)
		C(24)–S(6) (phenyl side)	1.712(6)

Ni1 is firmly bound by two chelating dithiolene ligands. The respective ligands' bond lengths here also exhibit an elongation of the C=C distances (C1–C4: 1.366(8) Å, C11–C14: 1.364(8) Å), while the C–S distances are significantly shortened on the phenyl sides (C4–S2: 1.711(6) Å and C14–S4: 1.729(6) Å) and more subtly on the ester sides (C1–S1: 1.745(6) Å and C11–S3: 1.749(6) Å). Overall a non-innocent effect is present here as well, but to a lesser extent overall. Interestingly, for all three dithiolene ligands, the phenyl sides exhibit the shorter C–S distances, suggesting that this is the sulfur atom which moves more electron density towards the nickel center, even though it is, in all cases, the ester side of the dithiolene ligand which engages in bridging binding modes. This un-symmetrical trend has already been observed in the ligand precursor, but to a much lesser extent. Possibly, some type of push-pull effect is present here in the unsymmetrically substituted dithiolenes, which is strengthened upon coordination due to the ligands' non-innocent behavior. In a nickel ion with a higher oxidation state, the non-innocent effect is presumed to be more prominent compared to more reduced, i.e., electron-rich, metal centers. In contrast to the recorded Ni–S bond lengths, which suggest that the outer nickel ions (Ni1) are smaller (hence, more oxidized), the structural data of the dithiolene ligands, therefore, support Ni2 being more oxidized with a stronger push of electron density from ligands to metal. This paradox can only be solved by involving additional analytical methods and observations; in particular, the magnetic properties are helpful in this respect (vide infra).

The unexpectedly formed tetrameric nickel complex **4** was further analyzed by a set of standard/routine characterization methods. The most prominent molecular ion peaks in its ESI-MS spectrum, which was recorded in methanolic solution and in the negative ion mode, can be assigned to monomeric bis-dithiolene nickel(III) species with charge ratios (m/e) of 533.9, 519.9, and 505.8. The oxidation of the nickel(II) centers to nickel(III) is not uncommon/unexpected, given that the solution of the probe needs to be handled in air prior to being injected into the instrument. The observed isotopic patterns thus belong to complex fragmentations and oxidation of the outer nickel centers (for the two lower mass signals) and to transformations of one or two of the ethyl esters to methyl ester functional groups in the methanol solvent (see the Supplementary Material, Figure S9). This strongly supports the notion that some of the nickel sulfur bonds in the complex are decidedly weak and most likely rather easily severed when dissolved in a coordinating solvent. From this behavior, we can therefore derive qualitative approximations of Ni–S bond strengths (in particular for bridging binding modes) from metrical parameters, i.e., it can be concluded whether they have to be presumed short enough to be stable, despite the presence of potentially competing coordinating ligands or whether they are too long and thus labile.

Reported Ni–S distances in FeNi and FeNiSe hydrogenases range from rather short (1.98 Å) to quite long (2.6 Å) [57–62]. It has to be taken into consideration, though, that the accuracy of such metrical parameters, particularly in proteins, greatly depends on the resolution of the gathered data. Naturally, the most recent published values tend to be the most accurate ones and they rest on the shorter side. In reports where the Ni–S distances are distinguished, apical and bridging ones are typically longer (up to 2.6 Å in comparison to shorter terminal ones of 2.2–2.3 Å), as expected. Bond lengths of 2.4 Å are long enough to suggest that they are labile and easily severed. The respective structural studies, however, are less recent. When compared to the observations of this study, the most recent published Ni–S bond lengths from an FeNiSe hydrogenase (1.98 Å, PDB code: 5JSH) strongly suggest that the respective interactions between the nickel center and sulfur donor atom are stable and have no tendency to dissociate [61]. Notably, in a recent comprehensive theoretical investigation [63] based on the most recent X-ray structure of an FeNiSe hydrogenase, the theoretical values were ca. 0.2 Å longer than the experimental ones and from a chemical point of view, the computed bond lengths were actually more reasonable for a Ni^{2+} or Ni^{3+} centre. However, these slightly longer computational bonds are decidedly shorter than the bonds in the Ni tetrameric structure of this study, which can apparently lead to a break-down of the complex in solution. This renders any potential mechanism in nickel-bearing hydrogenases less likely, which would involve replacing a Ni–S contact with an incoming donor atom.

Infrared spectra of **4** were recorded both in the solid-state (KBr matrix) and in solution (CH_3CN) (see the Supplementary Material, Figure S10). Assignment of the observed vibrational bands is based on a comparison with the ligand precursor's IR spectrum and on an extensive previous respective report [64]. Vibrational bands are indeed shifted, significantly in parts, in the solid-state spectrum compared to the solution spectrum. In particular, the Ni–S band (~550 cm^{-1}) and the C–S band (~1000 cm^{-1}) regions show fewer and less intense signals in solution than in the solid-state. This indicates considerable changes in the molecular and/or electronic structures upon dissolving, which is again in accordance with the proposed dissociation of the tetrameric species into dimers or monomers in solution.

The electronic absorption spectra of **4** were recorded in MeOH, CH_3CN, and water solvent (see the Supplementary Material, Figure S11). The broad low-energy near-IR band at 868 nm is very characteristic of monomeric bis-dithiolene nickel complexes with oxidation states of +3 or +4 and has been ascribed to transitions of sulfur lone pair electrons into mixed nickel-ligand molecular orbitals which are empty (OS +4) or singly occupied (OS +3), while Ni^{2+} bis-dithiolene complexes do not exhibit this transition because the respective receiving MO is already fully occupied [29,65–67]. Notably, this transition is missing in the aqueous spectrum, which means that either the tetrameric complex does not dissociate in water; it dissociates completely, even beyond the bis-dithiolene species to, for instance, mono-dithiolene species; or it is being reduced/not oxidized in aqueous solution. There is no reasonable explanation for why water as a medium would be better at protecting the tetrameric complex's fragments against oxidation by air or even induce reduction compared to acetonitrile or methanol. The most likely explanation is that water not only disintegrates the tetrameric complex into monomers by replacing the weaker bound sulfur donor atoms around the inner Ni2, but also interferes with the coordination shell of the outer Ni1, so that no bis-dithiolene complexes result from dissolving in water. The spectra in methanol and acetonitrile are slightly distinct, which suggests that interactions between solvent and dissolved species take place and result in subtle differences in the electronic structures of the mono-dithiolene fragments, which mix in the resulting solutions with the bis-dithiolene complexes.

The electrochemical properties of **4** were investigated by cyclic voltammetry (CV; referenced versus ferrocene, Fc/Fc^+, in CH_3CN solution with 0.1 M of $[^nBu_4N][PF_6]$ as supporting electrolyte; see the Supplementary Material, Figures S12 and S13). The voltammograms are relatively close to those of *bis*-dithiolene nickel complexes available in the literature [29,30], again supporting a dissociation of the tetrameric complex into monomers (or dimers). The two observed quasi-reversible redox transitions are, hence, tentatively ascribed to the couples $[Ni(ecpdt)_2]^{2-}/[Ni(ecpdt)_2]^-$ and $[Ni(ecpdt)_2]^-/[Ni(ecpdt)_2]^0$ at −1.1 and −0.1 V (vs. Fc/Fc^+), respectively. However, the evaluation here is impeded, due to the possible chemical dissociation equilibria and potentially manifold species, which may occur in solution.

Neves et al. have also investigated their related tetrameric nickel complex electrochemically and reported two redox transitions at −0.480 V and +0.286 V vs. Ag/AgCl [35], which roughly translates into values of −0.68 V and +0.08 V vs. Fc/Fc^+ [68]. These values are, firstly, more positive and their complex is thus easier to reduce. Secondly, the gap between the two redox events is smaller by ca. 0.24 V, suggesting a more significant involvement of the dithiolene ligand in the redox transitions than in the case of **4**. While we do not observe any further distinguishable signals, Neves et al. further reported a redox process in between at −0.095 V vs. Ag/AgCl (−0.298 vs. Fc/Fc^+) [35]. Overall, however, the electrochemical behavior of the two tetrameric complexes is very similar, particularly when considering that different solvents were used.

Lastly, in order to also unambiguously assign the oxidation states +2 and +3 to the appropriate nickel centers in **4**, an investigation of the magnetic properties of the complex was carried out with comparably simple/convenient methods. This was done in both the solid-state and in solution. For the solid-state investigation, a Magnetic Susceptibility Balance (MSB), also referred to as a Guy Balance, was used. Based on the measurement data, the magnetic moment was determined as μ_{eff} = 1.86 *BM* (Bohr magneton; see the Supplementary Material for calculation details) per non-dissociated tetrameric nickel complex. Therefore, **4** exhibits some paramagnetism, but much less than anticipated for the two unpaired electrons of the two Ni(III) centers, and only slightly more than the expected spin-only magnetism for a single unpaired electron of 1.73 BM. This clearly points toward significant antiferromagnetic coupling in the solid-state between the two metal centers and this can only take place if the respective atoms are close together. These observations, therefore, very clearly place the unpaired electrons on the two central nickel ions (Ni2), which are thus assigned the oxidation state +III. This is in accordance with the metrical structural ligand data and fits the results of the DFT study by Neves et al. on the related tetrameric complex perfectly well, including the predicted antiferromagnetic coupling [35]. Notably, Neves et al. did not have enough material to conduct a respective experiment with their complex. With our experimental data, we can confirm their computational results and vice versa, even though the complexes are not strictly identical.

The magnetic moment of **4** in solution was determined by the Evans method [69,70]. μ_{eff} of **4** was calculated to be 2.826 BM (see the Supplementary Material for details of measurements and calculations). The effective magnetism is, hence, in solution almost perfectly equal to the spin-only value for two unpaired electrons (2.828 BM) per tetrameric complex. This points towards two d^7/Ni(III) centers with one unpaired electron each, which are not antiferromagnetically coupled in solution, as opposed to what is observed in the solid-state. Again, the distinct magnetic properties in solution and in the solid-state firmly support the fragility of the central (long) Ni–S bonds and a dissociation of the tetramer upon interaction with a solvent.

3. Materials and Methods

3.1. Physical Measurements

NMR measurements were recorded on a Bruker Avance II-300 MHz instrument (Karlsruhe, Germany). All samples were dissolved in deuterated solvents, and chemical shifts (δ) are given in parts per million (ppm) using solvent signals as the reference ($CDCl_3$ 1H: δ = 7.24 ppm; 13 C: δ = 77.0 ppm) related to external tetramethylsilane (δ = 0 ppm). Coupling constants (*J*) are reported in Hertz (Hz), and splitting patterns are designated as s (singlet), d (doublet), t (triplet), q (quartet), quint (quintet), m (multiplet), and dd (doublet of doublet). The infrared spectra were recorded on a Perkin-Elmer Fourier-Transform Infrared (FT-IR) spectrophotometer in the range of 4000–400 cm^{-1} using KBr pellets (solid-state) or KBr windows and a concentrated CH_3CN solution of the analyte (in solution). Assignment of the bands was done with subjective appreciation: w = weak, m = medium, s = strong, vs = very strong, and br = broad. UV/vis spectra were recorded on a Shimadzu UV-3600PC spectrophotometer. Solutions with concentrations ranging from 10 to 1 μM were prepared for the measurements. All measurements were carried out at room temperature in quartz cuvettes (path length

= 1 cm). Elemental analyses (C, H, N, and S) were carried out with an Elementar Vario Micro Elemental Analyzer (Langenselbold, Germany). Mass spectra (APCI and ESI) were recorded with an Advion Expression CMS spectrometer (Ithaca, NY, USA) with a resolution of 0.5–0.7 m/e units (FWHM) at 1000 m/e units s^{-1} over the entire acquisition range. The magnetic moment experiment was performed with a Johnson-Matthey Mark I Magnetic Susceptibility Balance (Redwitz, Germany). Electrochemical measurements were carried out with an AUTOLAB PGSTAT12 potentiostat/galvanostat (Filderstadt, Germany) using a glassy carbon disc electrode with a reaction surface of 1 mm^2 as the working electrode. A platinum knob electrode (together with internal referencing versus ferrocene/ferrocenium (Fc/Fc+)) was used as a reference electrode and a platinum rod electrode was employed as an auxiliary electrode. All measurements were carried out inside a glove box and controlled with the NOVA software (2.0.1., Metrohm, Filderstadt, Germany). Tetrabutylammonium hexafluorophosphate (0.1 M; electrochemical grade from Fluka, München, Germany) was used as the electrolyte.

3.2. X-Ray Crystallography

Suitable single crystals of **3** and **4·2CH$_3$CN·2Et$_2$O** were mounted on a thin glass fiber coated with paraffin oil. X-ray single-crystal structural data were collected using a STOE-IPDS II diffractometer (Darmstadt, Germany) equipped with a normal-focus, 2.4 kW, sealed-tube X-ray source with graphite-monochromated Mo Kα radiation (λ = 0.71073 Å) at low temperature (170 K). The program XArea was used for the integration of diffraction profiles; numerical absorption corrections were carried out with the programs X-Shape and X-Red32, all from STOE© ((Darmstadt, Germany). The structures were solved by direct methods with SHELXT [71] and refined by full-matrix least-squares methods using SHELXL [72]. All calculations were carried out using the WinGX system, Ver 2018.3 [73]. All non-hydrogen atoms were refined anisotropically. The hydrogen atoms were refined isotropically on calculated positions using a riding model with their Uiso values constrained to 1.2 Ueq of their pivot atoms for methyl groups and 1.2 Ueq of their pivot atoms for all other groups. The refinement of the crystallographic data of **3** was unremarkable. In the refinement of **4·2CH$_3$CN·2Et$_2$O**, several disorder problems were encountered. The two solvent molecules coordinated to K1 exhibited thermal motion and the respective ellipsoids were comparably large. For acetonitrile, the three non-hydrogen atoms were constrained with SIMU and DELU. In addition, the ethyl groups of Et_2O were disordered over two positions each, which was treated with SADI, SIMU, and DELU constraints. Furthermore, two out of three ethyl groups of the dithiolene ethyl ester functionalities were disordered over two positions each. Again, this was treated with SADI, SIMU, and DELU constraints.

General crystallographic, crystal, and refinement data for **3** and **4·2CH$_3$CN·2Et$_2$O** are provided in the supplementary data file (see the Supplementary Material, Table S1). Crystallographic data were deposited with the Cambridge Crystallographic Data Centre, CCDC, 12 Union Road, Cambridge CB21EZ, UK. These data can be obtained free of charge upon quoting the depository numbers CCDC 1985587 (**3**) and 1985588 (**4·2CH$_3$CN·2Et$_2$O**) by FAX (+44-1223-336-033), email (deposit@ccdc.cam.ac.uk), or their web interface (at http://www.ccdc.cam.ac.uk).

3.3. Syntheses

3.3.1. Ethyl 2-bromo-3-oxo-3-phenylpropanoate (**1**)

Method (a). Ethyl benzoylacetate (10 g, 52 mmol) and *N*-bromosuccinimide (NBS) (9.26 g, 52 mmol) were triturated together in a porcelain mortar for 15 min. The resulting liquid paste was allowed to stand for 3 h and then washed three times with 100 mL H_2O in a separating funnel. The final crude product was distilled by Kugelrohr distillation to remove the unreacted starting material. Distillation was performed at 70 °C under a vacuum of 100 mmHg (b.p. 135 °C). The final product was a pale yellow oil that is a severe lachrymator. Yield: 6.34 g, 45%.

Method (b). NBS (9.26 g, 52 mmol) was added to ethyl benzoyl acetate (10 g, 52 mmol). The reaction mixture was refluxed one hour under microwave heating (magnetron power of 750 W until a temperature of 160 °C was reached). The dark brown reaction mixture was cooled slowly for 30 min and was distilled by Kugelrohr distillation. Yield: 10 g, 71%. ^{1}H NMR ($CDCl_3$, 300 MHz): δ, ppm = 7.94–8.05 (m, ortho-phenyl, 2H), 7.55–7.67 (m, para-phenyl, 1H), 7.43–7.54 (m, meta-phenyl, 2H), 5.73 (s, –CHBr, 1H), 4.27 (q, J = 6.9 Hz, $-CH_2$, 2H), 1.23 ppm (t, J = 7.2 Hz, $-CH_3$, 3H). ^{13}C NMR ($CDCl_3$, 75 MHz): δ, ppm = 187.9 (PhC=O), 164.8 (C=O), 134.0 (C-phenyl), 133.0 (C-phenyl), 128.8 (C-phenyl), 128.6 (C-phenyl), 62.9 (CH_2), 46.2 (CHBr), 13.5 (CH_3). APCI-MS (EI+): m/e calculated for $C_{11}H_{11}BrO_3$: 269.99, Found: 270.99 ($[M + H]^+$).

3.3.2. Ethyl 2-(isopropoxycarbonothioylthio)-3-oxo-3-phenylpropanoate (2)

Potassium o-isopropyl xanthate (3.34 g, 19.1 mmol) was added portion-wise to a solution of **1** (4.33 g, 16 mmol) dissolved in 60 mL acetone. The red mixture was heated at 50 °C for 1 h and then stirred overnight at RT. Precipitated white KBr was filtered off with suction employing Celite. The brown-yellow solution was concentrated on a rotary evaporator, and the red oil was dissolved in chloroform. The organic phase was washed with 10% HCl and water, and dried over anhydrous Na_2SO_4, and the solvent was evaporated on a rotary evaporator, resulting in a reddish-brown oil that was used without further purification. Yield: 4.4 g, 84%. APCI-MS (EI+): m/e calculated for $C_{12}H_{12}O_4S_2$: 284.02, Found: 284.65 ($[M + H]^+$).

3.3.3. 4-Ethylcarboxylate-5-phenyl-1,3-dithiole-2-one (3)

A total of 20 mL of sulfuric acid was slowly added to an ice-cold solution of **2** (4.09 g, 12.5 mmol) in 75 mL of Et_2O:DCM solvent mixture (1:1 ratio) and was stirred overnight at RT. The reaction was controlled by TLC (*n*-hexane/EtOAc). The solution was cooled down in an ice bath and slowly poured onto ice-cooled H_2O (200 mL). The solution was stirred for another hour and then extracted with CH_2Cl_2 (3 × 50 mL), dried over anhydrous Na_2SO_4, and concentrated on a rotary evaporator to produce a crude, dark red oil that was purified by flash column chromatography with gradient elution (5–15% EtOAc) to afford **3** as colorless crystals. Yield: 2.2 g, 63% (considering remaining 1/16 molecule of *n*-hexane and EtOAc per formula each, as supported by NMR and EA). ^{1}H NMR ($CDCl_3$, 300 MHz): δ, ppm = 7.31–7.54 (m, phenyl, 5H), 4.13 (q, J = 7.2 Hz, $-CH_2$, 2H), 1.12 ppm (t, J = 7.0 Hz, $-CH_3$, 3H). ^{13}C NMR ($CDCl_3$, 75 MHz): δ, ppm = 188.3 ($C{=}O_{dithiolene}$), 159.0 ($C{=}O_{ester}$), 145.5 ($Ph{-}C{=}C_{dithiolene}$), 131.0 (C-phenyl), 129.8 (C-phenyl), 128.9 (C-phenyl), 128.3 (C-phenyl), 119.9 ($COOEt{-}C{=}C_{dithiolene}$), 62.1 ($CH_2$), 13.7 ($CH_3$). FT-IR (KBr): (ν, cm^{-1}) = 3421 (br), 3005 (ws), 2980 (s), 1728 (s), 1695 (s), 1674–1658 (br), 1543 (s), 1489 (m), 1471(s), 1444 (m), 1390 (s), 1363 (w), 1263–1253 (br), 1205 (m), 1155 (m), 1109 (s), 1087 (w), 1066 (w), 1033 (m), 1020 (s), 948 (m), 912 (m), 893 (s), 867 (m), 825 (w), 802 (w), 761 (w), 748 (m), 727 (w), 690 (w), 628 (w), 588 (w), 591 (w), 491 (w), 466 (w), 445 (w). Elemental analysis for $C_{12}H_{10}O_3S_2{\cdot}1/16C_4H_8O{\cdot}1/16C_6H_{14}$ ($C_{12.625}H_{11.375}O_{3.063}S_2$; M = 277.23 g/mol) calc.: C, 54.70; H, 4.14; S, 23.13. Found: C, 55.30; H, 4.04; S, 22.73. APCI-MS (EI+): m/e calculated for $C_{12}H_{10}O_3S_2$: 266.01, Found: 266.94 ($[M + H]^+$). Fragmentation calc. for $C_{10}H_6O_3S_2$: 237.98, Found: 238.97 ($[M + H]^+$).

3.3.4. $K_2[Ni_4(ecpdt)_6]\cdot 2CH_3CN\cdot 2Et_2O$ (4·2CH$_3$CN·2Et$_2$O)

A total of 0.15 g of **3** (0.56 mmol) was suspended with 78 mg of KOH (1.4 mmol, 2.5 equiv.) in 10 mL of degassed methanol and stirred under nitrogen for 1 h at 50 °C, affording an orange solution. Next, 3 mL of a solution of $NiCl_2{\cdot}6H_2O$ (0.5 mmol, 0.12 g) in methanol was added, and the resulting reaction mixture was stirred for 30 min. Then, it was concentrated to dryness, affording a dark brown colored precipitate. Single crystals were obtained by recrystallization from a solvent mixture of CH_3CN:Et_2O (ratio 1:2) in a diethyl ether saturated atmosphere, while air was allowed to slowly diffuse in, providing molecular oxygen needed for the partial oxidation of nickel centers. Yield: 90 mg, 38%. FT-IR (KBr): ν, cm^{-1} = 3446 (br), 3055 (w), 2974 (s), 1668 (br), 1365–1473 (s), 1132–1303 (br), 985–1062 (m), 804 (w), 744 (w), 696 (w), 657 (w), 617 (m), 503 (m). ESI-MS (EI−): m/e calculated for $C_{20}H_{16}NiO_4S_4$ $[M]^-$ (*i*): 505.9, found: 505.0; calc. for $C_{21}H_{18}NiO_4S_4$ (*ii*): 519.9, found: 519.0, calc. for $C_{22}H_{20}NiO_4S_4$ (*iii*): 533.9, found: 533.0. *N.B.*: there appear to be signals in the mass spectrum supporting fragments much larger than monomers, but they are of a very low intensity. UV-Vis (CH_3CN): λ_{max} (ε/M^{-1} cm^{-1}), nm = 256,

315, 366, 476, 518, 868 (2000). Cyclic voltammetry (in CH_3CN, Bu_4NPF_6 vs. Fc/Fc+): $E_{1/2} = -1.1$ V ($E^2_{p,c}$), −0.1 V ($E^1_{p,c}$).

Note: parts of the study/experiments were included in the recent Ph.D. thesis of Mohsen Ahmadi [74].

4. Conclusions

In conclusion, an unexpected di-anionic, tetra-nuclear in-line, hetero-valent, centrosymmetric nickel dithiolene complex was synthesized, which had only one precedent in the literature. The employed dithiolene ligand was substituted unsymmetrically (Ph, –COOEt), probably inducing/experiencing a mild push-pull effect, which was intensified upon coordination. The compound's crystal structural data was evaluated in detail. Comparing the large range of observed Ni–S bond distances will facilitate the evaluation of respective bond strengths in nickel- and sulfur-dependent enzymes' active sites. The two distinct oxidation states +2 and +3 could be assigned unambiguously to the outer and inner nickel centers, respectively, by determining the complex's effective magnetism (μ_{eff}) and by consulting previously published computational results of a related compound. It was further established by several spectroscopic and other methods that the tetramer due to the comparably labile Ni–S bonds in its center dissociates in solution, either into identical dimers or, more likely, into two sets of distinct monomers. Previous computational results without an experimental backup regarding the assignment of oxidation states and antiferromagnetic coupling in the tetramer's solid-state could be verified experimentally (and vice versa) with this novel complex.

Supplementary Materials: The following are available online at http://www.mdpi.com/2304-6740/8/4/27/s1: Figure S1: ^{1}H NMR of **1**; Figure S2: ^{13}C NMR of **1**; Figure S3: APCI mass spectrum of **1**; Figure S4: APCI mass spectrum of **2**; Figure S5: ^{1}H NMR of **3** (*ecpdt*); Figure S6: ^{13}C NMR of **3**; Figure S7: APCI mass spectrum of **3**; Figure S8: FT-IR spectrum of **3**; Figure S9: The ESI(−)-MS of **4**; Figure S10: IR spectra of **4**; Figure S11: UV-vis-NIR spectra of **4** and extinction coefficient diagram; Figure S12: Cyclic voltammogram of **4**; Figure S13: Cyclic voltammogram of **4** at different scan rates; Figure S14: ^{1}H NMR for magnetic moment measurement of the Ni cluster; Figure S15: ^{1}H NMR for magnetic moment measurement of the standard; Table S1: General crystallographic data of **3** and **4·2CH$_3$CN·2Et$_2$O**; Figure S16: Molecular structure of the di-anion of **4·CH$_3$CN·Et$_2$O** shown with 50% probability thermal ellipsoids (the two potassium counter cations, disordered minor occupancies, H atoms, and lattice solvents are omitted for clarity); Table S2: Complete list of bond lengths [Å] and angles [°] for **4·2CH$_3$CN·2Et$_2$O**; Table S3: Hydrogen-bond geometry [Å and °] of **4·2CH$_3$CN·2Et$_2$O**; Table S4: Atomic coordinates and equivalent isotropic displacement parameters for **4·2CH$_3$CN·2Et$_2$O**; Table S5: Anisotropic displacement parameters (Å^2 × 10^3) for **4·2CH$_3$CN·2Et$_2$O**; Cif and checkcif files of **3** and **4·2CH$_3$CN·2Et$_2$O**.

Author Contributions: Synthesis, characterization, and manuscript draft, M.A.; additional investigation, validation, and manuscript draft, J.C.; crystallography, N.C. and C.S.; conceptualization, writing—review and editing, and resources, C.S.; all authors have read and agreed to the published version of the manuscript. All authors have read and agreed to the published version of the manuscript.

Funding: This research was funded by the DFG (Deutsche Forschungsgemeinschaft), grant number SCHU 1480/4-1, as part of the SPP 1927 (Priority Program) "Iron-Sulfur for life". The financial support is gratefully acknowledged.

Acknowledgments: The authors thank Marlen Redies and Gabriele Thede for carrying out the EA and NMR measurements.

Conflicts of Interest: The authors declare no conflicts of interest.

References

1. Schrauzer, G.N.; Mayweg, V.P. Preparation, Reactions, and Structure of Bisdithio-α-diketone Complexes of Nickel, Palladium, and Platinum. *J. Am. Chem. Soc.* **1965**, *87*, 1483. [CrossRef]
2. Schrauzer, G.N.; Mayweg, V. Reaction of Diphenylacetylene with Nickel Sulfides. *J. Am. Chem. Soc.* **1962**, *84*, 3221. [CrossRef]
3. Eisenberg, R.; Stiefel, E.I.; Rosenberg, R.C.; Gray, H.B. Six-Coordinate Trigonal-Prismatic Complexes of First-Row Transition Metals1. *J. Am. Chem. Soc.* **1966**, *88*, 2874–2876. [CrossRef]
4. Stiefel, E.I.; Eisenberg, R.; Rosenberg, R.C.; Gray, H.B. Characterization and Electronic Structures of Six-Coordinate Trigonal-Prismatic Complexes. *J. Am. Chem. Soc.* **1966**, *88*, 2956–2966. [CrossRef]

5. Balch, A.L.; Dance, I.G.; Holm, R.H. Characterization of dimeric dithiolene complexes. *J. Am. Chem. Soc.* **1968**, *90*, 1139–1145. [CrossRef]
6. Mueller-Westerhoff, U.T.; Vance, B.; Yoon, D.I. The synthesis of dithiolene dyes with strong near-IR absorption. *Tetrahedron* **1991**, *47*, 909–932. [CrossRef]
7. Winter, C.S.; Oliver, S.N.; Manning, R.J.; Rush, J.D.; Hill, C.A.S.; Underhill, A.E. Non-linear optical studies of nickel dithiolene complexes. *J. Mater. Chem.* **1992**, *2*, 443–447. [CrossRef]
8. Oku, H.; Ueyama, N.; Kondo, M.; Nakamura, A. Oxygen atom transfer systems in which the (*m*-oxo)dimolybdenum(V) complex formation does not occur: Syntheses, structures, and reactivities of monooxomolybdenum(IV) benzenedithiolato complexes as models of molybdenum oxidoreductases. *Inorg. Chem.* **1994**, *33*, 209–216. [CrossRef]
9. Davies, E.S.; Beddoes, R.L.; Collison, D.; Dinsmore, A.; Docrat, A.; Joule, J.A.; Wilson, C.R.; Garner, C.D. Synthesis of oxomolybdenum bis(dithiolene) complexes related to the cofactor of the oxomolybdoenzymes. *J. Chem. Soc. Dalton Trans.* **1997**, 3985–3995. [CrossRef]
10. Donahue, J.P.; Goldsmith, C.R.; Nadiminti, U.; Holm, R.H. Synthesis, Structures, and Reactivity of Bis(dithiolene)molybdenum(IV,VI) Complexes Related to the Active Sites of Molybdoenzymes. *J. Am. Chem. Soc.* **1998**, *120*, 12869–12881. [CrossRef]
11. Holm, R.H.; Solomon, E.I.; Majumdar, A.; Tenderholt, A. Comparative molecular chemistry of molybdenum and tungsten and its relation to hydroxylase and oxotransferase enzymes. *Coord. Chem. Rev.* **2011**, *255*, 993–1015. [CrossRef]
12. Periyasamy, G.; Burton, N.A.; Hillier, I.H.; Vincent, M.A.; Disley, H.; McMaster, J.; Garner, C.D. The dithiolene ligand—Innocent or 'non-innocent'? A theoretical and experimental study of some cobalt-dithiolene complexes. *Faraday Discuss.* **2007**, *135*, 469–488. [CrossRef]
13. Adams, H.; Morris, M.J.; Robertson, C.C.; Tunnicliffe, H.C.I. Synthesis of Mono- and Diiron Dithiolene Complexes as Hydrogenase Models by Dithiolene Transfer Reactions, Including the Crystal Structure of $[\{Ni(S_2C_2Ph_2)\}_6]$. *Organometallics* **2019**, *38*, 665–676. [CrossRef]
14. Maroney, M.J.; Ciurli, S. Nonredox Nickel Enzymes. *Chem. Rev.* **2014**, *114*, 4206–4228. [CrossRef]
15. Can, M.; Armstrong, F.A.; Ragsdale, S.W. Structure, Function, and Mechanism of the Nickel Metalloenzymes, CO Dehydrogenase, and Acetyl-CoA Synthase. *Chem. Rev.* **2014**, *114*, 4149–4174. [CrossRef]
16. Lubitz, W.; Ogata, H.; Rüdiger, O.; Reijerse, E. Hydrogenases. *Chem. Rev.* **2014**, *114*, 4081–4148. [CrossRef]
17. Van Gastel, M.; Shaw, J.L.; Blake, A.J.; Flores, M.; Schröder, M.; McMaster, J.; Lubitz, W. Electronic Structure of a Binuclear Nickel Complex of Relevance to [NiFe] Hydrogenase. *Inorg. Chem.* **2008**, *47*, 11688–11697. [CrossRef]
18. Choudhury, S.B.; Pressler, M.A.; Mirza, S.A.; Day, R.O.; Maroney, M.J. Structure and Redox Chemistry of Analogous Nickel Thiolato and Selenolato Complexes: Implications for the Nickel Sites in Hydrogenases. *Inorg. Chem.* **1994**, *33*, 4831–4839. [CrossRef]
19. Mondragón, A.; Flores-Alamo, M.; Martínez-Alanis, P.R.; Aullón, G.; Ugalde-Saldívar, V.M.; Castillo, I. Electrocatalytic Proton Reduction by Dimeric Nickel Complex of a Sterically Demanding Pincer-type NS2 Aminobis(thiophenolate) Ligand. *Inorg. Chem.* **2015**, *54*, 619–627. [CrossRef]
20. Song, L.-C.; Lu, Y.; Cao, M.; Yang, X.-Y. Reactions of dinuclear Ni2 complexes [Ni(RNPyS4)]2 (RNPyS4 = 2,6-bis(2-mercaptophenylthiomethyl)-4-R-pyridine) with Fe(CO)3(BDA) (BDA = benzylidene acetone) leading to heterodinuclear NiFe and mononuclear Fe complexes related to the active sites of [NiFe]- and [Fe]-hydrogenases. *RSC Adv.* **2016**, *6*, 39225–39233.
21. Schrauzer, G.N.; Mayweg, V.P.; Heinrich, W. Coordination Compounds with Delocalized Ground States. α-Dithiodiketone-Substituted Group VI Metal Carbonyls and Related Compounds. *J. Am. Chem. Soc.* **1966**, *88*, 5174–5179. [CrossRef]
22. Schrauzer, G.N. Coordination compounds with delocalized ground states. Transition metal derivatives of dithiodiketones and ethylene-1,2-dithiolates (metal dithienes). *Acc. Chem. Res.* **1969**, *2*, 72–80. [CrossRef]
23. Lim, B.S.; Donahue, J.P.; Holm, R.H. Synthesis and Structures of Bis(dithiolene)molybdenum Complexes Related to the Active Sites of the DMSO Reductase Enzyme Family. *Inorg. Chem.* **2000**, *39*, 263–273. [CrossRef] [PubMed]
24. Obanda, A.; Martinez, K.; Schmehl, R.H.; Mague, J.T.; Rubtsov, I.V.; MacMillan, S.N.; Lancaster, K.M.; Sproules, S.; Donahue, J.P. Expanding the Scope of Ligand Substitution from $[M(S_2C_2Ph_2)$ (M = Ni^{2+}, Pd^{2+},

Pt^{2+}) To Afford New Heteroleptic Dithiolene Complexes. *Inorg. Chem.* **2017**, *56*, 10257–10267. [CrossRef] [PubMed]
25. Goddard, C.A.; Holm, R.H. Synthesis and Reactivity Aspects of the Bis(dithiolene) Chalcogenide Series $[WIVQ(S_2C_2R_2)2]^{2-}$ (Q = O, S, Se). *Inorg. Chem.* **1999**, *38*, 5389–5398. [CrossRef]
26. Lim, B.S.; Fomitchev, D.V.; Holm, R.H. Nickel Dithiolenes Revisited: Structures and Electron Distribution from Density Functional Theory for the Three-Member Electron-Transfer Series $[Ni(S_2C_2Me_2)_2]^{0,1-,2-}$. *Inorg. Chem.* **2001**, *40*, 4257–4262. [CrossRef]
27. Ghosh, A.C.; Schulzke, C. Selectively detecting Hg^{2+}—A "mercury quick test" with bis-(coumarin–dithiolene) niccolate. *Inorg. Chim. Acta* **2016**, *445*, 149–154. [CrossRef]
28. Kean, C.L.; Pickup, P.G. A low band gap conjugated metallopolymer with nickel bis(dithiolene) crosslinks. *Chem. Commun.* **2001**, 815–816. [CrossRef]
29. Papavassiliou, G.C.; Anyfantis, G.C.; Mousdis, G.A. Neutral Metal 1,2-Dithiolenes: Preparations, Properties and Possible Applications of Unsymmetrical in Comparison to the Symmetrical. *Crystals* **2012**, *2*, 762–811. [CrossRef]
30. Anyfantis, G.C.; Papavassiliou, G.C.; Assimomytis, N.; Terzis, A.; Psycharis, V.; Raptopoulou, C.P.; Kyritsis, P.; Thoma, V.; Koutselas, I.B. Some unsymmetrical nickel 1,2-dithiolene complexes as candidate materials for optics and electronics. *Solid State Sci.* **2008**, *10*, 1729–1733. [CrossRef]
31. Cassoux, P. Molecular (super)conductors derived from bis-dithiolate metal complexes. *Coord. Chem. Rev.* **1999**, *185–186*, 213–232. [CrossRef]
32. Kato, R. Conducting Metal Dithiolene Complexes: Structural and Electronic Properties. *Chem. Rev.* **2004**, *104*, 5319–5346. [CrossRef] [PubMed]
33. Kusamoto, T.; Nishihara, H. Zero-, one- and two-dimensional bis(dithiolato)metal complexes with unique physical and chemical properties. *Coord. Chem. Rev.* **2019**, *380*, 419–439. [CrossRef]
34. Lieffrig, J.; Jeannin, O.; Auban-Senzier, P.; Fourmigué, M. Chiral Conducting Salts of Nickel Dithiolene Complexes. *Inorg. Chem.* **2012**, *51*, 7144–7152. [CrossRef]
35. Neves, A.I.S.; Santos, I.C.; Pereira, L.C.J.; Rovira, C.; Ruiz, E.; Belo, D.; Almeida, M. Ni-2,3-thiophenedithiolate Anions in New Architectures: An In-Line Mixed-Valence Ni Dithiolene (Ni4–S12) Cluster. *Eur. J. Inorg. Chem.* **2011**, *2011*, 4807–4815. [CrossRef]
36. Jeannin, O.; Clérac, R.; Fourmigué, M. Order−Disorder Transition Coupled with Magnetic Bistability in the Ferricinium Salt of a Radical Nickel Dithiolene Complex. *J. Am. Chem. Soc.* **2006**, *128*, 14649–14656. [CrossRef]
37. Deplano, P.; Pilia, L.; Espa, D.; Mercuri, M.L.; Serpe, A. Square-planar d8 metal mixed-ligand dithiolene complexes as second order nonlinear optical chromophores: Structure/property relationship. *Coord. Chem. Rev.* **2010**, *254*, 1434–1447. [CrossRef]
38. Basu, P.; Nigam, A.; Mogesa, B.; Denti, S.; Nemykin, V.N. Synthesis, characterization, spectroscopy, electronic and redox properties of a new nickel dithiolene system. *Inorg. Chim. Acta* **2010**, *363*, 2857–2864. [CrossRef]
39. Alves, H.; Simão, D.; Cordeiro Santos, I.; Gama, V.; Teives Henriques, R.; Novais, H.; Almeida, M. A Series of Transition Metal Bis(dicyanobenzenedithiolate) Complexes $[M(dcbdt)_2]$ (M = Fe, Co, Ni, Pd, Pt, Cu, Au and Zn). *Eur. J. Inorg. Chem.* **2004**, 1318–1329. [CrossRef]
40. Ghosh, A.C.; Weisz, K.; Schulzke, C. Selective Capture of Ni^{2+} Ions by Naphthalene- and Coumarin-Substituted Dithiolenes. *Eur. J. Inorg. Chem.* **2016**, *2016*, 208–218. [CrossRef]
41. Perochon, R.; Piekara-Sady, L.; Jurga, W.; Clerac, R.; Fourmigue, M. Amphiphilic paramagnetic neutral gold dithiolene complexes. *Dalton Trans.* **2009**, 3052–3061. [CrossRef]
42. Perochon, R.; Poriel, C.; Jeannin, O.; Piekara-Sady, L.; Fourmigué, M. Chiral, Neutral, and Paramagnetic Gold Dithiolene Complexes Derived from Camphorquinone. *Eur. J. Inorg. Chem.* **2009**, *2009*, 5413–5421. [CrossRef]
43. McLauchlan, C.C.; Ibers, J.A. Synthesis and Characterization of the Silver Maleonitrilediselenolates and Silver Maleonitriledithiolates $[K([2.2.2]\text{-cryptand})]_4[Ag_4(Se_2C_2(CN)_2)_4]$, $[Na([2.2.2]\text{-cryptand})]_4[Ag_4(S_2C_2(CN)_2)_4]\cdot 0.33MeCN$, $[NBu_4]_4[Ag_4(S_2C_2(CN)_2)_4]$, $[K([2.2.2]\text{-cryptand})]_3[Ag(Se_2C_2(CN)_2)_2]\cdot 2MeCN$, and $[Na([2.2.2]\text{-cryptand})]_3[Ag(S_2C_2(CN)_2)_2]$. *Inorg. Chem.* **2001**, *40*, 1809–1815.
44. Watanabe, E.; Fujiwara, M.; Yamaura, J.-I.; Kato, R. Synthesis and properties of novel donor-type metal–dithiolene complexes based on 5,6-dihydro-1,4-dioxine-2,3-dithiol (edo) ligand. *J. Mater. Chem.* **2001**, *11*, 2131–2141. [CrossRef]

45. Llusar, R.; Vicent, C. Trinuclear molybdenum cluster sulfides coordinated to dithiolene ligands and their use in the development of molecular conductors. *Coord. Chem. Rev.* **2010**, *254*, 1534–1548. [CrossRef]
46. Castillo, O.; Delgado, E.; Gómez-García, C.J.; Hernández, D.; Hernández, E.; Martín, A.; Martínez, J.I.; Zamora, F. Group 10 Metal Benzene-1,2-dithiolate Derivatives in the Synthesis of Coordination Polymers Containing Potassium Countercations. *Inorg. Chem.* **2017**, *56*, 11810–11818. [CrossRef]
47. Nicholson, J.R.; Christou, G.; Huffman, J.C.; Folting, K. The synthesis, structure and spectroscopic properties of the di- and tri-nuclear Ni(II) thiolate complexes. *Polyhedron* **1987**, *6*, 863–870. [CrossRef]
48. Breitzer, J.G.; Rauchfuss, T.B. Studies on α-$C_3S_5^{2-}$ (dmit^{2-}) and its dinuclear Ni(II) complex: Spectroscopic and structural characterization. *Polyhedron* **2000**, *19*, 1283–1291. [CrossRef]
49. Sheng, T.; Zhang, W.; Gao, X.; Lin, P. Two new nickel-dmit-based molecular conductors based on heteroleptic polymetallic complexes: Synthesis, structures and electrical properties. *Chem. Commun.* **1998**, 263–264. [CrossRef]
50. Ahmadi, M.; Fischer, C.; Ghosh, A.C.; Schulzke, C. An Asymmetrically Substituted Aliphatic Bis-Dithiolene Mono-Oxido Molybdenum(IV) Complex With Ester and Alcohol Functions as Structural and Functional Active Site Model of Molybdoenzymes. *Front. Chem.* **2019**, *7*, 486. [CrossRef]
51. Chrysochos, N.; Ahmadi, M.; Wahlefeld, S.; Rippers, Y.; Zebger, I.; Mroginski, M.A.; Schulzke, C. Comparison of molybdenum and rhenium oxo bis-pyrazine-dithiolene complexes—In search of an alternative metal centre for molybdenum cofactor models. *Dalton Trans.* **2019**, *48*, 2701–2714. [CrossRef]
52. Ghosh, A.C.; Samuel, P.P.; Schulzke, C. Synthesis, characterization and oxygen atom transfer reactivity of a pair of Mo(IV)O– and Mo(VI)O2– enedithiolate complexes—A look at both ends of the catalytic transformation. *Dalton Trans.* **2017**, *46*, 7523–7533. [CrossRef]
53. Holleman, A.; Wiberg, N.; Krieger-Hauwede, M. *Anorganische Chemie Band1: Grundlagen und Hauptgruppenelemente*, 103th ed.; Walter de Gruyter: Berlin, Germany; Boston, MA, USA, 2017; Volume 1.
54. Eisenberg, R.; Gray, H.B. Noninnocence in Metal Complexes: A Dithiolene Dawn. *Inorg. Chem.* **2011**, *50*, 9741–9751. [CrossRef]
55. Williams, R.; Billig, E.; Waters, J.H.; Gray, H.B. The Toluenedithiolate and Maleonitriledithiolate Square-Matrix Systems. *J. Am. Chem. Soc.* **1966**, *88*, 43–50. [CrossRef]
56. Yang, J.; Enemark, J.H.; Kirk, M.L. Metal–Dithiolene Bonding Contributions to Pyranopterin Molybdenum Enzyme Reactivity. *Inorganics* **2020**, *8*, 19. [CrossRef]
57. Volbeda, A.; Martin, L.; Cavazza, C.; Matho, M.; Faber, B.W.; Roseboom, W.; Albracht, S.P.J.; Garcin, E.; Rousset, M.; Fontecilla-Camps, J.C. Structural differences between the ready and unready oxidized states of [NiFe] hydrogenases. *J. Biol. Inorg. Chem.* **2005**, *10*, 239–249. [CrossRef]
58. Higuchi, Y.; Yagi, T.; Yasuoka, N. Unusual ligand structure in Ni–Fe active center and an additional Mg site in hydrogenase revealed by high resolution X-ray structure analysis. *Structure* **1997**, *5*, 1671–1680. [CrossRef]
59. Higuchi, Y.; Ogata, H.; Miki, K.; Yasuoka, N.; Yagi, T. Removal of the bridging ligand atom at the Ni–Fe active site of [NiFe] hydrogenase upon reduction with H2, as revealed by X-ray structure analysis at 1.4 Å resolution. *Structure* **1999**, *7*, 549–556. [CrossRef]
60. Ogata, H.; Kellers, P.; Lubitz, W. The Crystal Structure of the [NiFe] Hydrogenase from the Photosynthetic Bacterium Allochromatium vinosum: Characterization of the Oxidized Enzyme (Ni-A State). *J. Mol. Biol.* **2010**, *402*, 428–444. [CrossRef]
61. Marques, M.C.; Tapia, C.; Gutiérrez-Sanz, O.; Ramos, A.R.; Keller, K.L.; Wall, J.D.; De Lacey, A.L.; Matias, P.M.; Pereira, I.A.C. The direct role of selenocysteine in [NiFeSe] hydrogenase maturation and catalysis. *Nat. Chem. Biol.* **2017**, *13*, 544–550. [CrossRef]
62. Volbeda, A.; Charon, M.-H.; Piras, C.; Hatchikian, E.C.; Frey, M.; Fontecilla-Camps, J.C. Crystal structure of the nickel–iron hydrogenase from Desulfovibrio gigas. *Nature* **1995**, *373*, 580–587. [CrossRef]
63. Moubarak, S.; Elghobashi-Meinhardt, N.; Tombolelli, D.; Mroginski, M.A. Probing the Structure of [NiFeSe] Hydrogenase with QM/MM Computations. *Appl. Sci.* **2020**, *10*, 781. [CrossRef]
64. Schlaepfer, C.W.; Nakamoto, K. Infrared spectra and normal-coordinate analysis of 1,2-dithiolate complexes with nickel. *Inorg. Chem.* **1975**, *14*, 1338–1344. [CrossRef]
65. Bui, T.-T.; Thiebaut, O.; Grelet, E.; Achard, M.-F.; Garreau-de Bonneval, B.; Moineau-Chane Ching, K.I. Discotic Nickel Bis(dithiolene) Complexes—Synthesis, Optoelectrochemical and Mesomorphic Properties. *Eur. J. Inorg. Chem.* **2011**, *2011*, 2663–2676. [CrossRef]

66. Kirk, M.L.; McNaughton, R.L.; Helton, M.E. The Electronic Structure and Spectroscopy of Metallo-Dithiolene Complexes. In *Dithiolene Chemistry*; Stiefel, E.I., Ed.; John Wiley & Sons, Inc.: Hoboken, NJ, USA, 2004; Volume 52, pp. 111–212.
67. Ray, K.; Weyhermüller, T.; Neese, F.; Wieghardt, K. Electronic Structure of Square Planar Bis(benzene-1,2-dithiolato)metal Complexes [M(L)2]z (z = 2−, 1−, 0; M = Ni, Pd, Pt, Cu, Au): An Experimental, Density Functional, and Correlated ab Initio Study. *Inorg. Chem.* **2005**, *44*, 5345–5360. [CrossRef]
68. Izutsu, K. Potentiometry in Nonaqueous Solutions. In *Electrochemistry in Nonaqueous Solutions*; Wiley-VCH Verlag GmbH & Co. KGaA: Weinheim, Germany, 2009; Chapter 6; pp. 171–208. [CrossRef]
69. Evans, D.F. 400. The determination of the paramagnetic susceptibility of substances in solution by nuclear magnetic resonance. *J. Chem. Soc.* **1959**, 2003–2005. [CrossRef]
70. Evans, D.F.; James, T.A. Variable-temperature magnetic-susceptibility measurements of spin equilibria for iron(III) dithiocarbamates in solution. *J. Chem. Soc. Dalton Trans.* **1979**, 723–726. [CrossRef]
71. Sheldrick, G. SHELXT—Integrated space-group and crystal-structure determination. *Acta Crsyt. A* **2015**, *71*, 3–8. [CrossRef]
72. Sheldrick, G. Crystal structure refinement with SHELXL. *Acta Cryst. C* **2015**, *71*, 3–8. [CrossRef]
73. Farrugia, L. WinGX and ORTEP for Windows: An update. *J. Appl. Crystallogr.* **2012**, *45*, 849–854. [CrossRef]
74. Ahmadi, M. Phosphate Substituted Dithiolene Complexes as Models for the Active Site of Molybdenum Dependent Oxidoreductases. Universität Greifswald, Greifswald, Germany, 2019. Available online: https://epub.ub.uni-greifswald.de/frontdoor/index/index/start/1/rows/10/sortfield/score/sortorder/desc/searchtype/simple/query/Mohsen+Ahmadi/docId/3279 (accessed on 1 March 2020).

MDPI
St. Alban-Anlage 66
4052 Basel
Switzerland
Tel. +41 61 683 77 34
Fax +41 61 302 89 18
www.mdpi.com

Inorganics Editorial Office
E-mail: inorganics@mdpi.com
www.mdpi.com/journal/inorganics

www.ingramcontent.com/pod-product-compliance
Lightning Source LLC
LaVergne TN
LVHW070921170726
843515LV00005B/841

* 9 7 8 3 0 3 6 5 0 6 0 8 1 *